TECHNOLOGICAL ADVANCEMENTS IN SUSTAINABLE CROP IMPROVEMENT AND PRODUCTION

TECHNOLOGICAL ADVANCEMENTS IN SUSTAINABLE CROP IMPROVEMENT AND PRODUCTION

Devendra Kumar, PhD
Alok Kumar Singh, PhD
Nisha Malik, PhD
Editors

AAP | APPLE ACADEMIC PRESS

First edition published 2026

Apple Academic Press Inc.
1265 Goldenrod Circle, NE,
Palm Bay, FL 32905 USA

760 Laurentian Drive, Unit 19,
Burlington, ON L7N 0A4, Canada

CRC Press
2385 NW Executive Center Drive,
Suite 320, Boca Raton FL 33431

4 Park Square, Milton Park,
Abingdon, Oxon, OX14 4RN, UK

Library and Archives Canada Cataloguing in Publication

CIP data on file with Canada Library and Archives

Library of Congress Cataloging-in-Publication Data

CIP data on file with US Library of Congress

ISBN: 978-1-77964-314-8 (hbk)
ISBN: 978-1-77964-315-5 (ebk)
DOI: 10.1201/9781779643155

About the Editors

Dr. Devendra Kumar
Assistant Professor (Agricultural Biotechnology), College of Biotechnology, UP Pandit Deen Dayal Upadhyaya Pashu Chikitsa Vigyan Vishwavidyalaya Evam Go-Anusandhan Sansthan (DUVASU), Mathura, Uttar Pradesh, India

Devendra Kumar, PhD, is currently working as an Assistant Professor at the College of Biotechnology, DUVASU, Mathura, Uttar Pradesh, India. He had qualified ICAR-SRF, ICAR-ASRB NET, UPTET, CTET, REET, and OST AWES. With over five years of teaching and research experience, Dr. Kumar specializes in agricultural biotechnology. His current area of research is abiotic stress tolerance, hydroponics, plant tissue culture, gene expression, and plant breeding. His research portfolio includes three authored books, one edited book, seven book chapters, 14 research papers, three review articles, five lab manuals, and nine magazine articles. He has completed two university-funded research projects, holds three utility patents, and reviews manuscripts for national and international journals. Dr. Kumar has participated in over 30 seminars and completed more than 14 faculty development programs. His accolades include an ERDA Excellence in Research Award (2024), Subharti Young Achiever Award (2023), Young Scientist Award (2021), InSc Research Excellence Award (2021), and Young Achiever Award (2019). He is a lifetime member of several scientific organizations, including the Biotech Research Society, India (BRSI), Education Research & Development Association (ERDA), and Agricultural & Environmental Technology Development Society (AETDS). Beyond academics, he has organized over 10 conferences and workshops, delivered guest lectures, and conducted training sessions. Dr. Kumar's contributions highlight his dedication to agricultural biotechnology, education, and scientific advancement.

Dr. Alok Kumar Singh
Research Fellow, Division of Genomic Resources, ICAR–National Bureau of Plant Genetic Resources, New Delhi, India

Alok Kumar Singh, PhD, is a Research Fellow, in the Division of Genomic Resources at the ICAR-National Bureau of Plant Genetic Resources, New Delhi, India. He has published many books, book chapters, research articles, popular articles, and abstracts and has attended conferences, training sessions, and workshops of national and international repute. He has three years of research experience as a Junior Research Assistant in the Department of Agricultural Biotechnology, Sardar Vallabhbhai Patel University of Agriculture and Technology, Meerut, India. He has received various awards from prestigious societies for his outstanding contribution and recognition in the field of agricultural biotechnology. Dr. Singh is also a member of national and international agricultural professional societies. He was awarded his PhD in Plant Molecular Biology and Genetic Engineering from Acharya Narendra Deva University of Agriculture & Technology, Uttar Pradesh, India.

Dr. Nisha Malik
Assistant Professor, College of Biotechnology, U.P. Pandit Deen Dayal Upadhyaya Pashu Chikitsa Vigyan Vishwavidyalaya Evam Go-Anusandhan Sansthan (DUVASU), Mathura, Uttar Pradesh, India

Nisha Malik, PhD, is currently working as Assistant Professor in College of Biotechnology, DUVASU, Mathura, India. Dr. Malik has more than five years of teaching and research experience in the field of molecular biology and biotechnology, molecular markers, and plant tissue culture for crop improvement. Dr. Malik has published several research papers, review papers, book chapter, books, and popular articles with publishers of repute. She has received several scientific honors and appreciations from prestigious societies at conferences. She had organized several conferences and seminars and worked as an organizing committee member for hands-on training programs and workshops. Dr. Malik is also a member of national and international scientific professional societies. She was awarded a PhD in Agricultural Biotechnology from Sardar Vallabhbhai Patel University of Agriculture & Technology, Meerut, Uttar Pradesh, India.

Contents

Contributors

Faheem Ahamad
Department of Environmental Science, Keral Verma Subharti College of Science (KVSCOS), Swami Vivekanand Subharti University Meerut, Uttar Pradesh, India

Aftab Alam
Department of Mathematics, Keral Verma Subharti College of Science (KVSCOS), Swami Vivekanand Subharti University Meerut, Uttar Pradesh, India

Mohammad Irfan Ali
School of Allied Sciences, JECRC University, Jaipur, Rajasthan, India

S. Amulya
Division of Fruits and Horticultural Technology, ICAR-Indian Agricultural Research Institute, New Delhi, India

Kumar Anshuman
Kamla Nehru Institute of Physical and Social Sciences, Sultanpur, Uttar Pradesh, India

K. R. Anusha
Department of Vegetable Science, Punjab Agriculture University, Ludhiana, India

N. M. Anusha
Division of Fruits and Horticultural Technology, ICAR-Indian Agricultural Research Institute, New Delhi, India

K. R. Asha
Division of Agricultural Engineering, ICAR-Indian Agricultural Research Institute, New Delhi, India

Mainak Barman
Department of Genetics and Plant Breeding, Bidhan Chandra Krishi Viswavidyalaya, Mohanpur, Nadia, West Bengal, India

Partha Pratim Behera
Department of Plant Breeding and Genetics, Assam Agricultural University, Jorhat, Assam, India

Ankur Bhardwaj
Department of Biotechnology, Shri Vaishnav Institute of Sciences, Shri Vaishnav Vidyapeeth Vishwavidyalaya, Indore, Madhya Pradesh, India

Sweta Bhardwaj
Department of Food Technology, Faculty of Engineering and Technology (FET), Swami Vivekanand Subharti University, Meerut, Uttar Pradesh, India

Tanmaya Kumar Bhoi
Forest Protection Division, ICFRE-Arid Forest Research Institute (AFRI), Jodhpur, Rajasthan, India

Ramesh D, Bhusal
Department of Pharmaceutics, St. John Institute of Pharmacy and Research, Palghar, Maharashtra, India

M. R. Chandana
Division of Fruits and Horticultural Technology, ICAR-Indian Agricultural Research Institute, New Delhi, India

Machindra Chavan
Department of Pharmacognosy, Amrutvahini College of Pharmacy, Sangamner, Maharashtra, India

Arpita Das
Department of Genetics and Plant Breeding, Bidhan Chandra Krishi Viswavidyalaya, Mohanpur, Nadia, West Bengal, India

Sangeeta Dayal
Department of Biotechnology, Keral Verma Subharti College of Science (KVSCOS), Swami Vivekanand Subharti University Meerut, Uttar Pradesh, India

Md Faiyazuddin
School of Pharmacy, Al–Karim University, Katihar, Bihar, India

Amol D. Gholap
Department of Pharmaceutics, St. John Institute of Pharmacy and Research, Palghar, Maharashtra, India

Sanjeev Gupta
Crop Science Division, ICAR, Krishi Bhawan, New Delhi, India

Surendra Prakash Gupta
Department of Life Science, Shri Vaishnav Vidhyapeeth Viswavidhayalaya, Indore, Madhya Pradesh, India

Navnath T. Hatvate
Department of Pharmaceutical Technology, Institute of Chemical Technology, Marathwada Campus, Jalna, Maharashtra, India

Pankaj R. Khuspe
Department of Pharmaceutics, College of Pharmacy, Paniv, Solapur, Maharashtra, India

Devendra Kumar
Department of Plant Biotechnology, College of Biotechnology, U.P. Pt. Deen Dayal Upadhyaya Pashu Chikitsa Vigyan Vishwavidyalaya Evam Go-Anusandhan Sansthan (DUVASU), Mathura, Uttar Pradesh, India

Kapil Kumar
Department of Food Technology, Faculty of Engineering and Technology (FET), Swami Vivekanand Subharti University, Meerut, Uttar Pradesh, India

Krishna Kumar
Dr. Rajendra Prasad Central Agricultural University, Pusa, Bihar, India

Sandeep Kumar
Division of Agronomy, ICAR-Indian Agricultural Research Institute, New Delhi, India

Kamlesh Kumar Lakhena
Division of Agronomy, ICAR-Indian Agricultural Research Institute, New Delhi, India

Prasanta Kumar Majhi
Department of Plant Breeding and Genetics, Regional Research and Technology Transfer Station (RRTTS), Odisha University of Agriculture and Technology (OUAT), Keonjhar, Odisha, India

Sudipta Padhan
Department of Entomology, Lovely Professional University, Jalandhar, Punjab, India

Prachi Pattnaik
Department of Horticulture, Institute of Agricultural Sciences, Banaras Hindu University (BHU), Varanasi, Uttar Pradesh, India

Sarita, Pradhan
Department of Seed Science and Technology, Odisha University of Agriculture and Technology, Bhubaneswar, Odisha, India

H. R. Raghavendra
Division of Food Science and Postharvest Technology, ICAR-Indian Agricultural Research Institute, New Delhi, India

Soumyayan, Roy
Department of Genetics and Plant Breeding, Bidhan Chandra Krishi Viswavidyalaya, Mohanpur, Nadia, West Bengal, India

Mukesh Ruhela
Department of Environmental Engineering, Faculty of Engineering and Technology (FET), Swami Vivekanand Subharti University, Meerut, Uttar Pradesh, India

Sadikali F Sayyad
Department of Pharmaceutics, Amrutvahini College of Pharmacy, Sangamner, Maharashtra, India

Ipsita Samal
Department of Entomology, Faculty of Agriculture, Sri Sri University, Cuttack, Odisha, India

Subrat Kumar Senapati
Department of Floriculture and Landscaping, Odisha University of Agriculture and Technology (OUAT), Bhubaneswar, Odisha, India

Kripa Shankar
Division of Fruits and Horticultural Technology, ICAR-Indian Agricultural Research Institute, New Delhi, India

Navneet Sharma
Department of Biotechnology, School of Life Science & Technology, IIMT University, Meerut, Uttar Pradesh, India

Aalok Shiv
Division of Crop Improvement, ICAR-Indian Institute of Sugarcane Research, Lucknow, Uttar Pradesh, India

P. M. Sindhu
Division of Food Science and Postharvest Technology, ICAR-Indian Agricultural Research Institute, New Delhi, India

Sanjay Kumar, Singh
Dr. Rajendra Prasad Central Agricultural University, Pusa, Bihar, India

Ajit Kumar Singh
Assistant Professor, College of Horticulture and Forestry, at Acharya Narendra Deva University of Agriculture and Technology, Kumarganj , Ayodhya, Uttar Pradesh, India

Alok Kumar Singh
Division of Genomic Resources, ICAR–National Bureau of Plant Genetic Resources, New Delhi, India

Ujjwal Kumar Singh
Kamla Nehru Institute of Physical and Social Sciences, Sultanpur, Uttar Pradesh, India

Ankit Singh
Kamla Nehru Institute of Physical and Social Sciences, Sultanpur, Uttar Pradesh, India

Ashutosh Singh
Department of Biotechnology & Crop Improvement, College of Horticulture & Forestry, Rani Lakshmi Bai Central Agricultural University, Jhansi, Uttar Pradesh, India

Digvijay Singh
Dr. Rajendra Prasad Central Agricultural University, Pusa, Bihar, India

Harendra Singh
Dr. Rajendra Prasad Central Agricultural University, Pusa, Bihar, India

Rajendra K Singh
Integra Life Sciences, Princeton, NJ, USA

Yadvendra Pal Singh
Lovely Professional University, Phagwara, Punjab, India

Surabhi Singhal
Department of Botany, School of Life Science & Technology, Department of Botany, IIMT University, Meerut, Uttar Pradesh, India

Kavita Solanki
Dr. Rajendra Prasad Central Agricultural University, Pusa, Bihar, India

C. A. Sowndarya
Division of Computer Applications, ICAR-Indian Agricultural Statistics Research Institute, New Delhi, India

L.C. Sushmitha
Division of Vegetable Science, ICAR-Indian Agricultural Research Institute, New Delhi, India

Vatsala Tomar
Department of Botany, School of Life Science & Technology, IIMT University, Meerut, Uttar Pradesh, India

Md Jasim Uddin
Department of Pharmaceutical Technology, Faculty of Pharmacy, Universiti Malaya, Kuala Lumpur, Malaysia

Arvind Yadav
Department of Soil Science and Agricultural Chemistry, Banaras Hindu University (BHU), Varanasi, Uttar Pradesh, India

M. Yogananda
Division of Vegetable Science, ICAR-Indian Agricultural Research Institute, New Delhi, India

Abbreviations

1-MCP	methylcyclopropene
ACC	aminocyclopropane-1-carboxylic acid
ACO	ACC oxidase
ACS	ACC synthase
ADE	adenine base editor
AFLP	amplified fragment length polymorphism
ALS	acetolactate synthase
Blc	beta lycopene cyclase 1
Bt	*Bacillus thuringiensis*
CA	control atmosphere
CBE	cytosine base editor
CRISPR	clustered regularly interspaced short palindromic repeat
CRP	centromeric repeat probes
DArT	diversity array technology
DSB	double-strand breaks
dsDNA	double-stranded DNA
ESTs	expressed sequence tags
FFV	fresh fruit and vegetables
FISH	fluorescent *in-situ* hybridization
FM	functional markers
GAB	genomics-assisted breeding
GD	genetic diversity
GE	genome editing
GFP	green fluorescent protein
GISH	genomic *in-situ* hybridization
GMO	genetically modified organism
GT	gene targeting
GUS	β-glucuronidase
HDR	homology-directed repair
HR	homologous recombination
HWD	hot water dips
HWT	hot water treatment
IAA	indole-3-acetic acid
IoT	Internet of Things
KASP	kompetitive allele specific PCR
LD	linkage disequilibrium

LOD	logarithm of odds
LSP	locus specific probes
MAB	marker assisted breeding
MAP	modified atmosphere packaging
MAS	marker-assisted selection
MEG	maternally expressed genes
MMs	molecular markers
MNs	meganucleases
MWCNTs	multiwalled carbon nanotubes
NABIC	National Agricultural Biotechnology Information Centre
NGS	next-generation sequencing
NGT	new genomic technique
NHEJ	nonhomologous end joining
PAA	peroxyacetic acid
PAM	protospacer adjacent motif
PCR	polymerase chain reaction
PG	polygalacturonase
PM	physical map
PME	pectin methylesterase
PSY	phytoene synthase
QTL	quantitative trait loci
RAMP	randomly amplified microsatellite polymorphism
RAPD	random amplified polymorphic DNA
RFLP	restriction fragment length polymorphism
RH	radiation hybrid
RNAi	interference with RNA
RNP	ribonucleoprotein
SAM	S-adenosyl methionine
SCAR	sequence characterized amplified region
SGR1	stay-green 1
SNP	single nucleotide polymorphism
SRAP	sequence-related amplified polymorphism
SSN	sequence-specific nucleases
SSR	simple sequence repeat
STS	sequence-tagged sites
TALEN	transcription activator-like effector nucleases
TGA	triacylglycerol
TILLING	targeted induced local lesions in genome
TSS	total soluble solid content
VIGS	virus-induced gene silencing
WCP	whole chromosome probe
ZFN	zinc-finger nucleases

Acknowledgments

This comprehensive collection on recent crop improvement approaches and the integration of technology for sustainable crop improvement has been made possible through the dedicated efforts and unwavering support of numerous individuals and organizations. It is with heartfelt gratitude that we extend our appreciation to those who have contributed to the realization of this project.

First and foremost, we express our deepest gratitude to the esteemed authors who have shared their expertise, insights, and research findings. Your commitment to excellence and passion for advancing agricultural knowledge have enriched the content of this book and elevated its significance.

A special acknowledgment is extended to the reviewers whose meticulous assessments and constructive feedback have played a pivotal role in shaping the quality and credibility of the content. Your commitment to maintaining scholarly rigor has been instrumental in ensuring the book's academic integrity.

We extend our appreciation to the academic institutions, research organizations, and funding agencies that have supported the contributors in their research endeavors. Your commitment to fostering research and innovation in agriculture has had a profound impact on the development of this book.

The editorial and production team deserve sincere recognition for their diligence, professionalism, and tireless efforts in bringing this book to fruition. From manuscript preparation to the final stages of production, your attention to detail and commitment to excellence have been invaluable.

Our gratitude also extends to the individuals and organizations who provided resources, data, and permissions, facilitating the inclusion of diverse perspectives and enriching the depth of the book.

To our colleagues, friends, and families who provided encouragement and understanding during the demanding phases of this project, we express our heartfelt thanks. Your unwavering support has been a source of motivation and resilience.

Lastly, we extend our appreciation to the readers—students, researchers, and practitioners—who will engage with this book. It is our sincere hope that the insights and knowledge presented within these pages contribute to the advancement of agricultural practices and inspire further exploration in the field.

Collectively, these acknowledgments reflect the collaborative spirit that underpins the creation of this book. Each contributor, in their unique capacity, has played a vital role in shaping a publication that seeks to foster innovation, sustainability, and excellence in agriculture.

Thank you all for your contributions and support.

—Dr. Devendra Kumar
Editor

Preface

Agricultural biotechnology and crop production have become central to ensuring food security and sustainability in the face of global challenges such as climate change, population growth, and the increasing demand for natural resources. As we move into the 21st century, the necessity for innovative solutions to improve crop yields, resilience, and nutritional value has never been more pressing. This book, *Technological Advancements in Sustainable Crop Improvement and Production,* is a timely exploration of the advanced technologies and methodologies that are reshaping the landscape of modern agriculture.

The aim of this edited volume is to provide a comprehensive overview of the latest advancements in crop improvement and production technologies, focusing on the need for sustainability. It brings together a range of expertise from scholars and professionals across different facets of agricultural biotechnology, offering readers not only theoretical foundations but also practical applications for real-world challenges. Sustainable agriculture is at the heart of this work. As the world's population continues to grow, ensuring that we produce enough food without depleting natural resources or causing environmental harm is a monumental task. The integration of biotechnology in crop production offers numerous solutions, from developing crops resistant to abiotic stresses such as drought and salinity to enhancing crop nutritional content and postharvest quality.

The book is structured around key technological areas that are central to modern crop improvement. The first section delves into molecular markers and their role in identifying desirable traits for breeding superior crop varieties and their application in genome editing technologies. The book provides deeper insights into genome editing tools, tissue culture techniques, and nanotechnology. The exploration of recent advances in post-harvest technology underscores the book's commitment to covering the entire spectrum of agricultural advancements. In each chapter, the authors explore both the theoretical principles behind these technologies and their practical applications in agriculture. What sets this book apart is its holistic approach to agricultural biotechnology. While many books in the field focus on isolated techniques or technologies, this volume emphasizes their interconnectedness and their role in developing comprehensive, sustainable solutions to global agricultural challenges. The book integrates the latest research with practical examples, and applications, making it a valuable resource for researchers, practitioners, and students alike.

The content of this book is particularly relevant given the growing focus on sustainable agricultural practices that are eco-friendly, economically viable, and socially equitable. The book reflects the need to move beyond traditional agricultural methods and embrace the power of biotechnology to meet the demands of the future. Each chapter has been carefully curated to ensure that the content is up-to-date with the latest scientific developments, and it provides insights into the future of crop improvement and production technologies.

As editors, we believe that this book will serve as a significant resource for anyone involved in the fields of agricultural biotechnology, crop breeding, plant genetics, and sustainable agriculture. We have aimed to make this book accessible to a wide audience, including university professors, researchers, students, and professionals in the agricultural and biotechnology industries.

Finally, we would like to express our gratitude to all the contributors who have shared their expertise and time to make this book a comprehensive and authoritative resource. Their collective knowledge and commitment to advancing agricultural sciences have made this work possible. We also extend our thanks to Apple Academic Press for their support in bringing this volume to fruition.

It is our hope that this book will inspire further research, promote the adoption of sustainable agricultural practices, and contribute to the development of the next generation of crop improvement technologies. The future of agriculture lies in our ability to innovate, and this book serves as a testament to the power of scientific advancements in shaping a more sustainable and food-secure world.

—Dr. Devendra Kumar
Editor

Introduction

The world today faces unprecedented challenges in ensuring food security, environmental sustainability, and economic viability in the agricultural sector. With the global population expected to exceed 9 billion by 2050, the demand for food will increase drastically, placing immense pressure on agricultural systems to meet these needs. At the same time, climate change, soil degradation, and water scarcity are rapidly altering the landscape of global agriculture, making it more difficult to grow crops in traditional ways. In this context, the need for innovative solutions that increase crop productivity while minimizing environmental impact has never been more urgent.

Technological Advancements in Sustainable Crop Improvement and Crop Production is an edited volume that seeks to address these challenges through the lens of cutting-edge agricultural biotechnology. This book explores the latest technological advancements in the fields of crop improvement, molecular breeding, plant genetics, and sustainable agricultural practices. It brings together a wealth of knowledge from leading experts in the field, covering both fundamental principles and practical applications of modern agricultural techniques.

The primary goal of this book is to provide a comprehensive resource on the technologies and strategies that are revolutionizing crop improvement and production in the 21st century. While crop improvement has traditionally focused on enhancing yield and resistance to pests and diseases, the future of agriculture is increasingly shaped by the integration of biotechnology. These advances not only help increase productivity but also enable the development of crops that are more resilient to the stresses posed by climate change, more nutritious, and better suited for sustainable farming practices.

At the core of the book are discussions on key technologies such as molecular markers, genome editing, tissue culture, and nanotechnology. These techniques have already transformed how we breed and manage crops. Molecular markers, for example, allow for the identification of desirable traits at the genetic level, making it possible to accelerate the breeding process and produce crops with improved characteristics. Genome editing, particularly the revolutionary CRISPR/Cas9 technology, enables precise modifications of plant genomes, offering the potential to create crops with enhanced resistance to diseases, pests, and environmental stresses, all while maintaining or increasing yields.

The book is structured to guide readers through these innovations in a systematic manner. The first section focuses on molecular markers, detailing their applications in crop breeding and how they are used to select superior genotypes for enhanced performance. Subsequent chapters explore advanced techniques such as tissue culture, which enables the mass propagation of plants and nanotechnology, which holds immense potential in improving nutrient delivery, pest management, and postharvest preservation.

Another significant aspect of the book is its emphasis on sustainable agricultural practices. Sustainability is now a key factor in agricultural production, and this book provides detailed insights into how biotechnology can be leveraged to ensure long-term agricultural success without compromising the environment. The application of genome editing for sustainable crop development is explored in-depth, with a focus on how gene editing can create drought-tolerant, pest-resistant, and high-yielding varieties that contribute to sustainable agriculture.

In addition to providing a theoretical foundation, the book highlights real-world applications of these technologies in agriculture. Practical case studies and examples demonstrate how these technologies are already being implemented in various regions and industries, from large-scale commercial farming to smallholder agriculture. These examples not only showcase the potential of biotechnology but also provide a framework for understanding how such technologies can be scaled and adopted globally.

As agricultural research continues to evolve, the intersection of biotechnology, sustainability, and crop production becomes increasingly important. This book serves as a critical resource for anyone involved in agricultural science, from students and researchers to industry professionals and policymakers. By covering the latest research, providing practical applications, and offering insights into the future of crop improvement, this book provides a unique contribution to the field of agricultural biotechnology.

In conclusion, *Technological Advancements in Sustainable Crop Improvement and Production* is not just a compilation of technical information, but a call to action for the global agricultural community.

It emphasizes the need for innovation, sustainability, and collaboration to address the pressing challenges of food production and environmental conservation. As we move toward a future where agricultural systems must be more efficient, resilient, and sustainable, the technologies discussed in this book will play a crucial role in shaping the future of crop production, ensuring that we can feed the world while protecting the planet for generations to come.

Thank you all for your support.

—Dr. Devendra Kumar
Editor

CHAPTER 1

Markers for Crop Improvement

NAVNEET SHARMA,[1*] VATSALA TOMAR,[2] and SURABHI SINGHAL[2]

[1]*Department of Biotechnology, School of Life Science and Technology, IIMT University, Meerut, Uttar Pradesh, India*

[2]*Department of Botany, School of Life Science and Technology, IIMT University, Meerut, Uttar Pradesh, India*

*Corresponding author

ABSTRACT

Tremendous efforts have been made by the scientists for crop variety improvement. Several crop species like wheat, corn, and tomato have been used as models for research since these crops have significant important in food production. With the advent of molecular marker technology in the 1980s, the track of plant breeding research has changed. Different types of molecular markers have been developed during this time, and advancements in sequencing technology have geared up the research work in crop improvement. These sequencing technologies have also become a handy tool for identification of novel genetic markers for unstructured and complex populations. The DNA-based markers are widely used to reveal the DNA-based polymorphism. The type of markers, their advantages and disadvantages, and their applications in improvement of crop variety are discussed in this chapter.

1.1 INTRODUCTION

To investigate the diversity between various plant populations, different types of morphological and agronomical parameters have been successfully

Technological Advancements in Sustainable Crop Improvement and Production.
Devendra Kumar, Alok Kumar Singh, and Nisha Malik (Eds.)

DOI: 10.1201/9781779643155-1

used by plant scientists. With the advent of molecular marker techniques in the 1980s, a rapid increase in the knowledge about the genome sequence of plants and the genes controlling the physiological role of plants is observed (Cole, 2003). Different types of molecular markers have been developed during the last three decades and have revolutionized plant breeding research.

1.2 GENETIC MARKERS

The genetic marker is a gene or a specific DNA sequence. Genetic markers have two categories—classical marker and DNA or molecular marker (Kebriyaee et al., 2012). Types of classical markers are biochemical, cytological, and morphological. On the other hand, DNA markers are based on restriction endonuclease, fragmentation, and amplification. Major examples of DNA markers are restriction fragment length polymorphism (RFLP), simple sequence repeats (SSRs), diversity array technology (DArT), and amplified fragment length polymorphism (AFLP) (Jiang et al., 2013)

1.3 MOLECULAR MARKERS

Molecular markers are specific nucleotide sequences that can be investigated through polymorphism between the nucleotide sequences of different individuals. Main characteristic of an ideal DNA marker is evenly distributed throughout genome, co-dominant, highly reproducible, and should detect higher levels of polymorphism (Mondini et al., 2009).

1.3.1 CLASSIFICATION OF MOLECULAR MARKERS

Different types of DNA molecular markers have been developed on the basis of the following criteria (Table 1.1):

1. Mode of gene action
2. Mode of transmission
3. Method of detection

1.3.2 RESTRICTION FRAGMENT LENGTH POLYMORPHISM (RFLP)

RFLP is a molecular marker which is based on restriction enzyme combination and is specific to a single restriction enzyme. Different homologous

TABLE 1.1 Comparison of Important Characteristics of Most Commonly Used Molecular Markers

Characteristics	ISSR	Retrotransposons	SNP	DArT	RAPD	RFLP	AFLP	SSR
Sequencing	No	No	Yes	Yes	No	Yes	No	Yes
PCR requirement	Yes	Yes	Yes	No	Yes	No	Yes	Yes
Genome abundance	Medium	High	Very high	Very high	Very high	High	Very high	Medium
Required DNA (ng)	50	25–50	50	50–100	20	10000	500–1000	50
Cost	High	Cheapest	Variable	Cheapest	Less	High	High	High
Marker index	Medium	High	High	High	High	Low	Medium	Medium
Visualization	Agarose gel	Agarose gel	SNP-VISTA	Micro array	Agarose gel	Radioactive	Agarose gel	Agarose gel
Status	Present	Present	Present	Present	Past	Past	Past	Present
Reproducibility	Medium–High	High	High	High	High	High	Intermediate	High
Co-dominant/ Dominant	Dominant	Dominant	Co-dominant	Dominant	Dominant	Co-dominant	Dominant	Co-dominant
Required DNA quantity	Low	Low	Low	Low	Medium	High	Low	Low
Polymorphism level	High	High	High	High	Very high	Medium	High	High
Required DNA quality	Low	High	High	High	High	High	High	Low

DNA sequences can be detected by the presence of different length fragments after digestion of DNA sample with a specific restriction endonuclease.

Fragment of DNA samples are separated by gel electrophoresis and reveal a unique blotting pattern characteristic of a specific genotype. An RFLP probe is a labeled DNA sequence which hybridizes with one or more fragments of DNA on an electrophoresis gel.

1.3.2.1 DEVELOPING RFLP PROBES

The following steps are performed for development of RFLP probes:

1. Total DNA is digested with a restriction enzyme which is methylation sensitive (e.g., Psti). In Psti clones, expressed genes are not methylated.
2. The digested DNA fragments are separated on the basis of size, on a preparative agarose gel and the fragments of size 500–2000 bp are cut, eluted, and cloned with a plasmid vector (pUC 18).
3. Screening of digest in plasmid is done to check for inserts.
4. Southern blotting is developed, and the inserts can be probed with total DNA. It selects the clone that hybridizes to a single or low-copy sequence.

Probes are screened for RFLP using genomic DNA of different genotypes which are digested by restriction endonuclease. In a species with moderate-to-high polymorphism rates, more than one endonuclease is used, such as EcoRI. The restriction enzyme will not cut the DNA fragment even if a single based pair variation occurs. It will change the full sequence and will alter the restriction enzyme sequence also.

RFLP detects polymorphic sequences the restriction endonuclease digestion and DNA best hybridizations, and is very time-consuming and labor-intensive. Over the last few years, several novel genetic assays based on selective DNA amplification were developed with the help of polymerase chain reaction (PCR). These assays are easier for automation than conventional techniques. These assays are also simple to perform and are very helpful where genotype of a large number of individuals is to be determined. But due to the prerequisite of DNA sequence information, these assays got limited (Madhumati, 2014).

1.3.3 *RANDOM AMPLIFIED POLYMORPHIC DNA (RAPD)*

In the 1990s, two different laboratories, Williams et al. (1990) and Welsh & McClelland (1990) developed this procedure. RAPD detects nucleotide sequence polymorphism in DNA amplification-based assay using a single primer; this primer has an arbitrary nucleotide sequence.

1.3.3.1 PRINCIPLE OF RAPD

In this procedure, a single species of primer binds to the genomic DNA at two different sites on opposite strands. If the distances between these two sites are amplifiable, a discrete DNA product is produced through amplification cycles.

The presence of each amplified product gives an idea about the complete or partial nucleotide sequence similarity between the primer and genomic DNA. Each primer will detect the amplification sites at various positions on genome (Jones et al., 1997). This assay is an efficient way to screen the nucleotide sequence polymorphism between individuals (Figure 1.1).

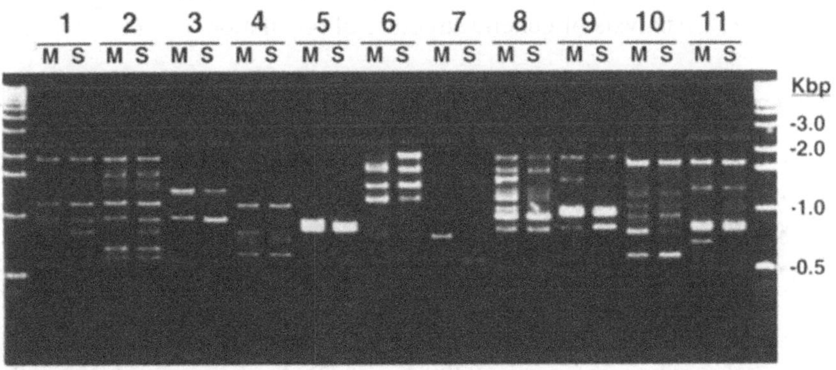

FIGURE 1.1 Effect of nucleotide substitutions in the primer. Genomic DNA from soybean G. max cv Bonus ("M") and G. soja PI81762 ("S") was amplified with primers, which were derived from a single primer sequence (primer 1= 5′TGGTCACTGA), by changing only a single nucleotide relative to the original primer (primer 2 = 5′-AGGTCACTGA, 3 = 5′-TCGTCACTGA, 4 = 5′-TGCTCACT'GA, 5 = 5′-TGGACACTGA, 6 = 5′-TGGTGACTGA 7 = 5′-TGGTCTCTGA, 8 = 5′-TGGTCAGTGA, 9 = 5′-TGGTCACAGA, 10 = 5′-TGGTCACTCA, 11 = 5′-TGGTCACTGT, progressively from 5′ to 3′. Molecular weight markers (kilobase pairs, kbp) are as indicated.

Source: Reprinted from Tingey et al. (1994)

As a result of DNA sequence difference in primer binding site between two individuals may result in, a failure of binding of primer hence a particular band will be absent among amplification products.

The quality and quantity of DNA, magnesium chloride concentration, PCR buffer, Taq polymerase, and annealing temperature are some important factors affecting the RAPD (Wolff et al., 1993).

1.3.3.2 APPLICATIONS OF RAPD

- Due to the need for small amount of DNA, it makes it possible to work RAPD with populations, which is inaccessible to RFLP analysis (e.g., single somatic embryo).
- It can be applied to a large variety of species where libraries are not available.
- RAPD markers are successfully used for the following applications:
    - Fingerprinting of an individual.
    - Mapping of trades.
    - Construction of genetic maps.
    - Targeting markers to specific regions of genome.
    - Analysis of genetic structure of population.

1.3.3.3 MAJOR ADVANTAGES OF RAPD

- There is no requirement for information about DNA sequence.
- Protocol of RAPD is relatively simple and easy to perform.
- A very small amount of DNA (in nanograms is required).
- Fluorescence is used in place of radioactivity, hence less chances of exposure.
- RAPD reaction producing five amplified bands; therefore, screen the polymorphism in 9200 bp of DNA sequence, while on the other hand, RFLP probe analyzes 12–24 nt of DNA sequence. Hence, rapid assay is a much more efficient marker in comparing DNA polymorphism.

1.3.4 AMPLIFIED FRAGMENT LENGTH POLYMORPHISM (AFLP)

To overcome the limitations of RFLP and RAPD, a new technique was developed named AFLP (Vos et al., 1995). This technique is the combination of previously discussed techniques like RFLP and RAPD (Lynch & Walsh, 1998).

In AFLP, total genomic DNA is digested with restriction endonuclease. This technique can be used for DNA of good quality as well as partially degraded.

1.3.4.1 STEPS OF AFLP

1. The first step of AFLP is digestion of total DNA with the help of a restriction enzyme. Subsequent to DNA digestion, ligation of oligonucleotide adapters is done.
2. In the second step, a specific primer set is used and PCR is performed where the primer binds to those fragments which have the same primer extension and nucleotides flanking the restriction sites. This type of amplification is termed as selective amplification.
3. Number of fragments that bind with primers and amplified are detected by polyacrylamide or agarose gel. The gel is stained with $AgNO_3$ and visualized by autoradiography. Normally, 50–100 fragments are amplified during this technique.

1.3.4.2 ADVANTAGE OF AFLP TECHNIQUE

- AFLP technique is very reliable and powerful due to stringent reaction conditions.
- The reliability of RFLP technique when combined with the power of PCR gives the best results, and the limitations are overcome in AFLP.
- Any quality of DNA can be used until unless the DNA does not contain any PCR or restriction enzyme inhibitor.

1.3.5 SIMPLE SEQUENCE REPEATS (SSRS) OR MICROSATELLITES

In the last few years, molecular markers have been very popular and extensively used for population genetics, marker-assisted selection and backcrosses, mapping of desired genes, and to find out the relationship between individuals (Tautz, 1989). Among available molecular markers, SSRs have gained importance. SSRs are tandem repeats of one to six nucleotides which are present abundantly in the genome of various individuals. Microsatellite can be mononucleotide (A) or GT, ATT, ATGC, and TAATC as *di*, *tri*, *tetra*, and *penta* nucleotide (Weber, 1990).

Besides distributed on genome, microsatellites are also present in mitochondria and chloroplasts (Rajendra Kumar et al., 2007). Some researchers have also revealed the presence of SSRs on expressed sequence tags (ESTs) and protein-coding genes. The lesser repetition of SSRs represents the high polymorphism. The higher polymorphism can be detected with the help of PCR (Kalia et al., 2011). The SSRs are widely utilized in plant genome research. SSRs can be developed from a genomic library or from enriched libraries for specific microsatellites. Searching public databases such as EMBL and Gene Bank could be helpful in searching microsatellites.

ETS databases are an important source of microsatellites information, as these can generate markers of interest directly. Procedure of development of SSR markers involved the formation of a library and detection of microsatellites. After this, primers are designed for favorable regions and PCR is performed. Banding pattern is analyzed for polymorphism study (Roder et al., 1998).

1.3.5.1 ADVANTAGES OF SSR MARKERS

SSR markers have become the choice of plant biotechnologists due to the following advantages:

1. Highly reproducible and greater relative abundance.
2. Hyper variability.
3. Extensive genome coverage.
4. Multiallelic nature.

1.3.6 SEQUENCE-RELATED AMPLIFIED POLYMORPHISM (SRAP)

AFLP is nowadays widely used for a number of applications due to its high multiplexing ratio. But the main disadvantage of AFLP is that it has multiple steps of DNA digestion, ligation, and amplification. All these steps make it difficult to optimize every step. In both RAPD and AFLP, markers need to be cloned in vectors, which require labor (Table 1.2).

TABLE 1.2 Advantages and Disadvantages of Different Genetic Markers

Markers	Advantages	Disadvantages
RAPD	Easy to use technique Less quantity of DNA is required	Requirement of highly purified DNA Dominant Not locus-specific Low reproducibility
RFLPs	Prior sequence information is not required Co-dominant	Expensive technique High quantity of pure DNA needed Time-consuming technique
AFLP	More informative technique Reliable technique High reproducibility technique	Large quantity of pure DNA needed Highly purified DNA is required Dominant marker
SRAP	Reliable and simple technique Easy isolation of bands	Moderate–high throughput ratio Dominant marker
SSRs	Co-dominant marker Less quantity of DNA is required High reproducibility	Costly technique Presence of more null alleles
ISSR	Prior sequence information is not required Highly polymorphic Simple and easy to handle technique	Very pure DNA is required low reproducibility Different-sized fragments are required
SNP	High reproducible cost-effective technique Widely distributed in genome Prior sequence information is not required Co-dominant marker	High developmental cost
Retrotransposons	High reproducibility Simple and easy to use Prior sequence information is not required	Dominant marker

TABLE 1.2 *(Continued)*

Markers	Advantages	Disadvantages
Isozymes	Co-dominant No need of specific instrument Easy to use	Less polymorphic Influenced by environmental factors
DArT	Prior sequence information is not required High throughput Highly polymorphic Cost-effective High reproducibility	High developmental cost Dominant marker
Morphological	Easy to use Cheaper Visually characterized	Less polymorphic Influenced by environment Influenced by plant growth stages

Source: Adapted with permission from Singh and Singh (2015).

To overcome this, Qiros et al. invented a new marker technique called sequence-related amplified polymorphism, which combines reliability, simplicity, and sequencing of selected bands.

SRAP is a simple technique with an efficient marker system which is widely used in map construction, genomic and cDNA fingerprinting (Salazar et al., 2014).

1.3.7 RANDOMLY AMPLIFIED MICROSATELLITE POLYMORPHISM (RAMP)

Previously discussed microsatellite markers show a great level of polymorphism, but the main drawback is labor input. On the other hand, RAPD markers, when used for crop improvement, are cost-effective, but the polymorphism detection is not as good as satellite markers.

To compensate for the weakness of these techniques, a new combined technique of RAPD and AFLP was suggested, named randomly amplified microsatellite polymorphism (RAMP) (Wu et al., 1994). This technique involves the SSR primer, which is radio-labeled and consists of 5′ anchor and 3′ repeats with ×33 p-dATP and used to amplify genome DNA. The resulting products are analyzed in denaturing polyacrylamide gel. PCR reaction was designed with high and low annealing temperatures.

RAMP markers show high polymorphism, are cost-effective, and have wide distribution in the genome. RAMP has been successfully applied in a number of plants for crop improvement through molecular characterization.

1.3.8 SINGLE NUCLEOTIDE POLYMORPHISM (SNP)

Even a single base pair change can lead to multiple aspects of marker research. Single base pair changes in the genome sequence are known as SNPs. A single unit nucleotide is done smallest unit of transfer of information. SNP may be transitions (G/A to C/T) or transversions (T/G, A/T, C/A, or C/G) on the basis of nucleotide change.

SNP is abundantly present in plants and animals with a frequency of 1 SNP in 100–300 base pairs. A large number of techniques exploring SNPs are used, but SNP and RFLP are the simplest and easiest. Identification of SNP is done with the analysis of sequence of data in databases (Xu, 2010). Different kinds of SNP genotyping assays are performed; out of them, invasive cleavage, primer extension, and oligonucleotide ligation are most important. Recent high-output methods with SNP are GBS, chip-based NGS, and allele-specific PCR, making SNP the marker of choice for plant researchers.

1.3.9 QTL

Most crops of interest are controlled by many genes on the same/different chromosomes. The regions on the chromosomes having genes for these qualitative traits are referred as QTL. QTL mapping is a method in which we use molecular markers to locate the gene controlling traits of interest. For QTL mapping, to diverse populations are selected (Angaji, 2009). Normally, 50–52 individuals are selected for QTL mapping, but for high resolutions, numbers can be increased.

1.3.9.1 MARKERS INVOLVED IN QTL MAPPING

Different types of markers like AFLP, SSR, ISSR, RFLP, ESTs, SNPs, and DArT have been used for the construction of linkage. Normally, 100–200 markers are used for linkage map construction; however, it depends on species genome size. The larger the species genome size, requires larger number of markers.

1.3.9.2 LINKAGE MAP CONSTRUCTION

It is a road map that describes the relative genetic distance and position between markers. Odd ratios are used for calculation of linkage between different markers. The most important method for QTL detection is single marker analysis, simple and composite internal analysis. Important statistical programs which facilitate the QTL mapping are PLAB QTL, QGENA, QTL Note, and Map Chart (Paterson, 1996).

1.3.9.3 FACTORS AFFECTING QTL DETECTION

Out of the factors affecting QTL detection, environmental factors, size of population, experimental error in phenotyping, and genetic properties of QTL are important ones. Environment affects the QTL detection. If QTL detection is done on the same size in different environments, it also gives an idea about the effect of environment. A large size of population facilitates the precise mapping and detection of QTL. Error in genotypic data and nonaccurate phenotypic data influence the distance between markers (Bernardo et al., 2015).

1.3.9.4 QTL CLONING

Molecular markers are used to detect QTL mapping by assigning the shortest genetic distance. These QTL-based mapping techniques provide precise results and are all less time-consuming; however, these techniques are costly as compared to map-based cloning. Nearly 3000–4000 plants are used as mapping population to achieve a higher level of recombination (Lukowitz et al., 2000). A candidate gene is a gene which is selected and sequenced for designing a specific primer for amplification.

1.3.9.5 ADVANTAGES AND DRAWBACKS OF QTL MAPPING

- Genes which are responsible for developing resistance against a specific disease are detected by QTL mapping.
- QTL is used to detect the gene of interest.
- The major drawbacks of QTL mapping are allele diversity and lower number of recombination in events (Price, 2006).
- Time-consuming is another drawback of QTL mapping (Neale & Savolainen, 2004).

1.3.10 MARKER-ASSISTED SELECTION (MAS)

MAS is a popular molecular breeding technique that overcomes the conventional breeding technique (Tabor et al., 2002). It helps in identification of suitable dominant or recessive alleles. Some important aspects of MAS are as follows:

- Marker-assisted recruitment selection
- Marker-assisted back crossing
- Genome selection
- Gene pyramiding

1.3.10.1 MARKERS-ASSISTED (MARS)

In this technique, molecular markers are applied to target all traits of interest at each generation. It is involved specifically for the improvement of the F2 population. This is a simple technique and does not require prior knowledge of QTL.

1.3.10.2 MARKER-ASSISTED (MARC)

The efficiency of back crossing was improved with the introduction of molecular markers for crop improvement. It is basically a back-crossing technique where molecular markers are used. The three levels of MARC are foreground selection, recombinant selection, and background selection (Eathington, 2007).

1.3.11 DIAGNOSTICS OR FUNCTIONAL MARKERS (FM)

FM is used to screen the large collection of germplasm in a short time and with high accuracy. With FM, any population can be screened as they are directly linked with the allele of interest (Andersen & Lu€bberstedt, 2003). FM fixes the alleles in the population more efficiently and in a balanced manner.

1.3.11.1 ADVANTAGE OF FM

- Most FM is applied for crop improvement by developing disease resistance.

- It provides an opportunity to select rare recombinant in a short time.
- For disease resistance and quality in wheat, 30 genes and more than 97 FM have been developed, and they are successfully used for crop improvement.

1.3.12 TARGETED INDUCED LOCAL LESIONS IN GENOME (TILLING)

The technique was developed by McCallum in the late 1990s while doing the characterization on Arabidopsis plants. It is a technique which is nontransgenic and successfully applied on all plants for crop improvement (McCallum et al., 2000). In 1st step of TILLING, a mutated population is generated with the use of mutagen ethyl methanesulfonate. Mutations in the targeted sequence are identified with the help of mass spectroscopy, high-performance liquid chromatography, and array-based technology. A Bioinformatics tool is PARSESNP (Project Aligned Related Sequences and Evaluate SNPs), is applied for identification of these mutants (Kurowska et al., 2011).

1.3.12.1 ADVANTAGES OF TILLING

- This technique is not affected by genome size
- It is a time-saving and low-cost technique
- It gives precise identification of new alleles

1.3.13 VIRUS-INDUCED GENE SILENCING (VIGS)

This technique uses a viral vector to exploit an RNA-mediated defense mechanism. RNA-based defense system is activated in a plant after viral infection. Viral RNA is replicated and produces dsRNA intermediate and siRNA, which base pair guides the RNAase complex. It is a technique which involves the plant defense system. In the last few years, VIGS has been applied by plant researcher for the gene of interest (Lu et al., 2003).

1.3.13.1 ADVANTAGES OF VIGS

- It is a very cost-effective and simple technique
- It can be used to identify the function loss of the gene of interest

1.3.14 RECENT ADVANCES IN MARKER TECHNOLOGY

Recent upgradations have been done in marker technology in the last few years. SNP marker becomes the marker of choice after NGS development and has been applied by a number of researchers on different crops for improvement. However, these markers were present in uniplex form, since there is a need to shift to multiplex system. KASP (kompetitive allele specific PCR) is the technology to convert uniplex to multiplex. KASP was first developed by KBioscience or LGC genomics to achieve in-house genotyping. This is a technique based on allele-specific oligo extension. KASP is done with 96,384 and 1536-well plates. The success rate of KASP is 98–100 %. It is being used in maize, wheat, rice, and a few other crops.

1.3.14.1 ADVANTAGES OF KASP

- It is a time-saving and low-cost technique.
- It is successfully applied in wheat to develop multiplex traits lined marker. Nearly 70 KASP assays have been done with wheat on various traits of interest.

1.3.15 GENOME SELECTION

It is a marker-assisted selection and was developed by Meuwissen et al. (2001); it is the technique that utilizes the high density of markers that are distinguished throughout the genome.

1.3.15.1 IMPORTANT STEPS OF GENOME SELECTION

- To develop a training population with diverse genotype.
- To perform phenotyping and genotyping of training population.
- To select the individual having superior GEBV.
- Progeny which are used in testing population are taken as input.
- Those individuals having maximum GEBV are again selected.

- The selected individuals are again used as parents of the next generation in the next experiment.

1.3.15.2 ADVANTAGES OF GENOME SELECTION

In this technique, genetic markers which have the ability to cover the whole genome are selected.

1.3.16 GENOME EDITING (CRISPR)

In the series of molecular markers described above, CRISPR is another genome editing technique which is applied successfully to various crops. It has also been helpful in some economically important crops like wheat, maize, Nicotiana tabacum, and Arabidopsis. It has gained popularity due to genome editing versatility and being easy to use. Two important parts of CRISPR technology are trans encoded CRISPR RNA (trac RNA) and CRISPR RNA (cr RNA), and two short RNAs which can cleave a particular target site with the involvement of Cas 9 endonuclease, a Cas protein which forms three types of CRISPR Cas systems: type I, II, and III.

CRISPR Cas system identifies and targets the pathogen genetic material by three steps: acquisition, expression, and interference. For acquisition of DNA fragment, a short stretch of conserved nucleotide (PAMs) is used as identification marks. The Cas protein moves toward a particular target area of foreign DNA and cleaves it by cr RNA. With this technique, the immunity is developed against pathogenic attack, and it has become a successful tool for crop improvement.

1.4 CONCLUSION

With the continuous development in plant research using molecular marker technology, various milestones have been achieved to improve crop quality and yield. Molecular marker technology has been developed from RFLP to SNPs and various array technology-based marker. With the advent of sequencing technologies, NGS platforms have been developed which are low-cost but give high-output. In spite of various developments in molecular genetic techniques, the goal of achievement is far away. The major drawbacks for this technique are inaccurate phenotyping and use of cameras, sensors, and computers. CRISPR technology has also played a very important role in plant molecular research of mostly economically important plants. We expect many more innovations in the coming years to make molecular marker technology more cost-effective, precise, and productive.

KEYWORDS

- **crop variety improvement**
- **molecular markers**
- **restriction fragment length polymorphism**
- **simple sequence repeats**
- **Taq polymerase**

REFERENCES

Andersen, J.R.; Lu€bberstedt, T. Functional markers in plants. *Trends Plant Sci.* 2003, 8(11), 554–560.

Angaji, S.A. QTL mapping: A few key points. *Int. J. Appl. Res. Nat. Prod.* 2009, 2(2), 1–3.

Bernardo, A.; Wang, S.; Amand, P.S.; et al. Using next generation sequencing for multiplexed trait-linked markers in wheat. *PLoS One* 2015, 10(12), e0143890.

Cole, C.T. Genetic variation in rare and common plants. *Annu. Rev. Ecol. Evol. Syst.* 2003, 34, 213–227.

Eathington, S.R.; Crosbie, T.M.; Edwards, M.D.; et al. Molecular markers in a commercial breeding program. *Crop. Sci.* 2007, 47(Suppl 3), S154–S163.

Jiang, G.L. Molecular markers and marker-assisted breeding in plants. In Andersen, S.B.; (Eds.), *Plant Breeding from Laboratories to fields*. InTech: Rijeka, 2013, pp. 45–83.

Jones, C.J.; Edwards, K.J.; Castaglione, S.; et al. Reproducibility testing of RAPD, AFLP and SSR markers in plants by a network of European laboratories. *Mol. Breed.* 1997, 3(5), 381–390.

Kalia, R.K.; Rai, M.K.; Kalia, S.; et al. Microsatellite markers: An overview of the recent progress in plants. *Euphytica.* 2011, 177(3), 309–334.

Kebriyaee, D.; Kordrostami, M.; Rezadoost, M.H.; et al. QTL analysis of agronomic traits in rice using SSR and AFLP markers. *Not. Sci. Biol.* 2012, 4(2), 116–123.

Kurowska, M.; Daszkowska-Golec, A.; Gruszka, D.; et al. TILLING—A shortcut in functional genomics. *J. Appl. Genet.* 2011, 52(4), 371–390.

Lu, R.; Malcuit, I.; Moffett, P.; et al. High throughput virus-induced gene silencing implicates heat shock protein 90 in plant disease resistance. *Embo. J.* 2003, 22(21), 5690–5699.

Lukowitz, W.; Gillmor, C.S.; Scheible, W.R. Positional cloning in Arabidopsis. Why it feels good to have a genome initiative working for you. *Plant Physiol.* 2000, 123(3), 795–806.

Lynch, M.; Walsh, B. *Genetics and Analysis of Quantitative Traits*. Sinauer: Sunderland (MA), 1998.

Madhumati, B. Potential and application of molecular markers techniques for plant genome analysis. *Int. J. Pure App. Biosci.* 2014, 2(1), 169–188.

McCallum, C.M.; Comai, L.; Greene, E.A.; et al. Targeting induced local lesions in genomes (TILLING) for plant functional genomics. *Plant Physiol.* 2000, 123(2), 439–442.

Meuwissen, T.H.; Hayes, B.J.; Goddard, M.E. Prediction of total genetic value using genome-wide dense marker maps. *Genetics* 2001, 157(4), 1819–1829.

Mondini, L.; Noorani, A.; Pagnotta, M.A. Assessing plant genetic diversity by molecular tools. *Diversity* 2009, 1(1), 19–35.

Neale, D.B.; Savolainen, O. Association genetics of complex traits in conifers. *Trends Plant Sci.* 2004, 9(7), 325–330.

Paterson, A.H. Making genetic maps. In Paterson, A.H.; (Eds.), *Genome Mapping in Plants*. R.G. Landes Company: Austin (TX), 1996, pp. 23–39.

Price, A.H. Believe it or not, QTLs are accurate! *Trends Plant Sci.* 2006, 11(5), 213–216.

Rajendra Kumar, P.; Biswal, A.K.; Balachandran, S.M.; et al. Simple sequence repeats in organellar genomes of rice: Frequency and distribution in genic and intergenic regions. *Bioinformatics* 2007, 23(1), 1–4.

Roder, M.S.; Korzun, V.; Wendehake, K.; et al. A microsatellite map of wheat. *Genetics* 1998, 149(4), 2007–2023.

Salazar, J.A.; Rasouli, M.; Moghaddam, R.F.; et al. Low-cost strategies for development of molecular markers linked to agronomic traits in Prunus. *Agric. Sci.* 2014, 5(05), 430–439.

Singh, B.D.; Singh, A.K. *Marker-Assisted Plant Breeding: Principles and Practices*. Springer, New Delhi, 2015.

Tingey, S.V.; Rafalski, J.A.; Hanafey, M.K. Genetic analysis with RAPD markers. In *Plant Molecular Biology*. Du Plant Agricultural Products, E.I. du Pont de Nemours & Company: Wilmington, Delaware, 1994, H 81, p. 491.

Tabor, H.K.; Risch, N.J.; Myers, R.M. Candidate-gene approaches for studying complex genetic traits: Practical considerations. *Nat. Rev. Genet.* 2002, 3(5), 391–397.

Tautz, D. Hypervariability of simple sequences as a general source for polymorphic DNA markers. *Nucleic Acids Res.* 1989, 17(16), 6463–6471

Vos, P.; Hogers, R.; Bleeker, M.; et al. AFLP: A new technique for DNA fingerprinting. *Nucleic Acids Res.* 1995, 23(21), 4407–4414.

Weber, J.L. Informativeness of human (dC-dA) n (dG-dT) n polymorphisms. *Genomics* 1990, 7(4), 524–530.

Welsh, J.; McClelland, M. Fingerprinting genomes using PCR with arbitrary primers. *Nucleic Acids Res.* 1990, 18(24), 7213–7218.

Williams, J.G.; Kubelik, A.R.; Livak, K.J.; et al. DNA polymorphisms amplified by arbitrary primers are useful as genetic markers. *Nucleic Acids Res.* 1990, 18(22), 6531–6535

Wolff, K.; Schoen, E.D.; Peters-Van Rijn, J. Optimizing the generation of random amplified polymorphic DNAs in chrysanthemum. *Theor. Appl. Genet.* 1993, 86(8), 1033–1037.
Wu, K.S.; Jones, R.; Danneberger, L.; Scolnik, P.A. Detection of microsatellite polymorphisms without cloning. *Nucleic Acids Res.* 1994, 22(15), 3257–3258.
Xu, Y. *Molecular Plant Breeding*. CABI: Wallingford, 2010.

CHAPTER 2

Molecular Markers: Emerging Strategy for Crop Improvement

ANKUR BHARDWAJ,[1*] MOHAMMAD IRFAN ALI,[2] and SURENDRA PRAKASH GUPTA[3]

[1]*Department of Biotechnology, Shri Vaishnav Institute of Sciences, Shri Vaishnav Vidyapeeth Vishwavidyalaya, Indore, Madhya Pradesh, India*

[2]*School of Allied Sciences, JECRC University, Jaipur, Rajasthan, India*

[3]*Department of Life Science, Shri Vaishnav Vidhyapeeth Viswavidhayalaya, Indore, Madhya Pradesh, India*

*Corresponding author

ABSTRACT

Molecular markers (MMs) are extensively used in crop development due to their simplicity, repeatability, and precise placement. Simple sequence repeat and single nucleotide polymorphism (SNP) markers are the most commonly employed in plant breeding programs. Recent research has significantly advanced our understanding of DNA markers, benefiting plant breeding, genetics, genomic selection, and genome editing. This chapter provides an overview of current advancements in MMs and their applications in crop improvement and plant breeding. MMs allow plant breeders to select superior plant progenies without environmental influences, enhancing the efficiency of breeding programs. The use of MMs has increased with advancements in DNA analysis, particularly next-generation sequencing technology, which has made SNPs a valuable resource. These markers are essential for constructing saturated linkage maps, crucial for marker-assisted selection, map-based gene cloning, and

Technological Advancements in Sustainable Crop Improvement and Production.
Devendra Kumar, Alok Kumar Singh, and Nisha Malik (Eds.)

DOI: 10.1201/9781779643155-2

quantitative trait loci mapping. The chapter covers various types of plant MMs, focusing on recent developments in crop enhancement technology. It highlights examples of genetic diversity in agricultural plants, which enhance plant breeding techniques. By leveraging MMs, plant breeders can improve crop traits more effectively, contributing to sustainable agricultural practices. Overall, MMs play a pivotal role in modern plant breeding, offering precise tools for genetic improvement and crop development.

2.1 INTRODUCTION

A DNA sequence that is known to be on a chromosome and connected to a specific characteristic is described as a molecular marker (MM). It can be characterized as a variance that could result from a mutation or change in the detectable genomic locus. Recently, MMs have gained more interest as they are highly discoverable in DNA polymorphism, required for crop improvement approaches (Musunuru, 2022). Selecting plants with superior potential qualities to serve as parents for the following generation is the primary goal of a plant breeder.

When used to measure genetic variation among populations, allozymes (first markers) bring molecular genetics and ecology together (DeWoody et al., 2021). Although they have not been as prominent as DNA markers recently, allozymes have continued to be used in research to investigate animal and plant populations since their debut in the 1960s. Allozymes are co-dominant by nature, cost-effective, and do not require DNA order information (Singh et al., 2021). They do, however, give conservative estimates of genetic diversity (GD) since nonsynonymous changes in protein-coding genes account for all of their variability.

Advancements in MM technology have been seen during the 1980s, and this has brought about many changes in the fate of plant breeding techniques and crop improvement (Mao et al., 2013). A thorough perception of MMs and their technology has been perceived in the field of genetics and molecular breeding, along with genetic selection and editing (Nadeem et al., 2018). Many high-throughput genotyping technology and phenotyping platforms, such as GWAS, were analyzed to specifically explore the genetic constitution of plant traits (Leng et al., 2017; Xiao et al., 2017). The identification of GD, FG, and the advancement of breeding technologies all benefit greatly from the use of sophisticated MM methods. The new era has changed from the conventional use of MM to the latest genome editing system CRISPR/Cas9 (Dheer et al., 2020).

Conventional methods of breeding basically relied upon genetic traits and selection methods to produce cultivars of breeder's interest (Gupta et al., 2013). Undoubtedly, through these methods, many hybrid varieties were produced. But the recent advances in MM technology have a more powerful impact.

Since most important crop species have a substantial number of MMs that are closely connected to disease resistance features, the MMs are currently routinely employed in crop breeding programs to track loci and targeted genomic regions (Kim et al., 2018; Miah et al., 2013; Ramakrishnan et al., 2015). The majority of MMs have been recovered from libraries of randomly amplified polymerase chain reaction (PCR) fragments or from genomic DNA libraries RAPF. MMs are essential for mapping genes of interest, marker-assisted breeding (MAB), and cloning genes by utilizing mapping-based cloning strategies (Miah et al., 2013; Singh et al., 2015). Additional applications for MMs include phylogenetic research, germplasm characterization, and backcrossing gene insertions (Gupta et al., 2002; Zhang et al., 2021b). Microsatellites have become the standard marker of choice for plant breeding applications among the several classes of now available markers (Darkwa et al., 2020; Kesawat et al., 2009). While both amplified fragment length polymorphisms (AFLPs) and microsatellites are effective in identifying polymorphism, microsatellite-based techniques are easier to automate (Avvaru et al., 2020). Furthermore, AFLP analysis is not always simple since, especially when utilizing large genomic templates, ostensibly single bands may really be made up of many fragments (Bangarwa et al., 2021). In addition, MMs can be used to determine the proportion of an introgressed genome, trait-related genes, and to select genes during multiple rounds of introgression (Kumar et al., 2020; Zhang et al., 2021a).

For a given DNA sequence in a gene pool, the MMs identify polymorphism between distinct genotypes or allelic forms of a gene and disclose mutations or genetic variants (Jiang, 2015).

Using PCR or southern blotting method followed by electrophoresis (like PAGE, AGE, CE, etc.), the polymorphism in DNA patterns could be known as per product band size. Several new polymorphism detection systems have been developed, like array chip technique, which uses DNA hybridization with labelled nucleotides and modern sequencing techniques (Nag, 2022). Ideal DNA markers for effective usage in MAB should fulfill the following requirements:

- High polymorphic level
- Even division across the whole genome

- Co-dominance
- Distinct allelic features
- No pleiotropy
- Cost efficient
- Easy detection and automation
- High availability (unrestricted use)
- Polyploidy in genome
- No damaging impact on phenotype

The correlation between an inheritable disease and its genetic etiology can be investigated using genetic markers (Gautam, 2022). Genetic markers remain nearby on the chromosomes that are inherited together. This property of genetic marker enables to ascertain the gene's exact inheritance pattern, which has not yet been precisely located (Molfino, 2004).

The genetic differences revealed by MMs could be visualized by using AGE, radioactive, or colorimetric probes. When DNA markers show variations across members of the same or distinct species, they are very helpful and referred to as polymorphic (pm) markers. The monomorphic (mm) markers do not differentiate between genotypes. Polymorphic markers that can discriminate between homo and heterozygotes are also described as co-dominant or dominant markers (Maurye et al., 2014).

2.1.1 CO-DOMINANT MARKERS

Each co-dominant marker in a diploid species identifies two alleles in heterozygous and one allele in homozygous individuals. One of the most crucial characteristics of co-dominant markers is their capacity to differentiate between homozygotes and heterozygotes (Eduardo et al., 2015). When two alleles co-occur in one individual, the marker is said to be co-dominant since both alleles express the marker. Co-dominant markers could distinguish heterozygotes among homozygotes. On the other hand, the band profiles of these markers are graded according to whether or not specific-sized fragments are present (Amiteye, 2021).

Considering Hardy–Weinberg law for calculating allele frequency from the percentage of people with missing phenotype, one can only gain an approximated allele frequency. For predominantly self-fertilizing species (discarding heterozygosity), allele frequencies could be equal to observed band frequencies (Kumar et al., 2009). The information obtained using co-dominant markers assay is usually lower compared to other multilocus techniques. Some examples of co-dominant markers are–restriction fragment length polymorphism (RFLP), sequence characterized amplified region (SCAR), simple sequence repeat (SSR), CpSSR, SAMPL, SSCP, CAPS, single nucleotide polymorphism (SNP), sequence-tagged sites (STS), and RAMP (McGregor et al., 2000).

2.1.2 DOMINANT MARKERS

Dominant markers are the genetic markers that represent a single allele even when multiple alleles are co-occurring in an individual (Vieira et al., 2016). A dominant marker will only show the dominant gene either in parents or its offspring. Dominant marker results in a loss of information concerning the genotype of an individual. As a result, such markers provide less information about the parental individuals. The loss of information will be reflected in a higher variance for results obtained from dominant loci compared with co-dominant loci in a given population (Adhikari et al., 2017). Some examples of dominant markers are—random amplified polymorphic DNA (RAPD), AFLP, ISSR, SRAP, DArT, and EST.

2.2 MMS-LINKAGE AND MAPPING

MMs can examine allelic polymorphisms that exist between the nucleotide sequences of different plant species. The basis for polymorphisms within a particular population is the result of nucleotide sequence deletions, mutations, insertions, duplications, and translocations; these events, however, have little effect on the functioning of individual genes.

MMs are phenotypically neutral and are located at specified sites across the genome. They are employed to determine the location of a certain gene or the character's inheritance. For detection of markers, either restriction enzymes or PCR or their combination are used to generate/amplify the DNA sequences that are linked to a heritable trait such as yield or disease resistance. With the introduction of marker-assisted selection (MAS), breeding populations now have access to a new tool for more precise and beneficial selection. The process of analyzing MM marker data to produce linkage maps is the same as it is for genes that affect phenotypic variation (Kumar et al., 2021). DNA markers allow geneticists to identify many differences in genomic sequences between parents and track the inheritance of several marker loci in a single cross (Table 2.1).

TABLE 2.1 Important Characteristics of the Most Significant MMs (Kesawat et al., 2009; Miah et al., 2013; Nadeem et al., 2018)

S. No.	Characteristics	Types of MMs							
		RFLP	**RAPD**	**AFLP**	**ISSR**	**SSR**	**SNP**	**DArt**	**Retero-transposons**
1	Co-dominant/Dominant	CoD	D	D	D	CoD	CoD	D	D
2	Reproducibility	H	H	M	M	H	H	H	H
3	Polymorphism type	M	VH	H	H	H	H	H	H
4	DNA quality required	H	H	H	L	L	H	H	H
5	DNA quantity required	H	L	L	L	L	L	L	L
6	Marker index	L	H	M	M	M	H	H	H
7	Genomic richness	H	VH	VH	M	M	VH	VH	H
8	Price	H	L	H	H	H	H	C	C
9	Sequencing	Y	N	N	N	Y	Y	Y	N
10	PCR requirements	N	Y	Y	Y	Y	Y	N	Y
11	Type of electrophoresis	Radioactive	Agarose gel	Agarose gel	Agarose gel	Agarose gel	SNP-Vista	Microarray	Agarose gel

Note: H: High; VH: Very High; L: Low; M: Medium; Y: Yes; N: No; C: Cheap; CoD: Co-dominant; D: Dominant.

2.2.1 MMS-LINKAGE

The biological process of genetic recombination takes place in the parental gametes during meiosis, is the basis of linkage. When two chromosomes, one from each parent, align in the first metaphase of a parental gamete, chromosomal segment is frequently exchanged in a crossover event, the site of which is assumed to be random. This chromosomal recombination results in linkage disequilibrium (LD), which would be inherited from either parent alone (Bangarwa et al., 2021). Thus, each LD inherits a unique set of chromosomes that are recombinants of the parents. Linkage analysis depicts about marker alleles traveling in proximity on the same chromosome or are farther away or on different chromosomes (Phan et al., 2021; Sharanabasappa et al., 2022). Linked genes are those that are positioned on a chromosome so that there is less than a 50% chance of crossing between them during meiosis. Even though recombination rates vary throughout the genome, this idea has produced a useful biological model for linkage research (Shipilina et al., 2022).

2.2.2 MMS-MAPPING

The linear arrangement of MMs in a particular genome is reflected in genetic linkage (gl) maps. By tracking the segregation of MMs within a population and organizing them in a linear order according to pair-wise recombination rates, linkage maps are created. For a mapping population, the occurrence of numerous polymorphisms throughout the entire genome is a highly desired trait (Kulwal and Singh, 2022). For molecular mapping, recombinant inbred lines are the most commonly utilized populations (Arafa et al., 2022; Zhou et al., 2022).

Genomic complexity of heighted plants, heterozygosity among clones, higher self-incompatibility, and absence of phenotypic genetic markers make formation of genetic linkage (gl) maps a tedious task (Cropano et al., 2021). In outcrossing tall species, the pseudo-testcross method is thought to be an effective way to create linkage maps (Cropano et al., 2021; Yu et al., 2022). Through pseudo-testcross technique mapping population is created by hybridizing two unrelated, vastly heterozygous parents that generate a group of F1 progeny (Song and Li, 2021). In majority of these groups, significant amounts of marker discrimination are seen. However, because markers can be dominant or codominant, one or both parents may be heterozygous at a particular locus, and the linkage phase of marker alleles is typically unknown, creating genetic linkage (gl) maps, which is challenging (Shimizu et al., 2020; Wang et al., 2020). When compared to other important crop species, the genomic information for many crop species is extremely restricted because the majority of these species are polyploids or outcrossing (Pasquali et al., 2022; Shahabzadeh et al., 2020).

Finding a marker that co-segregates with the desired gene will allow linkage analysis to follow the gene within a family without requiring knowledge of the mutation. The genetic marker has to co-segregate with the relevant gene, meaning it needs to be present in family members who are afflicted but absent in those who are not. Linkage analysis was the primary technique used in the days before quick sequence analysis to determine homo or heterozygosity (Bailey-Wilson et al., 2011; Bishop, 2002).

2.3 MICROSATELLITE MAP

Recently, several microsatellite and STS markers were developed for many plant species (Saha et al., 2009). Microsatellites are the most widely used type of MMs, as PCR can easily amplify them with a higher extent of allelic difference at each locus (Miah et al., 2013). Compared to other genetic markers, microsatellites or SSR markers are highly variable, species-specific, widely dispersed, and co-dominant in nature (Zhang et al., 2014). These factors have made them more and more significant genetic markers in crop improvement initiatives. For crop breeders, the emergence of novel pest and disease biotypes and climate-related pressures provides significant obstacles (Chukwu et al., 2019; Zaghum et al., 2022). Thus, MAS may be utilized to track the presence or absence of genes in breeding populations in conjunction with traditional breeding techniques. Using marker-assisted backcross breeding, for instance, major genes with noteworthy biological effects have been included into a number of widely cultivated rice types. Through the application of MAS techniques and accurately mapped, cost-effective microsatellite markers, breeders may create high-yield rice cultivars that are resistant to blasts (Boopathi, 2020; Hajiaqatabar et al., 2019; Parihar, 2022).

Microsatellite markers have been a popular tool for screening GD and characterizing it in cereal crops over the past few years (Gupta et al., 2003; Miah et al., 2013). Many microsatellite markers that are closely associated with certain qualitative genes have been created using publicly accessible datasets (www.gramene.org) (Fjellstrom et al., 2004; Hasan et al., 2015). Moreover, the use of MAS and MAB (backcross) breeding might be accelerated with the advent of molecular techniques that effectively detect novel resistant genes, potentially leading to a significant improvement in contemporary cultivars.

2.4 LINKAGE DISEQUILIBRIUM (LD)

A sensitive measure of the population genetic factors that shape a genome is the nonrandom combination of allelic variants at various locations (Slatkin, 2008). The LD is being used more and more finely by evolutionary biologists and geneticists to map genes linked with quantitative traits and acquired genetic disorders (Mackay, 2001; Slatkin, 2008).

The LD is important in evolutionary biology and human genetics, as so many factors affect and are affected by it. In addition to providing historical context, LD limits possible responses to both artificial and natural selection (Gaitán-Espitia et al., 2021; Slatkin, 2008). Natural selection, gene switching, mutation, along with other processes that lead to the evolution of gene frequencies, are all reflected in the history of LD in a particular genomic area. Recombination frequencies determine how these variables impact LD within a genome or among a specific loci pair (Jasielczuk et al., 2020; Yoshida et al., 2019). The advanced population genetics concept of LD is commonly utilized to map genes and shed light on the origins and evolution of species (Rahman et al., 2022).

The "haplotype sets" were a surprising discovery with great practical importance for mapping of inherited diseases (Matias et al., 2017; Wu et al., 2022). However, fewer SNPs may need to be tested. The "haplotype set" discovery shows that LD typically extends over considerably greater chromosomal distances, as well as suggests that analyzing one SNP inside each set for significant relationship with a character may be adequate for determining linkage with SNP within that set (Matias et al., 2017; Wu et al., 2021).

2.5 MOLECULAR (GENETIC) MARKERS IN PLANT BREEDING PROGRAMS AND CROP IMPROVEMENT

Plant breeders can choose individual plants using the MM technique by considering their genotype, or marker pattern, as opposed to their phenotype, or observable qualities (Singh and Singh, 2015a). Genetic markers, which are closely connected to the target gene and function as flags, are significant advancements in the practice of plant breeding (Ragimekula et al., 2013). Some of the frequently used genetic markers are RFLP, AFLP, SSRs, SNPs, and DArT markers. Nonetheless, a range of MMs were created, and sequencing tools development has accelerated crop enhancement initiatives.

A relatively recent method of traditional breeding called MAS avoids and improves the phenotypic selection criteria by selecting the required genes (Hospital, 2009). The MMs do not bear environmental impact and have no effect on crop plants they observed in any growth stages of any respective plant. Since a range of MMs and genetic maps are available, MAS is possible for both quantitative trait loci (QTLs) and features managed by substantial quality. The efficacy of a given MM depends on its ability to reveal polymorphism, allowing for sorting between different MM alleles (Ghafoor and McPhee, 2012).

Crop breeding system improvement using MMs depends on a few factors.

- A genetic map linked with QTLs of relevance in relevant crop plants and an MM associated with it.
- The significant genes are strongly correlated with MMs.
- Adequate recombination in between the rest of the genome and the MMs linked to desired features.
- An efficient time and money approach for analyzing many diverse individuals.

Various genomic events, like deletions, insertions, gene mutation, etc., produce polymorphic variances in the overall population, but they have little effect upon the way gene functions. Several types of MMs exist that are used currently as per convenience (Table 2.2), nonetheless, an ideal DNA marker ought to be evenly distributed and co-dominant. and should have the capacity to recognize the significant level of polymorphism.

TABLE 2.2 A Summarized Lists of MMs Used in Crop Improvement Programs

S. No.	MM	Abbreviation	S. No.	MM	Abbreviation
1.	Allele Specific Associated Primers	ASAP	17.	Random Amplified Microsatellites	RAM
2.	Allele Specific Oligo	ASO	18.	Random Amplified Polymorphic DNA	RAPD
3.	Allele Specific Polymerase Chain Reaction	AS-PCR	19.	Restriction Fragment Length Polymorphism	RFPL
4.	Amplified Fragment Length Polymorphism	AFLP	20.	Short Tandem Repeats	STR
5.	Anchored Microsatellite Primed PCR	AMP-PCR	21.	Selective Amplification of Microsatellite Polymorphic Loci	SAMPL
6.	Anchored Simple Sequence Repeats	ASSR	22.	Sequence Characterized Amplified Regions	SCAR
7.	Arbitrarily Primed PCR	AP-PCR	23.	Sequence Tagged Microsatellite Site	STMS
8.	Cleaved Amplified Polymorphic Sequence	CAPS	24.	Sequence-Tagged Site	STS
9.	Diversity Arrays Technology	DArT	25.	Simple Sequence Length Polymorphism	SSLP
10.	DNA Amplification Fingerprinting	DAF	26.	Simple Sequence Repeats	SSR
11.	Expressed Sequence Tags	ESTs	27.	Single Nucleotide Polymorphism	SNP
12.	Inter-Simple Sequence Repeats	ISSR	28.	Single Primer Amplification Reactions	SPAR
13.	Microsatellite Primed PCR	MP-PCR	29.	Single Stranded Conformational Polymorphism	SSCP
14.	Multiplexed Allele-Specific Diagnostic Assay	MASDA	30.	Variable Number Tandem Repeats	VNTR
15.	Random Amplified Microsatellite Polymorphisms	RAMP	31.	CAAT-box derived polymorphism	CBDP
16.	Start Codon Targeted Polymorphism	ScoT	32.	Inter-Retrotransposon Amplified Polymorphism	IRAP

2.6 DESIGN AND DEVELOPMENT OF MM BASED MAPPING

The sequential assigning of loci at a specific place within a chromosome is known as molecular mapping. These positions/distances are calculated on the basis of frequency of chromosomal crossing over taking place during meiosis (Williams, 2019). Dense genetic marker maps are accessible for many crop species, and next-generation sequencing (NGS) technologies are easing genetic maps construction for other species, too. Stress (biotic and abiotic) related gene mapping requires the use of genetic maps. Integrating genetic mapping and disease gene mapping with next-generation sequencing has proven to be a powerful strategy in crop genetic research (Ouyang et al., 2022).

Genetic mapping has seen a rise in the usage of PCR-based MMs due to their simplicity of production from minuscule DNA amounts. Earlier AFLP technique was in extensive use due to high efficiency, but SSRs were then preferred because they were reported at high frequencies in plant genomes exhibiting high mutation rates, and good analytical resolution. Genetic maps offer insights into different regions of the chromosome, for instance, the genetic-physical distance proportion, diverse recombination phenomenon, indicating euchromatic and heterochromatic regions in the genome (Pyeritz et al., 2018).

The formation of genetic markers and selection of mapping population are the initial steps to create a genetic map. To map gene order, markers' "co-segregation" patterns are analyzed. The preciseness of genetic maps is largely dependent on the quantity of genetic markers present and the total number of mapping populations. The two factors are interlinked in a larger mapping population that could increase the "resolution" of constructed maps, avoiding maps being "saturated" (Quezada et al., 2021; Ward et al., 2013).

Any sequence characteristic that can be reliably labelled among two parents could be used as a genetic marker. Within this sense, "traits" that reliably discriminate between two parents are a representation of genes. The actual gene loci are then encircled in an area between the two closest neighboring markers, and their link with another genomic marker is computed as though they were common markers. Once a single causal locus has been found, the same procedure is repeated by examining additional markers that target that area. This technique, called "positional cloning," is widely utilized in the investigation of several plant species (Gallavotti and Whipple, 2015; Jaganathan et al., 2020; Kalia et al., 2011).

2.6.1 GENETIC LINKAGE MAPPING

The assessment of co-segregation patterns among several polymorphic MMs assessed for numerous individuals from a biparental cross is the basis for linkage or genetic mapping, in plants (Oraguzie and Wilcox, 2007). Statistical stepwise methods provide the foundation for creating a genetic map: first grouping of markers on linkage groups, then setting order of markers, and finally distance establishing between identified markers (Cheema and Dicks, 2009). The application of genetic maps is the identification of genomic regions linked to quantitative traits termed as QTL mapping. The generation of map-based cloning of critical genes influencing significant agronomic factors can be facilitated by genetic maps. Additional uses include helping with genome sequence assembly, deciphering the structure of the genome and traits, and comprehending the evolutionary links between and among species.

The first genetic map was constructed in 1913 using phenotyping markers in a segregating *Drosophila melanogaster* population (Singh and Singh, 2015b). In order to perform genetic linkage mapping, two phenotypically different parents with at least one characteristic of interest's must be crossed to create segregating mapping populations. (Scott et al., 2020).

Genetic mapping is closely allied with crop breeding and certainly, many populations bred for crops have had their genetic makeup documented (Allen, 2020; Rafalski, 2010). The elite parental crop plants are crossed to create numerous breeding populations, though, and these parents arc frequently closely related. Consequently, populations created by crossing parents that are most divergent at the genetic level are typically used to create reference maps, making it possible to uncover and characterize a large number of DNA markers at a low cost (Begna et al., 2021; Doelle et al., 2009).

Breeders find great benefit from highly saturated genetic linkage maps, which are also a prerequisite for many scientific applications, including potential gene discovery, mapping QTL, marker-trait relationships, developing MMs for MAS, and comparative genetic research. High-resolution genetic linkage map creation and QTL, which offer potent tools for genetic analysis of economic characteristics, are based on MMs (Rani et al., 2023). Genetic investigations regarding quantitative characteristics in important crops are presently the focus of research in the AM area. The 1980s saw the introduction of MM technology, which altered the nemesis of plant breeding (Nadeem et al., 2018).

Genetic mapping measures the relative distances between DNA markers along an organism's chromosomes using the Mendelian principles of recombination. Advances in genome editing and MAS plant breeding strategies have allowed for greater comprehension of the diverse pool of crops and a more thorough understanding of MMs (Bhatia et al., 2022; Parihar, 2022) (Table 2.3).

TABLE 2.3 Summarized List of Online Software/Tools for Molecular Maker-Based Genetic Linkage Mapping (Cheema and Dicks, 2009; Duran et al., 2010; Varshney et al., 2021)

S.No.	Name of Tool	Usage
1	LDAssoc	Interactively visualize association *p*-value results and linkage disequilibrium patterns for a genomic region of interest
2	LDexpress	Search for a list of variants in gene expression in multiple tissue types
3	LDhap	Depict haplotype frequencies among query variants
4	LDmatrix	Create an interactive heat map matrix of pairwise linkage disequilibrium statistics
5	LDpair	Inspect allelic correlation variants with high LD
6	LDpop	Investigate allele frequencies and linkage disequilibrium patterns across 1000G populations
7	LDproxy	Explore substituted and functional variants for query variant
8	LDtrait	Search if variants are being associated with any trait or disorder
9	SNPchip	Find commercial genotyping platforms for variants
10	SNPclip	Prune variants by LD
11	MAPmaker	This approach combines E–M algorithm for recombination estimation with HMM for calculating expected inter-marker recombination outcomes, for huge datasets
12	CarthaGe'ne	It is a hybrid mapping program that uses robust algorithms and a quick E–M method to find maximum likelihood consensus maps for numerous populations
13	Neighbor mapping (NM)	NM approach develops genetic map, employing a template of two-point distances
14	MapManager QTX	Carry out marker grouping and sorting, further refined using rippling for map construction
15	RECORD software	Construct complicated maps (500 markers/lg)
16	AntMap	AntMap uses a Travelling Salesmen Problem-based search algorithm to find the marker order, minimizing total adjacent recombination fractions

TABLE 2.3 *(Continued)*

S.No.	Name of Tool	Usage
17	MadMapper	It was developed for analyzing high-throughput MM datasets, specifically for recombinant inbred lines of sixth or higher generation
18	MSTmap	Uses notions from graph speculation and ML, swiftly estimates highly dense maps
19	THREaD Mapper	Calculate marker positions geometrically

Maps created using AFLPs, RAPDs, or ISSRs have finite transferability across populations and species since the sequence knowledge of each marker mostly defines it. Unless proved by hybridization experiments, bands with the same size amplified across populations or species may not always have the same sequence (Begna et al., 2021). In contrast, a better tool for finding genes or QTLs linked to economically significant characteristics is now available to researchers with the creation of high-density molecular maps constructed using modern MMs methods. Additionally, markers that have been mapped in one population can be utilized as primers or probes to characterize additional populations of the same species (Begna et al., 2021; N. Hasan et al., 2021).

Genetic maps with good genome coverage and confidence in locus order require a higher number of DNA markers, large numbers of individuals from a population. Mapping studies have stressed the need for multiple markers or mapping populations to boost map accuracy (Crane and Crane, 2005; Jeuken et al., 2001; Sharma et al., 2002b). The methods for detection and genetic analysis through widely used markers have become automated and much faster over some years (Guichoux et al., 2011). The ability to analyze many marker loci simultaneously using multiplex PCR is one example of how marker analysis has become more efficient. PCR products up to nine different primer pairs with nonoverlapping sorts of allele sizes can be multiplexed and run high-throughput DNA sequencing machines (Begna et al., 2021). New types of high-throughput marker systems, such as SNPs, play an important role in construction of high-density maps (Begna et al., 2021). SNP markers are anticipated to have significant influence on upcoming mapping research due to SNPs quantity and advancements in high-throughput SNP detection techniques (Koebner and Summers, 2003).

Numerous genetic maps of commercially important crops, such as rice, wheat, sunflower, pea, cotton, soybean, sorghum, cowpea, tobacco, and many more, have been created using DNA markers (Bhatia and Bajwa, 2022; Diaz et al., 2011). In constructing genetic maps, it is not always possible to have the identical map distance plus marker organization. With the advent of several high-throughput approaches for the physical mapping of chromosomes, such problems can be resolved. Generally, genome mapping is a widely applicable approach for probing the genetic data of any specified crop. The need to create high-yielding cultivars with greater resilience to biotic and abiotic stress has arisen because of the fast expansion in the global population.

A precise correlation between crop genetics and crop phenotype is required. Thus, by figuring out the sequence and distance between genetic characteristics related to genetic deviation or polymorphisms, genetic mapping may be utilized to pinpoint the attributes that are advantageous for raising production. Many important genes could be placed in genetic maps using MMs, which can subsequently be used as genetic markers themselves.

2.6.1.1 RFLP

RFLP marker is based on the distinct hybridization of probe DNA to DNA fragments in a sample of enzymatically digested DNA. The resulting DNA fragments are size separated on gel electrophoresis, transferred to nylon films (Southern blotting), and ultimately hybridized with labelled probe to envision DNA polymorphisms. These outcomes of RFLP are immense in number of segments with various lengths.

Plant breeding that incorporates the RFLP approach has the potential to:

1. Quicken the transfer of beneficial genes across cultivars.
2. Permit the introduction of new genes from closely similar wild species.
3. Provide the ability to analyze complicated polygenic features as collections of individual Mendelian variables; and
4. Determine the genetic connections among agricultural plants that are sexually incompatible.

The significance of RFLPs in plant breeding has been covered by Prince and Tanksley, who focused on labelling genes, map-assisted cloning, genetic distances and variation determination, and relative genome mapping (Prince and Tanksley, 1992). An RFLP map may be used to locate genes in relation to DNA markers, which serves as a foundation for cloning the genes via chromosomal walking. Three disease-resistant (*Pto, Mi,* and *Tm2a*) genes that have been identified on the tomato RFLP map are presently the targets of this methodology (Gebhardt, 2023).

2.6.1.2 RANDOM AMPLIFIED POLYMORPHIC DNA (RAPD)

One of the extremely prevalent MM in mapping of plants genome is RAPD, a method that has been widely used to study genetic relationships and GD within or between populations of various crops, varieties, marking, and polymorphism of species. In this method, PCR is commonly carried out with random primers of length 10–11 nucleotides. Use of RAPD method is especially percept plant quantitative traits attributes to make genetic map. The PCR amplified products are visualized through agarose gel electrophoresis. This reaction is carried out by using a single primer that anneals with the target DNA at forward and reverse positions as per Chargaff's rule. The use of RAPD is suitable for anonymous genomes through DNA profiling with limited quantities of sample and higher efficiency with low cost. The leading obstacle of RAPD is low reproducibility; hence, highly regulated experimental measures are required because of sensitivity and elevated reproducibility.

2.6.1.3 AMPLIFIED FRAGMENT LENGTH POLYMORPHISM (AFLP)

RFLP and RAPD are used to create the AFLP marker method. The process of digesting DNA is followed by PCR amplification. The AFLPs are highly polymorphic and mostly heritable, like RFLP. The method entails ligating two adapters, restriction-digesting DNA with two distinct enzymes, amplifying selected restriction fragments, and analyzing amplified fragments on a gel.

AFLP is a robust technique to analyze the bi-allelic data that makes AFLP well appropriate for high-throughput analysis. Moreover, fluorescent-labelled primers have distinct advantages for scoring automated fragments through computerized analysis.

Recent studies have revealed the purpose of AFLP markers for the evaluation of genetic variations in many plant species, like *Cymbopogon* species (Adhikari et al., 2022), *Ficus carica* (Baraket et al., 2009), Cumin

(*Cuminum cyminum L.*) (Archangi et al., 2019), and Wolfberry (*Lycium* spp.). However, more genetic information is required for the explication of GD. Recently, a combination of complementary DNA markers from two or more techniques has been adopted to uncover new genetic information to predict GD. The AFLP markers serve as a basis for improving genes and selective breeding of the genome assembly, and are also useful for exploring and improving a diverse set of germplasm. AFLP has been considered as a co-dominant marker for constructing genetic maps in many crop plants, including tea (Chang et al., 2017), wheatgrass, pear, and sugarcane (Andru et al., 2011).

2.6.1.4 SIMPLE SEQUENCE REPEAT (SSR) MARKER

SSRs are tandem repeats of 1–6 nucleotide base pairs that provide relatively comprehensive genomic coverage of various taxa. Appropriate primers (usually 22–27 bp) can be developed for amplifying the microsatellite regions through PCR if the flanking sequences of the microsatellites have been determined. By building a genomic library, screening it by using an artificially labelled oligonucleotide repeat, followed by sequencing the clones, flanking sequences of microsatellites can be determined. An alternative method of identifying microsatellites is to search microsatellite sequence databases, from which neighboring primers can be derived. Many agronomic traits show quantitative inheritance, and the genes governing these traits were quantified using QTL tools.

SSR, as codominant markers, exhibit good reproducibility; can identify homozygosity, stable amplification, and high polymorphism; and can provide advantages in analyzing results, constructing genetic maps of plants (Chen et al., 2014; Padmakar et al., 2015) and germplasm conservation (Luo et al., 2018).

The SSRs develop in the plant genome either by sliding of ssDNA, dsDNA recombination or transpositioning of transposable elements. Mainly, the sequences which are flanking the SSR regions are conserved and used in constructing primers. Recent studies on microsatellites have suggested the progress of linkage maps, GD analysis, and fingerprinting owing to their richness, great polymorphic nature, co-dominance and reproducibility in various crop plants, like sweet potato, in cashew, sesame (*Sesamum indicum* L.*)*, Moroccan saffron, and common bean (*Phaseolus vulgaris* L.*)*.

2.6.1.5 STS (SEQUENCE-TAGGED SITES) AND SCAR (SEQUENCE CHARACTERIZED AMPLIFIED REGION) MARKERS

A single small DNA sequence with known base-pair sequences and position within the genome is called an STS marker. Particular primers make it simple to identify the STS marker, which is helpful for creating a basic genomic map. A web-based database has been developed to enable searches for STS markers that can be used in agricultural systems using NGS technologies. The database of STS markers can be retrieved freely through NABIC (National Agricultural Biotechnology Information Centre) portal (http://nabic.rda.go.kr/Molecularmarker). Numerous useful functions, including comprehensive viewer, downloading function, and marker searches, are available for free with the NABIC database, with no license fee (Kim et al., 2015).

RAPD-derived SCAR markers are monologs that allow amplification of polymorphic DNA that represents the development of co-dominant markers from dominant markers, to prevent prolonged RAPD reactions (Li et al., 2010). SCAR markers are derived by cloning and sequencing the two ends of RAPD markers (e.g., a RAPD band present in disease-resistant lines but absent in susceptible lines), and longer primers are designed (22–24 bp) for specific amplification at a specific locus.

SCAR markers are effective in discriminating alleles when the desirable morphological and phenotypic characters are indistinguishable at initial developmental stage. SCAR markers have also been used in other crop improvement areas, like detection of disease-resistant loci in tomatoes, presence of lignotuber organs in *Eucalyptus*, etc.

Applications of SCAR markers are in gene mapping studies, MAS, and distinguishing between sexual forms of related species such as *C. papaya*, *Salix viminalis,* and *Gracilaria changii* (Alstrom-Rapaport, 1998).

2.6.1.6 SINGLE NUCLEOTIDE POLYMORPHISM (SNP)

SNPs are extremely common and can be found in either coding or noncoding DNA (intergenic region) with different frequencies distributed throughout the genome in various plant species. SNP marker is only a base change in the DNA sequence at which different nucleotides occur in different individuals of populations. SNPs are distributed in plant ranges between one SNP in every 100–300 bp. SNPs refer to sequence polymorphism at the genomic level, which results from single-nucleotide variation, including single-base deletion, insertion, transition, and transversion. They are an attractive tool for genetic research and molecular breeding studies, including mapping, MAS breeding, QTL mapping, comparative genomics, LD, and map-based cloning.

Sobrino and colleagues classified SNP genotyping assays into four groups based on molecular mechanism: allele-specific hybridization, invasive cleavage, oligonucleotide ligation, and primer extension (Sobrino et al., 2005).

Recently, Yang and co-workers have built a 2032 germplasm vegetables SNP database (VegSNPDB) for 16 vegetables (Yang et al., 2022). Chromosomal number, location, SNP type, individual, and heterozygosity values may all be used to search for SNPs. Since SNPs may adjust splicing options, affect transcription factor binding capability, change protein structure and function, and control noncoding RNA, they are strongly associated with phenotypic diversity and immunity to disease. (Guo et al., 2018).

2.6.2 MOLECULAR PHYSICAL MAPPING

Employing DNA markers, the physical mapping approach is used to determine the orientation along with physical distance among DNA bp (Brown, 2002). The physical map (PM) is highly accurate in identifying DNA bp sequence. Larger DNA segments are assembled from DNA fragments and DNA markers using PM. The overlapping areas of DNA fragments can be used by researchers to deduce the bp positions of DNA.

A few techniques for determining the gene's location include somatic cell hybridization, radiation hybrid (RH), and IH (Womack, 2020). There are various physical mapping methods available for studying genomes of various sizes and obtaining various degrees of precision. The three broadly used approaches to construct a PM are fluorescent *in situ* hybridization (FISH), restriction site mapping, and sequencing via clones (Jiang and Gill, 2006).

As a typical method used in genomic research, physical mapping aims to gather a whole genome sequence in order to infer any relationships within targeted DNA sequences with phenotypic features (Hasan et al., 2021). It is feasible to cure genetic disorders by prophylaxis or by creating new medicines if the exact gene locations that govern different phenotypes have been identified.

2.6.2.1 THREE BROAD CATEGORIES APPLY TO PHYSICAL MAPS

- **Chromosomal or cytogenetic maps:** When stained and seen under a microscope, a chromosome appears as a cytogenetic map. Different sections of bright and dark bands give each of the chromosomes a distinct look that is particularly significant. With the use of this characteristic, chromosomes can be examined in a clinical test called a karyotype, which enables researchers to check for chromosomal abnormalities.
- **Radiation hybrid (RH) maps:** This class of physical mapping is the most competent method for long-range gene mapping using both polymorphic and nonpolymorphic markers and has several advantages. Nonpolymorphic markers are informative, making them useful for anchoring clones that normally do not contain polymorphic markers. RH maps provide high map resolution for a relatively small number of hybrids.
- **Sequence maps:** Sequence mapping combines gene mapping with whole genome sequencing to increase identification of the mutated section of chromosome. This method allows the evaluation of a wide range of MMs in a particular plant population. More importantly, it depicts the different cross-sectional aspects, number of recombination, and the range (bp) of sequence necessary for successful map construction.

The resolution, or the capacity to measure the distance between closely spaced alleles, varies among various kinds of maps.

2.6.3 TECHNIQUES USED FOR PHYSICAL MAPPING

2.6.3.1 FISH

The FISH procedure makes it possible to locate a particular DNA sequence or a whole chromosome within a cell. It is applied to identify genetic disorders, GM, chromosomal abnormalities identification, and to study the comparison among the gene chromosomal arrangement of allied species. FISH involves DNA double helix unwinding and binding of fluorescent-labeled probes with target DNA, which can be visualized under a fluorescent microscope (Shakoori, 2017). A chromosome may be seen or visualized when a fluorescent tag-bound probe gets attached to it (Veselinyová et al., 2021).

Three distinct kinds of probes are used by scientists to carry out FISH, and each has a unique purpose:

1. **Locus specific probes (LSP)**: The LSP attaches themselves to a certain chromosomal region. When a small portion of a gene is isolated by researchers and wish to find out whether gene location on either chromosome or multiple copies of a gene are present or not in a given genome, they can employ this kind of probe (Breen et al., 2001).
2. **Centromeric repeat probes (CRP)**: Sequences that repeat in the center of chromosome is used to create CRPs. These probes are used by researchers to assess the appropriate number of chromosomes. Additionally, these probes can be used together with "LSPs" to ascertain whether a plant lacks any DNA segment from a certain chromosome (Shakoori, 2017).
3. **Whole chromosome probe (WCP):** The WCPs are sets of shorter probes, which attach to a distinct sequence across a chromosome's span. Researchers may identify each chromosome with distinct color by using several probes labelled with a variety of fluorescent dyes. A spectrum karyotype is a full-color map of a chromosome that is produced as a result. When analyzing chromosomal anomalies, WCPs are especially helpful (Liu et al., 2020).

2.6.3.2 GENOMIC IN SITU HYBRIDIZATION (GISH)

Genomic *in situ* hybridization (GISH) employs whole genomic DNA as a probe to identify the source of hybrid genomes or chromatin. This method was used to identify the barley chromosomes in wheat hybrids (Gupta and Vasistha, 2018). The method FISH, which was first used in molecular cytogenetics about 30 years ago, is modified into GISH (Khrustaleva et al., 2019). Ever since, GISH has developed into one of the most effective methods for analyzing naturally occurring polyploids, hybrid plants, and their offspring through backcrossing, GC, IGR, chromosomal integration, and recombination maps (Jiang and Gill, 2006). In interspecific plant hybrids, GISH can distinguish the parental genomes without knowing the sequence. The use of GISH was successful in characterizing the chromosomal makeup of tomato somatic hybrids (De Vries et al., 1992; Khrustaleva et al., 2019; Kudryavtseva et al., 2019).

2.6.3.3 RESTRICTION MAPPING (RM)

Restriction mapping, maps the unknown DNA segment by slicing it into pieces and then locating the breakpoints utilizing restriction enzymes is the basis of this technique (Prasad et al., 2017). This approach is frequently used to create a restriction map that entails three distinct digestions of the unknown DNA. The third piece of DNA sample is double-digested with two (previously used) restriction enzymes at the same time after two sections of the DNA sample have been separately digested with two restriction enzymes. The sizes of the DNA fragments in each digestion sample are then recorded after separation using gel electrophoresis. The overall length of each digested piece will be the same. However, because the locations of each individual DNA fragment's restriction sites determine its length, the lengths of the fragments may be used to map each restriction site. For accurate site mapping, the information from the double-digestion is especially helpful. A restriction map is the final representation of the DNA segment that depicts the locations of restriction sites (Kilian et al., 2012). Adding up the sizes of different fragments in each lane is the initial step, once electrophoresis is complete. Each digest's pieces should amount to the same size as the original fragment, and the size of each individual fragment should match that of the original fragment. If the sum of the fragment sizes is incorrect, two issues are probable. One instance could have resulted in some of the smaller particles slipping off the gel's end (if the gel is left on for too long, this commonly happens). Another probable cause of inaccuracy is that the gel was not thick enough to distinguish between fragments of similar sizes. Now, this results in the failure to separate the pieces that were close enough in size. Placing the restriction endonuclease sites on the original DNA fragment at locations that would meet the fragment sizes produced by all three digests would allow one to determine the position of the sites if all the digests result in fragments that add up (Kilian et al., 2012; Prasad et al., 2017).

2.6.3.4 CHROMOSOME WALKING

Chromosome walking procedure is used in modified cloning and sequencing methods in plants to clone a gene (of interest) from its acknowledged closest markers (Bhatia et al., 2015). It is used to locate, isolate, and clone a specific sequence that is present close to the gene that is required to be mapped. The majority of genomic initiatives use libraries of large pieces, primarily bacterial artificial chromosome libraries. The library is first screened with the necessary probe to find the desired colony and choose a specific clone. Following screening, the clone and probe are overlapped, and overlapping fragments are mapped (Bhatia et al., 2015; Reeve, 2014).

2.6.3.5 CHROMOSOME JUMPING

Chromosome jumping is a peculiar form of chromosomal walking in which the 5′ and 3′ ends of a DNA sequence are only cloned, without an intermediate DNA segment, by using the breakpoints of chromosomal rearrangements as guideposts. Chromosomal jumping is a technique for physically mapping organism genomes. This method was developed to get over a chromosomal walking hurdle that emerged when repeating DNA sections were discovered in the cloning process (Sharma et al., 2002a). As a result, the method of chromosome jumping may be thought of as a special case of chromosomal walking. When compared to chromosomal walking, it is a quicker process and allows for the bypassing of repetitive DNA segments that are less likely to be cloned. Chromosomal jumping reduces the distance between the target gene and the known markers that can be used to map the genome. The first step in creating a chromosome jumping tool is cutting a specific DNA using specific restriction endonucleases, followed by ligating the DNA pieces into circularized loops. The circularized loops are then sequenced using a primer created from an already known sequence. This primer makes it possible to jump and sequence in a different way. Because of this, it may skip through repeated DNA sequences and quickly navigate the chromosome in search of the target gene (Stewart Jr, 2016).

2.6.3.6 RADIATION HYBRID (RH) MAPPING

The most effective technique for creating long-range genomic maps employing both polymorphic and nonpolymorphic markers is RH mapping. A block of genes or chromosomal regions is said to be connected if they do not assort separately during segregation. This serves as the basis for creating a genomic map. Recombination between polymorphic loci and the possibility of recombination between two sites on a chromosome are often influenced by the physical proximity of these regions. Consequently, genetic linkage is observed as a measure of physical separation. The frequency of recombination, however, varies over the whole genome because it is controlled by chromosomal structure, which changes with chromosome length. RHs provide high resolution map using less than 100 hybrids. Along with its effectiveness, RH mapping also has the advantage of facilitating the integration of physical and genetic maps.

Although a suitable gene arrangement has a maximum number of hybrid markers. The computational approach made it easier by providing most hybrids with several chromosomal marker groups and providing an estimate of separation between the markers. Slonim and co-workers used RHMAPPER to arrange the markers and assess their distance (Slonim et al., 1996; Stein, 1995). RHMAPPER was designed to accommodate large numbers of markers and was used at the MIT Centre for Genome Research (MCRG) to build a whole genome map (Hudson et al., 1995). The construction of PMs depicts the real position of MMs and/or genes on the chromosome(s), and the actual distance in nucleotides between them has become more significant due to genomics research.

2.6.3.7 QUANTITATIVE TRAIT LOCI (QTL) MAPPING

Mapping of QTL is carried out among biparental and natural progenies populations in which relationships are unknown. The ability of QTL detection is affected by allele frequencies, the makeup, and significance of a mapping population. Biparental progenies mapping has been recognized to be helpful in breeding programs (Morrell et al., 2011). Without using data from the linkage map's genetic distances, single-marker analysis is performed for QTL mapping in biparental populations to identify QTL that correlate to the differences between the phenotypes of various genotypic groups. Quantitative characteristics can be mapped more precisely by analyzing numerous QTLs at once because it is widely known that they are often influenced by multiple loci. In developing QTL mapping populations, novel improved marker technologies are frequently employed (Rani et al., 2023).

For QTL mapping, just a connection between a genetic marker and a quantitative characteristic is required. For instance, if every single corn plant, regardless of its height, carries a certain allele of a genetic marker, there is a very high possibility that a QTL for tallness in the population of corn plants is associated with this allele (Dhingani et al., 2015). The genetic variation of a quantitative feature is determined by the interactions of several genes, known as QTL. For the discovery of QTLs of agronomic significance and their use in crop growth, it is important to map these QTLs using MMs within the genome of crop species. It is fundamental to create a linkage map utilizing genotyping information from any chosen mapping population before beginning any QTL research. A segregating mapping population can have both recombinant and parental genetic constitution. The rate of recombinant genotype recurrence must be known in order to calculate recombination frequencies, which are then used to calculate the genetic distance among markers.

Two frequently used mapping functions that convert recombination frequency into cM distance are the KMF, which assumes how recombination events impact the probability of closest recombination incidents, and the HMF, which presumes there is no disruption between crossover occurrences (Vinod, 2011; Kosambi, 2016). To determine the association between markers, researchers frequently employ odds ratio (i.e., probability of linkage against no linkage). This ratio can also be stated more easily as its logarithm, which is known as a logarithm of odds (LOD) value or LOD score (Gopinath, 2020; Risch, 1992). Linkage maps are frequently built with LOD values >3. To identify connection across a wider area or to add more markers to maps created at higher LOD levels, LOD values may be reduced (Collard et al., 2005; Ma et al., 2010). Linkage groups, which stand in for chromosomal segments or complete chromosomes, are collections of linked markers.

2.7 MARKER-ASSISTED SELECTION (MAS)

MAS is a method using DNA markers to pick out agriculturally significant traits for crop breeding. MAS technique is used to increase the efficacy of breeding program by selection of desired (traits) qualities. Numerous studies have been conducted on the application of MAS for various attributes and crop species. It is simple to employ MAS to manipulate basic/qualitative features, and it has been well documented. (Ashraf et al., 2012).

DNA-based MMs that are strongly linked to genes or QTLs in the desired crops that confer tolerance to different types of abiotic stressors and resistance to biotic stress factors have improved as a result of recent technological advancements. The procedure of MAS uses the technique of gene pyramiding to accelerate the progress of tolerant/resistant cultivars in a few feasible generations; it is possible to transfer many genes with prospective traits into a single genotype (Hasan et al., 2021)

The most cutting-edge way for transferring a desired gene to a chosen crop plant with the needed combination is now MM methods. A unique method to increase crop yield is the development of breeding programs utilizing the most widely used molecular approaches (Das et al., 2017).

For decades, MAS has been crucial in maximizing the potential of contemporary genomic resources, identifying, and using allelic diversity for crop and germplasm improvement. Future sustainability of genomics-assisted breeding (GAB) will depend on a number of novel strategies that expedite the focused manipulation of allelic variation for the generation of novel diversity and enable their quick and effective integration into crop improvement programs. Future crop design will need the use of GAB that optimize crop genomes by accumulating good alleles and eliminating harmful alleles (Varshney et al., 2021).

2.8 PROSPECTS MMS IN CROP IMPROVEMENT

In today's world, crop production and its improvement methods are affected a lot by the population explosion, changed and diversified climatic conditions; therefore, there is a diehard demand for novel crop design methods and sustainable utilization procedures. Staple food crops genome sequence information has opened newer pathways for discovery and utilization of desirable genetic marker information for plant breeders or farmers. Conventional plant breeders are always in search of potential selection criteria for their breeding programs, aiming at improving overall quality or stress (biotic and abiotic) tolerance of crops. With the advances and new discoveries in the MM technology and plant genomics, metabolomics, nearby decades, have observed a remarkable development in novel techniques for crop improvement. Recently, the discovery of CRISPR-Cas9 has opened up a new door toward genome editing and crop improvement procedures. CRISPR-Cas9 mediated genome editing enables the plant biotechnologist to alter plant yield, survival pathways, and improve pest and stress resistance. These new techniques are highly functional in the determination of genetic and functional genomics, especially in the development of crop improvement and breeding technologies. The availability of different cost-effective genomic markers as discussed here (RFLP, RAPD, SNPs, etc.) and more effective agronomical development are expected to speed up and improve the production of complex traits of different crops in the coming years.

KEYWORDS

- **crop improvement**
- **marker**
- **linkage**
- **plant breeding**
- **polymorphism**

REFERENCES

Adhikari, S.; Biswas, A.; Saha, S.; Bandyopadhyay, T.K.; Ghosh, P. AFLP-based assessment of genetic variation in certain Indian elite cultivars of *Cymbopogon species*. *J. Appl. Res. Med. Aromatic Plants*. 2022, 29, 100372.

Adhikari, S.; Saha, S.; Biswas, A.; Rana, T.S.; Bandyopadhyay, T.K.; Ghosh, P. Application of molecular markers in plant genome analysis: A review. *The Nucleus* 2017, 60(3), 283–297.

Allen, F.L. Usefulness of plant genome mapping to plant breeding. In: *Plant Genome Analysis*. CRC Press, 2020, pp. 11–18.

Alstrom-Rapaport, C. Identification of a RAPD marker linked to sex determination in the basket willow (Salix viminalis L.). *J. Heredity*. 1998, 1, 44–49.

Amiteye, S. Basic concepts and methodologies of DNA marker systems in plant molecular breeding. *Heliyon* 2021, 7(10), e08093.

Andru, S.; Pan, Y.B.; Thongthawee, S.; Burner, D.M.; Kimbeng, C.A. Genetic analysis of the sugarcane (*Saccharum spp.*) cultivar 'LCP 85–384'. I. Linkage mapping using AFLP, SSR, and TRAP markers. *Theo. Appl. Gen*. 2011, 1, 77–93.

Arafa, R.A.; Prohens, J.; Solberg, S.Ø.; Plazas, M.; Rakh, M. Breeding and genome mapping for resistance to biotic stress in eggplant. In: *Genomic Designing for Biotic Stress Resistant Vegetable Crops*. Springer: Cham, 2022, p. 147–187.

Archangi, A.; Heidari, B.; Mohammadi-Nejad, G. Association between seed yield-related traits and cDNA-AFLP markers in cumin (*Cuminum cyminum*) under drought and irrigation regimes. *Indus. Crop. Prod.* 2019, 133, 276–283.

Ashraf, M.; Akram, N.A.; Foolad, M.R. Marker-assisted selection in plant breeding for salinity tolerance. In: *Plant Salt Tolerance*. Humana Press: Totowa, NJ, 2012, pp. 305–333.

Avvaru, A.K.; Sharma, D.; Verma, A.; Mishra, R.K.; Sowpati, D.T. MSDB: A comprehensive, annotated database of microsatellites. *Nucleic Acids Res.* 2020, 48(D1), D155-D159.

Bailey-Wilson, J.E.; Wilson, A.F. Linkage analysis in the next-generation sequencing era. *Human Heredity*, 2011, 72(4), 228–236.

Bangarwa, S.K.; Balwan, K.L.S.; Choudhary, M.; Choudhary, M.K. An introduction to DNA-markers and their role in crop improvement. *Pharma J.* 2021, 10(7), 638–643.

Baraket, G.; Chatti, K.; Saddoud, O.; Mars, M.; Marrakchi, M.; Trifi, M.; Salhi-Hannachi, A. Genetic analysis of Tunisian fig (*Ficus carica L.*) cultivars using amplified fragment length polymorphism (AFLP) markers. *Sci. Horti.* 2009, 4, 487–492.

Begna, T.; Yesuf, H.; Abdurezake, M.; Eshetu, G. Genetic mapping in crop plants. *Open J. Plant Sci.* 2021, 6(1), 019–026.

Bhatia, D.; Bajwa, G.S. Molecular marker techniques and recent advancements. In: *Genotyping by Sequencing for Crop Improvement,* John Wiley and Sons: Hoboken, NJ, USA, 2022, pp. 1–21.

Bhatia, S.; Sharma, K.; Dahiya, R.; Bera, T. *Modern Applications of Plant Biotechnology in Pharmaceutical Sciences*. Academic Press: Cambridge, MA, USA, Elsevier, 2015, pp. 31–107, London, UK.

Bishop, D.V. The role of genes in the etiology of specific language impairment. *J. Commun. Disord.* 2002, 35(4), 311–328.

Boopathi, N.M. Marker-assisted selection (MAS). In: *Genetic Mapping and Marker Assisted Selection.* Springer: Singapore, 2020, pp. 343–388.

Breen, M.; Jouquand, S.; Renier, C.; Mellersh, C.S.; Hitte, C.; Holmes, N.G.; Galibert, F. Chromosome-specific single-locus FISH probes allow anchorage of an 1800-marker integrated radiation-hybrid/linkage map of the domestic dog genome to all chromosomes. *Genome Res.* 2001, 11(10), 1784–1795.

Brown, T.A. Mapping genomes. 2nd edn., In: *Genomes.* Wiley-Liss, 2002.

Chang, Y.; Oh, E.U.; Lee, M.S.; Kim, H.B.; Moon, D.G.; Song, K.J. Construction of a genetic linkage map based on RAPD, AFLP, and SSR markers for tea plant (*Camellia sinensis*). *Euphytica*, 2017, 8.

Cheema, J.; Dicks, J. Computational approaches and software tools for genetic linkage map estimation in plants. *Brief. Bioinform.* 2009, 10(6), 595–608.

Chen, H.; Song, Y.; Li, L.T.; Khan, M.A.; Li, X.G.; Korban, S.S.; Wu, J.; Zhang, S.L.Construction of a high-density simple sequence repeat consensus genetic map for pear (*Pyrus spp.*). *Plant Mol. Biol. Rep.* 2014, 2, 316–325.

Chukwu, S.C.; Rafii, M.Y.; Ramlee, S.I.; Ismail, S.I.; Oladosu, Y.; Okporie, E.; Jalloh, M. Marker-assisted selection and gene pyramiding for resistance to bacterial leaf blight disease of rice (*Oryza sativa L.*). *Biotechnol. Biotechnol. Equip.* 2019, 33(1), 440–455.

Collard, B.C.Y.; Grams, R.A.; Bovill, W.D.; Percy, C.D.; Jolley, R.; Lehmensiek, A.; Sutherland, M.W. Development of molecular markers for crown rot resistance in wheat: Mapping of QTLs for seedling resistance in a '2-49'×'Janz'population. *Plant Breed.* 2005, 124(6), 532–537.

Crane, C.F.; Crane, Y.M. A nearest-neighboring-end algorithm for genetic mapping. *Bioinfo*. 2005, 21(8), 1579–1591.

Cropano, C.; Place, I.; Manzanares, C.; Do, J.; Canto, T.L.; Studer, B.; Thorogood, D. Characterization and practical use of self-compatibility in outcrossing grass species. *Ann. Bot.* 2021, 127(7), 841–852.

Darkwa, K.; Olasanmi, B.; Asiedu, R. Asfaw, A. Review of empirical and emerging breeding methods and tools for yam (*Dioscorea spp.*) improvement: Status and prospects. *Plant Breed.* 2020, 139(3), 474–497.

Das, G.; Patra, J.K.; Baek, K.H. Insight into MAS: A molecular tool for development of stress resistant and quality of rice through gene stacking. *Front. Plant Sci.* 2017, 8, 985.

De Vries, J.N.; Wietsma, W.A.; De Vries, T. Introgression of leaf blight resistance from *Allium roylei* Stearn into onion (A. cepa L.). *Euphytica.* 1992, 62(2), 127–133.

DeWoody, J.A.; Harder, A.M.; Mathur, S.; Willoughby, J.R. The long-standing significance of genetic diversity in conservation. *Mol. Ecol.* 2021, 30(17), 4147–4154.

Dheer, P.; Rautela, I.; Sharma, V.; Dhiman, M.; Sharma, A.; Sharma, N.; Sharma, M.D. Evolution in crop improvement approaches and future prospects of molecular markers to CRISPR/Cas9 system. *Gene.* 2020, 753, 144795–144812.

Dhingani, R.M.; Umrania, V.V.; Tomar, R.S.; Parakhia, M.V.; Golakiya, B. Introduction to QTL mapping in plants. *Ann. Plant Sci.* 2015, 4(04), 1072–1079.

Diaz, A.; Fergany, M.; Formisano, G.; Ziarsolo, P.; Blanca, J.; Fei, Z.; Monforte, A.J. A consensus linkage map for molecular markers and Quantitative Trait Loci associated with economically important traits in melon (*Cucumis melo* L.). *BMC Plant Biol.* 2011, 11(1), 1–14.

Doelle, H.W.; Rokem, J.S.; Berovic, M. *Biotechnology-Volume VIII: Fundamentals in Biotechnology.* EOLSS Publications, 2009, Vol. 8.

Duran, C.; Eales, D.; Marshall, D.; Imelfort, M.; Stiller, J.; Berkman, P.J.; Batley, J.J.G. Future tools for association mapping in crop plants. *Genome.* 2010, 53(11), 1017–1023.

Eduardo, I.; Picañol, R.; Rojas, E.; Batlle, I.; Howad, W.; Aranzana, M.J.; Arús, P.J.E. Mapping of a major gene for the slow ripening character in peach: Co-location with the maturity date gene and development of a candidate gene-based diagnostic marker for its selection. *Euphytica.* 2015, 205(2), 627–636.

Fjellstrom, R.; Conaway-Bormans, C.A.; McClung, A.M.; Marchetti, M.A.; Shank, A.R.; Park, W.D. Genomics, molecular genetics and biotechnology. *Crop Sci.* 2004, 44, 1790–1798.

Gaitán-Espitia, J.D.; Hobday, A.J. Evolutionary principles and genetic considerations for guiding conservation interventions under climate change. Glob. *Change Biol.* 2021, 27(3), 475–488.

Gallavotti, A.; Whipple, C.J. Positional cloning in maize (*Zea mays* subsp. mays, Poaceae). *Appl. Plant Sci.* 2015, 3(1), 1400092.

Gautam, A. Genome mapping. In: *DNA and RNA Isolation Techniques for Non-Experts.* Springer Nature, 2022, pp. 171–178.

Gebhardt, C. (2023). A physical map of traits of agronomic importance based on potato and tomato genome sequences. *Front. Genetics*, 14, 1197206.

Ghafoor, A.; McPhee, K.J. Marker assisted selection (MAS) for developing powdery mildew resistant pea cultivars. *Euphytica*. 2012, 186(3), 593–607.

Gopinath, I. QTL mapping for cold tolerance with specific reference to rice. *J. Pharmacogn. Phytochem.* 2020, 9(3), 146–153.

Guichoux, E.; Lagache, L.; Wagner, S.; Chaumeil, P.; Léger, P.; Lepais, O.; Salin, F.J. Current trends in microsatellite genotyping. *Mol. Ecol. Res*. 2011, 11(4), 591–611.

Guo, L.; Wang, X.; Zhao, M.; Huang, C.; Li, C.; Li, D.; Yang, C.J.; York, A.M.; Xue, W.; Xu, G.; Liang, Y.; Chen, Q.; Doebley, J.F.; Tian, F. Stepwise cis-Regulatory changes in ZCN8 contribute to maize flowering-time adaptation. *Curr. Biol*. 2018, 18, 3005–3015

Gupta, P.; Rustgi, S.; Sharma, S.; Singh, R.; Kumar, N.; Balyan, H. Transferable EST-SSR markers for the study of polymorphism and genetic diversity in bread wheat. *Mol. Genet. Genom.* 2003, 270(4), 315–323.

Gupta, P.K.; Balyan, H.S.; Varshney, R.K.; Gill, K.S. Development and use of molecular markers for crop improvement. *Plant Breed.* 2013, 132, 431–432.

Gupta, P.K.; Varshney, R.K.; Prasad, M. Molecular markers: Principles and methodology. In: *Molecular Techniques in Crop Improvement.* Springer: Dordrecht, 2002, pp. 9–54.

Gupta, P.K.; Vasistha, N.K. Wheat cytogenetics and cytogenomics: The present status. *Nucleus*. 2018, 61(3), 195–212.

Hajiaqatabar, A.; Kiani, G.; Kazemitabar, S.; Alavi, M. Marker aided selection (MAS) for qualitative traits in segregating F2 populations of rice. *Genetika. J.* 2019, 51(3), 1031–1038.

Hasan, N.; Choudhary, S.; Naaz, N.; Sharma, N.; Laskar, R.A. Recent advancements in molecular marker-assisted selection and applications in plant breeding programmes. *J. Genet. Eng. Biotechnol.* 2021, 19(1), 1–26.

Hasan, N.A.; Rafu, M.Y.; Rahim, H.A.; Ali, N.S.; Mazlan, N.; Abdullah, S. Introgression of Pi-kh resistance gene into a malaysian cultivar, MR264 using marker-assisted backcrossing (MABC). *Int. J. Agric. Biol.* 2015, 17(6), 1–12.

Hospital, F.J.G. (2009). Challenges for effective marker-assisted selection in plants. *Genetica*. 136(2), 303–310.

Hudson, T.J.; Stein, L.D.; Gerety, S.S.; Ma, J.; Castle, A.B.; Silva, J.; Xu, S.H. An STS-based map of the human genome. *Genetica.* 1995, 270(5244), 1945–1954.

Jaganathan, D.; Bohra, A.; Thudi, M.; Varshney, R.K. Fine mapping and gene cloning in the post-NGS era: Advances and prospects. *Theor. Appl. Genet.* 2020, 133(5), 1791–1810.

Jasielczuk, I.; Gurgul, A.; Szmatoła, T.; Semik-Gurgul, E.; Pawlina-Tyszko, K.; Szyndler-Nędza, M.; Bugno-Poniewierska, M. Comparison of linkage disequilibrium, effective population size and haplotype blocks in Polish Landrace and Polish native pig populations. *Livest. Sci.* 2020, 231, 103887.

Jeuken, M.; Van Wijk, R.; Peleman, J.; Lindhout, P. An integrated interspecific AFLP map of lettuce (Lactuca) based on two L. sativa× L. saligna F2 populations. *Theor. Appl. Genet.* 2001, 103(4), 638–647.

Jiang, G.L. Molecular marker-assisted breeding: A plant breeder's review. In: Al-Khayri, J.; Jain, S.; Johns, D.; (Eds.), *Advances in Plant Breeding Strategies: Breeding, Biotechnology and Molecular Tools*. Springer: Cham, Switzerland, 2015.

Jiang, J.; Gill, B.S. Current status and the future of fluorescence in situ hybridization (FISH) in plant genome research. *Genome*. 2006, 49(9), 1057–1068.

Kalia, R.K.; Rai, M.K.; Kalia, S.; Singh, R.; Dhawan, A. Microsatellite markers: An overview of the recent progress in plants. *Euphytica*. 2011, 177(3), 309–334.

Kesawat, M.S.; Das Kumar, B.J. Molecular markers: It's application in crop improvement. *J. Crop. Sci. Biotechnol.* 2009, 12(4), 169–181.

Khrustaleva, L.; Mardini, M.; Kudryavtseva, N.; Alizhanova, R.; Romanov, D.; Sokolov, P.; Monakhos, G. The power of genomic in situ hybridization (GISH) in interspecific breeding of bulb onion (*Allium cepa* L.) resistant to downy mildew (Peronospora destructor [Berk.] Casp.). *Plant. J.* 2019, 8(2), 36–48.

Kilian, A.; Wenzl, P.; Huttner, E.; Carling, J.; Xia, L.; Blois, H.; Uszynski, G. Diversity arrays technology: A generic genome profiling technology on open platforms. In: *Data Production and Analysis in Population Genomics*. Humana Press: Totowa, New Jersey, 2012, pp. 67–89.

Kim, B.; Hwang, I.S.; Lee, H.J.; Lee, J.M.; Seo, E.; Choi, D. Identification of a molecular marker tightly linked to bacterial wilt resistance in tomato by genome-wide SNP analysis. *Theor. Appl. Genet.* 2018, 131(5), 1017–1030.

Kim, C.K.; Lee, G.S.; Mo, J.S.; Bae, S.H.; Lee, T.H. Molecular marker database for efficient use in agricultural breeding programs. *Bioinfo*. 2015, 9, 444–446.

Koebner, R.M.; Summers, R.W. 21st century wheat breeding: Plot selection or plate detection? *Trend. Biotechnol.* 2003, 21(2), 59–63.

Kosambi, D.D. *The Estimation of Map Distances from Recombination Values.* Ch. 15, In: Ramaswamy, R.; (Eds.), Springer India: New Delhi, 2016, pp. 125–131.

Kudryavtseva, N.; Havey, M.J.; Black, L.; Hanson, P.; Sokolov, P.; Odintsov, S.; Khrustaleva, L. Cytological evaluations of advanced generations of interspecific hybrids between *Allium cepa* and *Allium fistulosum* showing resistance to *Stemphylium vesicarium*. *Gene. J.* 2019, 10(3), 195–210.

Kulwal, P.L.; Singh, R. (2022). Biparental crossing and QTL mapping for validation of genome-wide association studies. *Methods Mol. Biol.* 2022, 2481, 273–285.

Kumar, P.; Gupta, V.; Misra, A.; Modi, D.; Pandey, B. Potential of molecular markers in plant biotechnology. *Plant. Omics* 2009, 2(4), 141–162.

Kumar, S.; Hilario, E.; Deng, C.H.; Molloy, C. Turbocharging introgression breeding of perennial fruit crops: A case study on apple. *Hortic. Res.* 2020, 7, 1–15.

Kumar, S.; Kumar, M.; Mir, R.R.; Kumar, R.; Kumar, S. Advances in molecular markers and their use in genetic improvement of wheat. In: *Physiological, Molecular, and Genetic Perspectives of Wheat Improvement.* Springer: Cham, 2021, pp. 139–174.

Leng, P.F.; Lübberstedt, T.; Xu, M.L. Genomics-assisted breeding–a revolutionary strategy for crop improvement. *J. Integr. Agric.* 2017, 16(12), 2674–2685.

Li, S.F.; Tang, S.J.; Cai, W.Q. RAPD-SCAR markers for genetically improved NEW GIFT Nile Tilapia (Oreochromis niloticus niloticus L.) and their application in strain identification. *Zoological Res*. 2010, 2, 147–154.

Liu, X.; Sun, S.; Wu, Y.; Zhou, Y.; Gu, S.; Yu, H.; Liu, B.J. Dual-color oligo-FISH can reveal chromosomal variations and evolution in *Oryza* species. *Plant J.* 2020, 01(1), 112–121.

Luo, C.; Chen, D.; Cheng, X.; Liu, H.; Li, Y.; Huang, C. SSR analysis of genetic relationship and classification in chrysanthemum germplasm collection. *Horti. Plant J.* 2018, 2, 73–82.

Ma, J.; Li, H.; Zhang, C.; Yang, X.; Liu, Y.; Yan, G. Identification and validation of a major QTL conferring crown rot resistance in hexaploid wheat. *Theor. Appl. Genet.* 2010, 120(6), 1119–1128.

Mackay, T.F. Quantitative trait loci in Drosophila. *Nat. Rev. Genet.* 2001, 2(1), 11–20.

Mao, Y.; Zhang, H.; Xu, N.; Zhang, B.; Gou, F.; Zhu, J.K. Application of the CRISPR–Cas system for efficient genome engineering in plants. *Mol. Plant.* 2013, 6(6), 2008–2011.

Matias, F.I.; Galli, G.; Correia Granato, I.S.; Fritsche-Neto, R. Genomic prediction of autogamous and allogamous plants by SNPs and haplotypes. *Crop Sci.* 2017, 57(6), 2951–2958.

Maurye, P.; Basu, A.; Gupta, A.J. Simple and cost effective apparatus for silver staining of polyacrylamide gels with sequential reagents addition and real time monitoring. *J. Biosci. Bioeng.* 2014, 117(6), 769–774.

McGregor, C.; Lambert, C.; Greyling, M.; Louw, J.; Warnich, L. A comparative assessment of DNA fingerprinting techniques (RAPD, ISSR, AFLP and SSR) in tetraploid potato (*Solanum tuberosum L.*) germplasm. *Euphytica.* 2000, 113(2), 135–144.

Miah, G.; Rafii, M.Y.; Ismail, M.R.; Puteh, A.B.; Rahim, H.A.; Islam, K.N.; Latif, M.A. A review of microsatellite markers and their applications in rice breeding programs to improve blast disease resistance. *Int. J. Mol. Sci.* 2013, 14(11), 22499–22528.

Molfino, N.A. Genetics of COPD. *Chest.* 2004, 125(5), 29–40.

Morrell, P.L.; Buckler, E.S.; Ross-Ibarra, J. Crop genomics: Advances and applications. *Nat. Rev. Gen.* 2011, 2, 85–96.

Musunuru, K.J.T. Moving toward genome-editing therapies for cardiovascular diseases. *J. Clin. Investig.* 2022, 132(1), 1–15.

Nadeem, M.A.; Nawaz, M.A.; Shahid, M.Q.; Doğan, Y.; Comertpay, G.; Yıldız, M. DNA molecular markers in plant breeding: Current status and recent advancements in genomic selection and genome editing. *Biotechnol. Biotechnol. Equip.* 2018, 32(2), 261–285.

Nag, A. Understanding the environment and sustainability with molecular approaches. In: *Biotechnological Innovations for Environmental Bioremediation.* Springer: Singapore, 2022, pp. 895–914.

Oraguzie, N.C.; Wilcox, P.L. An overview of association mapping. *Journal Association Mapping in Plants,* 2007, pp. 1–9.

Ouyang, W.; Chen, L.; Ma, J.; Liu, X.; Chen, H.; Yang, H.; Chen, S. Identification of quantitative trait locus and candidate genes for drought tolerance in a soybean recombinant inbred line population. *Int. J. Mol. Sci.* 2022, 23(18), 108–128.

Padmakar, B.; Kanupriya, C.; Latha, P.M.; Prashant, K.S.; Dinesh, M.R.; Sailaja, D.; Aswath, C. Development of SRAP and SSR marker-based genetic linkage maps of guava (*Psidium guajava L.*). *Scientia Hort.* 2015, 158–165.

Parihar, A. Molecular breeding and marker-assisted selection for crop improvement. In: *Plant Genomics for Sustainable Agriculture.* Springer: Cham, 2022, pp. 129–164.

Pasquali, E.; Palumbo, F.; Barcaccia, G. Assessment of the genetic distinctiveness and uniformity of pre-basic seed stocks of Italian ryegrass varieties. *Gene.* 2022, 13(11), 2097.

Phan, H.T.; Furuki, E.; Hunziker, L.; Rybak, K.; Tan, K.C. GWAS analysis reveals distinct pathogenicity profiles of Australian Parastagonospora nodorum isolates and identification of marker-trait-associations to *septoria nodorum* blotch. *Sci. Rep.* 2021, 11(1), 1–14.

Prasad, B.D.; Sahni, S.; Kumar, P.; Siddiqui, M.W. *Plant Biotechnology.* In: *Principles, Techniques, and Applications*, CRC Press: London, 2017, pp. 45–95, Vol. 1.

Prince, J.P.; Tanksley, S.D. (1992). Restriction fragment length polymorphisms in plant breeding and genetics. *Proceedings of the Royal Society of Edinburgh, Section B: Biol. Sci.* 99(3–4), 23–29.

Pyeritz, R.E.; Korf, B.R.; Grody, W.W. *Emery and Rimoin's Principles and Practice of Medical Genetics and Genomics. Foundations*; Academic Press: Cambridge, MA, USA; 2018.

Quezada, M.; Amadeu, R.R.; Vignale, B.; Cabrera, D.; Pritsch, C.; Garcia, A.A. Construction of a high-density genetic map of *Acca sellowiana* (Berg.) Burret, an outcrossing species, based on two connected mapping populations. *Front. Plant Sci.* 2021, 12, 626811.

Rafalski, J.A. Association genetics in crop improvement. *Curr. Opin. Plant Biol.* 2010, 13(2), 174–180.

Ragimekula, N.; Varadarajula, N.N.; Mallapuram, S.P.; Gangimeni, G.; Reddy, R.K. Kondreddy, H.R. Marker assisted selection in disease resistance breeding. *J. Plant Breed.Gen.* 2013, 1(2), 90–109.

Rahman, M.; Hoque, A.; Roy, J. Linkage disequilibrium and population structure in a core collection of Brassica napus (L.). *PLoS One.* 2022, 17(3), e0250310.

Ramakrishnan, A.P.; Ritland, C.E.; Blas Sevillano, R.H.; Riseman, A. Review of potato molecular markers to enhance trait selection. *Am. J. Potato Res.* 2015, 92(4), 455–472.

Rani, K.; Kumar, M.; Razzaq, A.; Ajay, B.; Kona, P.; Bera, S.K.; Wani, S.H. Recent advances in molecular marker technology for QTL mapping in plants. QTL Mapp. *Crop Improve.* 2023, 1–15.

Reeve, E.C. *Encyclopedia of Genetics*. Routledge: New York, NY, USA, 2014.

Risch, N. Genetic linkage: Interpreting LOD scores. Science 1992, 255(5046), 803–804.

Saha, M.C.; Hopkins, A.A.; Wang, Z.Y. Genomic tools for improvement. *Tall Fescue for the Twenty-first Century* 2009, 53, 377–394.

Scott, M.F. Ladejobi, O.; Amer, S. Bentley, A.R.; Biernaskie, J.; Boden, S.A.; Filippi, C.V. Multi-parent populations in crops: A toolbox integrating genomics and genetic mapping with breeding. *Heredity J.* 2020, 125(6), 396–416.

Shahabzadeh, Z.; Mohammadi, R.; Darvishzadeh, R.; Jaffari, M. Genetic structure and diversity analysis of tall fescue populations by EST-SSR and ISSR markers. *Mol. Biol. Rep.* 2020, 47(1), 655–669.

Shakoori, A.R. Fluorescence in situ hybridization (FISH) and its applications. In: *Chromosome Structure and Aberrations*. Springer: New Delhi, 2017, pp. 343–367.

Sharanabasappa, B.Y.; Kumari, V.; Sharma, R.; Punia, S.S. Quantitative Trait Locus (QTL) mapping in crop improvement. In: *Biotechnology and Crop Improvement*. CRC Press, 2022, pp. 227–236.

Sharma, H.; Crouch, J.; Sharma, K.; Seetharama, N.; Hash, C. Applications of biotechnology for crop improvement: Prospects and constraints. *Plant Sci.* 2002a, 163(3), 381-395.

Sharma, R.; Aggarwal, R.A.; Kumar, R.; Mohapatra, T.; Sharma, R. Construction of an RAPD linkage map and localization of QTLs for oleic acid level using recombinant inbreds in mustard (*Brassica juncea*). *Genome.* 2002b, 45(3), 467–472.

Shimizu, T.; Aka Kacar, Y.; Cristofani-Yaly, M.; Curtolo, M.; Machado, M.A. Markers, maps, and marker-assisted selection. In: *The Citrus Genome*. Springer: Cham, 2020, pp. 107–139.

Shipilina, D.; Näsvall, K.; Höök, L.; Vila, R.; Talavera, G.; Backström, N. Linkage mapping and genome annotation give novel insights into gene family expansions and regional recombination rate variation in the painted lady (*Vanessa cardui*) butterfly. *Genomic J.* 2022, 114(6), 110481.

Singh, B.; Singh, A.K. *Marker-Assisted Plant Breeding: Principles and Practices,* 1st edn., Springer India: Delhi, India, 2015a; ISBN 978-81-322-2316-0.

Singh, B.; Singh, A. Linkage mapping of molecular markers and oligogenes. In: *Marker-Assisted Plant Breeding: Principles and Practices*. Springer: New Delhi, 2015b, pp. 151–183.

Singh, H.P.; Raigar, O.P.; Chahota, R.K. Estimation of genetic diversity and its exploitation in plant breeding. *J. Bot. Rev.* 2021, 1–23.

Singh, S.P.; Nigam, A.; Singh, R.K. Significant role of molecular markers in sugarcane improvement. *Agrica. J.* 2015, 4(2), 67–78.

Slatkin, M. Linkage disequilibrium—Understanding the evolutionary past and mapping the medical future. *Nat. Rev. Genet.* 2008, 9(6), 477–485.

Slonim, D.; Stein, L.; Kruglyak, L.; Lander, E. RHMAPPER: An interactive computer package for constructing radiation hybrids maps. *Comput. Biol.* 1996, 26–42.

Sobrino, B.; Brión, M.; Carracedo, A. (2005). SNPs in forensic genetics: A review on SNP typing methodologies. *Forensic Sci. Int. 154(2–3*), 181–194.

Song, Z.; Li, X. Recent advances in molecular marker-assisted breeding for quality improvement of traditional Chinese medicine. *Curr. Pharm. Biotechnol.* 2021, 22(6), 867–875.

Stein, L. RHMAPPER. unpublished software, Whitehead Institute/MIT Center for Genome Research, 1995. http://www. genome. wi. mit. edu/ftp/pub/software/rhmapper/.

Stewart Jr, Neal, C.; (Eds.) *Plant Biotechnology and Genetics: Principles, Techniques, and Applications.* John Wiley and Sons, 2016.

Varshney, R.K.; Bohra, A.; Yu, J.; Graner, A.; Zhang, Q.; Sorrells, M.E. Designing future crops: Genomics-assisted breeding comes of age. *Trend. Plant Sci.* 2021, 26(6), 631–649.

Veselinyová, D.; Mašlanková, J.; Kalinová, K.; Mičková, H.; Mareková, M.; Rabajdová, M. Selected in situ hybridization methods: Principles and application. *J. Mol.* 2021, 26(13), 3874.

Vieira, M.L.C.; Santini, L.; Diniz, A.L.; Munhoz, C.D.F. Microsatellite markers: What they mean and why they are so useful. *Genet. Mol. Biol.* 2016, 39, 312–328.

Vinod, K. Kosambi and the genetic mapping function. *Reson. J.* 2011, 16(6), 540–550.

Ward, J.A.; Bhangoo, J.; Fernández-Fernández, F.; Moore, P.; Swanson, J.; Viola, R.; Sargent, D.J. Saturated linkage map construction in *Rubus idaeus* using genotyping by sequencing and genome-independent imputation. *BMC Genet.* 2013, 14(1), 1–14.

Wang, Y.; Jia, H.M.; Shen, Y.T.; Zhao, H.B.; Yang, Q.S.; Zhu, C.Q.; Jiao, Y. Construction of an anchoring SSR marker genetic linkage map and detection of a sex-linked region in two dioecious populations of red bayberry. *Hortic. Res.* 2020, 7–21.

Williams, K.L. Gene mapping. In: *Encyclopedia of Bioinformatics and Computational Biology*. Academic Press, 2019, pp. 242–250.

Womack, J.E. Strategies and technologies for comparative gene mapping. In: *Gene-Mapping Techniques and Applications*. CRC Press, 2020, pp. 3–20.

Wu, X.; Jiang, W.; Fragoso, C.; Huang, J.; Zhou, G.; Zhao, H.; Dellaporta, S. Causal haplotype block identification in plant genome-wide association studies. *bioRxiv.* 2021, 1–20.

Wu, X.; Jiang, W.; Fragoso, C.; Huang, J.; Zhou, G.; Zhao, H.; Dellaporta, S. Prioritized candidate causal haplotype blocks in plant genome-wide association studies. *PLoS Genet.* 2022, 18(10), e1010437.

Xiao, Y.; Liu, H.; Wu, L.; Warburton, M.; Yan, J. Genome-wide association studies in maize: Praise and stargaze. *J. Mol. Plant.* 2017, 10(3), 359–374.

Yang, J.; Zhang, J.; Du, H.; Zhao, H.; Li, H.; Xu, Y.; Wen, C. The vegetable SNP database: An integrated resource for plant breeders and scientists. *Genomics* 2022, 114(3), 110348.

Yoshida, G.M.; Barria, A.; Correa, K.; Cáceres, G.; Jedlicki, A.; Cadiz, M.I.; Yáñez, J.M. Genome-wide patterns of population structure and linkage disequilibrium in farmed Nile tilapia (*Oreochromis niloticus*). *Front. Genet.* 2019, 10, 745–767.

Yu, X.; Lara, N.A.; Carbajal, E.M.; Milla-Lewis, S.R. QTL mapping of morphological characteristics that correlated to drought tolerance in *St. Augustinegrass. PLoS One* 2022, 17(5), e0268004.

Zaghum, M.J.; Ali, K.; Teng, S. Integrated genetic and omics approaches for the regulation of nutritional activities in rice (*Oryza sativa L.*). *J. Agric.* 2022, 12(11), 1757.

Zhang, F.; Shi, Y.; Ali, J.; Xu, J.; Li, Z. Breeding by selective introgression: Theory, practices, and lessons learned from rice. *Crop J.* 2021a, 9(3), 646–657.

Zhang, S.; Tang, C.; Zhao, Q.; Li, J.; Yang, L.; Qie, L.; Zhao, M. Development of highly polymorphic simple sequence repeat markers using genome-wide microsatellite variant analysis in Foxtail millet (*Setaria italica (L.) P. Beauv.*). *BMC Genome* 2014, 15(1), 1–9.

Zhang, X.; He, Q.; Zhang, W.; Shu, F.; Wang, W.; He, Z.; Deng, H. Genetic relationships and identification of core germplasm among rice photoperiod-and thermo-sensitive genic male sterile lines. *BMC Plant Biol.* 2021b, 21(1), 1–13.

Zhou, J.; Gali, K.K.; Jha, A.B. Tar'an, B.; Warkentin, T.D. Identification of quantitative trait loci associated with seed protein concentration in a pea recombinant inbred line population. *J. Gene.* 2022, 13(9), 1531.

CHAPTER 3

Application of Genome Editing in Modern Crop Technology

SURENDRA PRAKASH GUPTA,[1*] ANKUR BHARDWAJ,[2] and MOHAMMAD IRFAN ALI[3]

[1]*Department of Life Science, Shri Vaishnav Vidhyapeeth Viswavidhayalaya, Indore, Madhya Pradesh, India*

[2]*Department of Biotechnology, Shri Vaishnav Institute of Sciences, Shri Vaishnav Vidyapeeth Vishwavidyalaya, Indore, Madhya Pradesh, India*

[3]*School of Allied Sciences, JECRC University, Jaipur, Rajasthan, India*

*Corresponding author

ABSTRACT

A cutting-edge technique in agriculture called genome editing (GE) is now widely applied to plants to improve crop quality and characterize gene function. GE has allowed gene engineering to find extensive use in a variety of industries, including agriculture, industry, and medicine. It is feasible to precisely insert or delete a specific DNA segment at a precise location utilizing GE. Research in this field is becoming more and more competitive as it aims to improve crop productivity and quality in many different ways. This field has advanced the working models of pathological processes over the course of the year, expanding our potential to explain the contribution of genomics. Zinc-finger nucleases, transcription activator-like effector nucleases, and clustered regularly interspaced short palindromic repeat–Cas-associated nucleases are examples of novel GE tools that are being developed with increased research effort. These instruments have shown promise in the modification of several crops with important compositional features.

Technological Advancements in Sustainable Crop Improvement and Production.
Devendra Kumar, Alok Kumar Singh, and Nisha Malik (Eds.)

DOI: 10.1201/9781779643155-3

3.1 INTRODUCTION

Modern agriculture would face major challenges from the anticipated 10 billion people globally by 2050, who would demand crops with higher yields, better quality, and noticeably lower input requirements (Tilman et al., 2011). While genetic variation has emerged as the most widely used method for plant development, getting from the initial stages of screening genotypes and phenotypes to the first crosses into commercial varieties is a labor-intensive process that usually takes several years. In order to take advantage of naturally occurring or artificially induced genetic variations, conventional breeding techniques have been used more frequently in the past 10 years. This has led to the need for thorough characterization of multigenerational offspring that are the product of many genetic crosses. Contrary, the best varieties can be improved thanks to GE tools, which make it possible to alter genomes quickly, precisely, and predictably. Modern breeding strategies have increased agricultural productivity, and other advanced techniques have been introduced as a result to ensure global food security. Crops must be developed with increased adaptability, high productivity, and resilience against a variety of biotic and abiotic stresses because they have an impact on both crop productivity and plant quality. The standard techniques of genetic engineering entail laborious processes.

With the addition of next-generation sequencing and the recent development of GE tools that allow targeted mutagenesis, this technology is being utilized more and more in the practice of cutting-edge OMICS approaches. This approach has expanded the possibilities for precision genome engineering. In order to better understand gene functions and enhance agricultural traits, GE methods have been employed to a range of crops. Comparatively speaking to traditional mutation breeding, GE is a less risky method for producing improved plant varieties by changing the functions of genes. Furthermore, plant varieties with difficult-to-breed complex genomes can benefit directly from the application of GE technology. Furthermore, plants possessing intricate genomes or those challenging to cultivate through conventional methods stand to gain directly from the utilization of GE technologies. Almost any location in the plant genome can now be accurately and reliably edited with highly versatile GE tools, broadening the range of potential applications to include molecular plant breeding and functional genomics. Plant scientists can now manipulate desired genes in crop plants thanks to a variety of GE tools, including meganucleases (MNs), zinc-finger nucleases (ZFNs), and transcription activator-like effector nucleases (TALENs).

3.2 HISTORY OF GENOME-/GENE-EDITING TECHNOLOGY

In order to improve yield and quality, a multitude of studies have looked into the development of GE technologies in recent years (Chen and Gao, 2013). These technologies allow for precise and effective targeted modification of the genomes of different crops. Genetic variations known as mutations happen on their own during evolution as a result of replication errors or damage to DNA. The term "mutagenesis" refers to this process (Hidema et al., 2005). Breeding for any desired trait is now feasible thanks to the accessibility of genome sequences for numerous crops and recent developments in plant GE techniques.

This has been of interest for a while; numerous interesting GE approaches have been developed as a consequence of recent molecular biology developments that have made it possible to precisely edit and cause site-specific mutations in many genomes (Brooks et al., 2014). A sequence-specific DNA binding domain is fused to a nonspecific nuclease domain in engineered nucleases (GE). Double-strand breaks (DSBs) caused by particular engineered nucleases in double-stranded DNA (dsDNA) are the main mechanism underlying GE. Either nonhomologous end joining (NHEJ) or homology-dependent recombination (HDR) is used to repair the dsDNA break. With minimal editing frequencies, precise point mutations, deletions, or gene knock-ins that are beneficial for plant breeding can now be produced thanks to HDR (Veillet et al., 2019).

3.3 RECENT NOVEL TECHNICAL BREAKTHROUGHS

In recent literature, several strategies have been put forth where crop improvement through the use of GE has been greatly aided by the introduction of beneficial foreign genes or the suppression of endogenous gene expression in crop plants. Abiotic stress tolerance, insect resistance, disease resistance, herbicide tolerance, and improved nutrition are just a few of the advantageous traits that genetically modified plants display. New developments are mainly reported by Kumar et al. (2020) on the widespread usage of transgenic plants with foreign genes and include issues about their potential toxicity and allergenicity to humans as well as potential hazards to the environment.

Investigations like these have demonstrated that GE is being done with recently created instruments. Kaur et al. (2021) provided an overview of the uses of the clustered regularly interspersed short palindromic repeats (CRISPR)/Cas9 GE technology in decorative, fruit, and vegetable plants. Li and coworkers (2021) described the most recent advancements in wheat functional genomics technologies, including fast breeding, synthetic biology, high-throughput genotyping and phenotyping, and GWAS. With the elite winter wheat cultivar Zhangmai 7698 (ZM), they have recently developed an effective multiplexing strategy that produces multiple mature single chimeric guide RNAs (sgRNAs) at the same time using a polycistronic tRNA strategy. Ul Haq et al. (2022) looked at the growth of targeted mutagenesis using CRISPR/Cas9 in a variety of forage crops, such as *Panicum virgatum, Medicago truncatula, Hordeum vulgare, Sorghum bicolor,* and *Alfalfa*. He et al. (2021) gave a succinct overview of the use of CRISPR-based GE technology to create a genome-wide mutant library in a variety of oil crop varieties, as well as its possible applications and future developments in oil crops. Hossain et al. (2022), in order to increase crops' resistance to abiotic stress, CRISPR–Cas GE technology was updated in terms of concept, mechanism, and application.

Talakayala et al. (2022) addressed the use of CRISPR GE tool to cultivate plant disease resistance against bacteria, fungi, and viruses. Numerous disease-prone, recessive, negatively regulated genes and transcription factors have been discovered in host plants; these may contribute to the plants' ability to interact with pathogens in a way that is mutually beneficial. These genes have been altered in a variety of plants, comprising rice, wheat, cucumbers, cassava, cocoa, grapes, citrus fruits, apples, and bananas, to increase their resistance to disease. Only modest advancements in genetically modified technology have resulted from consumer dissatisfaction in certain parts of the world. Therefore, during the primary or early stages of GE development, it would be advantageous to concentrate on nonvector integration of foreign DNA.

3.4 APPLIED GE TOOLS

Over the past 10 years, GE tools have experienced rapid advancements and are commonly utilized by research teams. Their widespread use has enormous implications for basic research in the life sciences, medicine, and agriculture. In recent years, even more diverse, efficient, and easier-to-implement gene editing tools are expected to be developed.

1. ZFNs
2. TALENs
3. CRISPR/Cas systems
4. Sequence-specific nucleases (SSNs)
5. Double-stranded breaks (DSBs)
6. MN

3.4.1 ZFNS

ZFNs are synthetic restriction enzymes used in genetic engineering (GE) that can specifically target DNA cleavage domains and cause DSBs in the genome. Cellular DNA repair processes, which produce site-directed mutagenesis or targeted gene stacking and/or replacement, depending on the presence or absence of DNA, are mediated by ZFN-induced DSBs (Carroll, 2011). These chimeric nucleases are composed of nonspecific cleavage domain chimeric fusion structure fused to sequence-specific DNA binding domains of between Cys2-His2 zinc-finger DNA-binding domains attached with a FokI nuclease. As a dimer, the FokI nuclease binds 24–36 nucleotides upstream and downstream of the interface through pairs of zinc finger domains, which increases specificity. Its enzymatic activity depends on the dimerization of the FokI domain, which is made up of four to six 30-amino acid domains that can bind extended nucleotide segments and give the DNA binding domain as a whole specificity (Davies et al., 2017). Bonawitz and others (2008) showed the targeted integration of biolistically supplied DNA into fatty acid desaturase via HDR or NHEJ, and reported the targeted integration of multiple transgenes into a single soybean locus using a ZFN, where embryogenic cells are found in tissue cultures.

3.4.2 TALENS

TALENs have been the subject of numerous investigations as a potentially revolutionary tool for GE in an extensive range of organisms and cell types. This has a straightforward design with excellent precision and fission activity efficiency. TALENs induce site-specific chromosomal DSBs that dramatically boost the efficiency of desired gene expression. At particular points in the genome, TALENs function as programmable DNA scissors and create DSBs. Homologous recombination (HR) and NHEJ are two of the repair mechanisms that the cell uses to fix the transcription activator-like effector fusions that cause DSBs. Because of this, they offer a practical technique for modifying particular regions of the genome, including insertions, deletions, rearrangements, and mutations. Neural replacements of targeted genes or native GE are hampered by HR repair at DSBs.

Li et al. (2016) used donor DNA and ballistic delivery in the Oryza (rice) calli-TALENs construct to cause double point mutations in the rice acetolactate synthase gene (OsALS) and produce herbicide-resistant rice lines. To increased seed storability and viability, Ma et al. (2015a) performed site-directed mutagenesis in rice using TALENs nucleases. The seed longevity is impacted by Lipoxygenases particularly Lox3 activity. Lipoxygenases acted on polyunsaturated fatty acids by catalyzing the dioxygenation process on *cis,cis*-1,4-pentadiene to form hydroperoxide radical. The activity of lipoxygenases, specifically Lox3, affects the seedling longevity. By catalyzing the dioxygenation of *cis,cis*-1,4-pentadiene to form hydroperoxide radicals, lipoxygenases acted on polyunsaturated fatty acids. These radicals have a substantial effect on the viability and longevity of seeds. The findings indicate that rice can successfully undergo TALEN-mediated GE and offer valuable insights for future GE efforts using alternative nuclease-based GE platforms.

3.4.3 CRISPR/CAS SYSTEMS

The advent of RNA-guided nucleases, CRISPR and their cooperation with CRISPR-associated (Cas) genes has marked their application as the most reliable and powerful tools for GE and engineering. During an analysis of genes associated in phosphate metabolism, It was initially discovered as CRISPR–Cas system in the *Escherichia coli* genome, which enabled researchers to precisely and efficiently manipulate specific genomic elements.

The cellular DNA repair process that follows produces the intended insertions, deletions, or substitutions at the target sites and makes it easier to understand how target genes function in health and in biology. Target sequences that match CRISPR RNA (crRNA) and a protospacer-adjacent motif that is situated downstream of the target sequences are necessary for the specificity of CRISPR/Cas9-mediated DNA cleavage (Zhang et al., 2014). As an illustration, consider how simple it is to use CRISPR systems to effectively target several genes at once. Using CRISPR systems, a number of tactics and toolkits have been created to accomplish multiplex targeting (Cao et al., 2018). Numerous studies that have been reviewed specifically highlight the capabilities of CRISPR/Cas9 as a tool for plant genome engineering and outline the advantages and disadvantages of the technology (Bortesi and Fischer, 2015; Samanta et al., 2016; Zhang et al., 2017).

A key opportunity for enhancing drought tolerance is to comprehend the underlying physiological, biochemical, genetic, and environmental mechanisms as well as how they interact. Just lately, Bapela et al. (2022) showed how to use the CRISPR–Cas system to create wheat varieties that are resistant to drought and talked about the importance of innate genetic variation and environmental adaptation. Numerous nucleases have previously been used to demonstrate site-directed mutagenesis in soybean (Glycine max) through NHEJ-mediated repair, as well as homology-directed repair (HDR)-mediated integration of a single transgene into endogenous target soybean loci using CRISPR/Cas9 (Bonawitz et al., 2008).

3.4.4 SSNS

In order to facilitate targeted genomic modifications like gene mutations, insertions, replacements, or chromosomal rearrangements, GE of SSNs causes DNA DSBs at particular genomic locations. These DSBs are then primarily repaired by HR or NHEJ pathways. The creation of artificial SSNs opened the door to precise gene editing in several organisms, including plants. SSNs generate a DSB at the sequence-specific site of DNA. When an adequate donor template is accessible, these DSBs have been almost exclusively repaired employing error-prone NHEJ and very infrequently utilizing error-free HDR (Miki et al., 2021). Gene targeting (GT) associated with the integration or replacement of a particular sequence has been accomplished through the use of repair donor templates combined with SSN combinations. On the other hand, the efficiency is actually very low in some higher systems. Consequently, by enabling the insertion of precise point mutations and new gene functions, or the integration of foreign gene functions in a predetermined manner, genome modification through SSNs-mediated HDR for GT could offer an unprecedented ability to accelerate plant breeding.

3.4.5 DOUBLE-STRANDED BREAKS (DSBS)

Genetic diversity is produced via meiotic recombination, which also promotes proper chromosome segregation. DSBs that are artificially created in chromosomal DNA by DNA topoisomerase leads recombination. In order to replace targeted genes and introduce desired DNA elements at specific gene loci for crop improvement, crRNA generates DSBs with long 5'-protruding ends. This may make it easier to pair and insert repair templates through HDR. Crossovers (COs) are formed as a result of certain DSBs during meiotic recombination. The repair of certain DNA DSBs results in the creation of COs events. In the majority of organisms, such as plants, there are significantly more DSBs than there are CO events. The formation of COs is affected by the DNA sequence and epigenetic features of chromosomes, along with broader mechanisms that regulate the dynamics of genetic recombination (Xue et al., 2018).

3.4.6 MNS

MNs or homing endonucleases are highly sequence-specific enzymes that induce HR discovered initially in yeast cells, later on found in mammalian cells, mice, plants, Drosophila, trypanosomes, prokaryotes (*E. coli*), archaea, and unicellular eukaryotes (Paques and Duchateau, 2007). Consequently, these enzymes are useful models for the development and commercialization of genomic research tools dedicated to in cellulo genome customization, including knock-in, knock out and modulation of gene expression. In order to understand protein–DNA interaction and mechanism of DNA cleavage, several MNs have been improved by X-ray studies (Arnould et al., 2011).

3.5 PRECISE GE IN PLANTS

GE technologies offer potential for enhancing crops by utilizing plant genes to make precise changes to DNA sequences. This process involves using specific nucleases to create DNA DSBs, which then prompt repair mechanisms such as nonhomologous end joining or HDR to make targeted modifications to plant genomes.

Numerous modifications have been made to plant cell DNA by researchers, including the deletion of genes or chromosome segments, the insertion of foreign DNA into particular genomic regions, and the creation of nucleotide substitutions that alter a protein's amino acid sequence. Such customized DNA sequence alterations are made possible by SSNs that induce DSBs in the genomic locus. HR or nonhomologous end joining can be used to repair the breaks and produce the desired sequence change (Voytas, 2013). We highlight some of the significant advancements that have occurred recently.

3.5.1 IMPROVING SINGLE-GENE TRAITS

Traits controlled by a single gene are known as single-gene traits. GE technologies are valuable tools for enhancing agronomic traits by altering gene mutations. For instance, pin genes in wheat play a vital role in determining grain hardness (Nadolska-Orczyk et al., 2009).

The evolution of high-yielding cereal varieties has traditionally involved adding a single important gene into the genome. While conventional breeding methods have been effective in manipulating traits like Rht-B1 and Rht-D1, CRISPR-based approaches offer a faster and less labor-intensive alternative. This approach has led to the creation of new cereal varieties that are shorter, more resistant to stem lodging, and better able to withstand nitrogen-based fertilizers (Youssefian et al., 1992).

3.5.2 ENGINEERING COMPLEX TRAITS

Complex genetic networks control a wide range of crucial crop agronomic properties. Large gene families control the complex plant features that adapt to environmental changes. For instance, the shortage of nitrogen in the soil influences the rates and patterns of gene expression, which has an impact on traits like root development, growth rate, and flowering time. For the development of crops, genetic polymorphisms within coding sequences have been extensively used. Introducing complex traits that are unattainable through conventional breeding methods is one of the major benefits of using GE tools. One of the most crucial methods for creating double haploids in breeding is the haploid induction process.

3.6 APPLICATION OF GE TO CROP BREEDING

The basic goal of crop breeding is to develop agricultural varieties that produce high-quality, high-yielding crops that are tolerant to biotic and abiotic stresses. This has rapidly become a highly significant subject area to create recent crop varieties with necessitated agronomic traits by utilizing efficient targeted genome modifications instead of traditional breeding methods. This is affected according to several stages of crossing, selection, and testing during the breeding process, as well as the duration of the growing season and the maturity stage of the plants. Additionally, the targeted effect on the genome is not feasible using conventional physical and chemical mutation techniques. However, a recent phase in plant breeding has initialized with the growth of the modern DNA-technology techniques like GE. In a broad range of plant species, genome-editing methods have been employed to elucidate gene functions and enhance agricultural variants. Target mutations are routinely used to alleviate such features, which result in the activity of the regulatory genes controlling the development of undesirable traits being suppressed (Zhang et al., 2014)

Among the most valuable qualities of agricultural crops from an economic perspective is productivity. GE technologies are a potentially useful tool for quickly and precisely mutating target genes in this sense (Jansing et al., 2019). The most crucial factors influencing yield are multiple blooms, early flowering, a particular growth pattern, fruit size, and the quantity of compartments. Ripeness, color, and metabolites are the three main factors used to evaluate fruit quality and nutritional value. Removing ("turning off") yield-detrimental genes is the most efficient way to boost yield with GE technologies (Sedeek et al., 2019). The floral repressor Self Pruning 5G, for instance, grows compactly and determinately when CRISPR/Cas9 is used to modify it. This has resulted in "doubly determinate" plants that yield earlier than expected.

Biodiversity has been revolutionized by GE. GE opens up new avenues for crop development when combined with simplified regulatory system and the possibility to create transgene-free variants (Nadakuduti et al., 2018).

In recent past, techniques for genome engineering have developed swiftly and are now often employed by investigators. Their broad use would significantly influence agriculture, medicine, and basic life science research. In the upcoming years, it is conceivable that even more gene-editing technologies will be created that are versatile, effective, and simple to use. Because genome-editing technologies have already been discussed in depth elsewhere (Hua et al., 2019; Komor et al., 2017), we have briefly summarized the various GE techniques and emphasis on how they are being used to improve crops.

ZFNs are applied to increase the efficiency of transgenic products by introducing transgenes into specific genomic locations and subsequently adding or removing transgenes from already-existing transgenic events. They can also be used to create new genetic variations by deleting or editing endogenous nucleotide sequences (Davies et al., 2017; Negi et al., 2022; Novak, 2019).

TALENs are chimeric proteins with two functional domains: a nuclease domain and a DNA-recognition transcription activator-like effector and has been implemented on various plant species like *Arabidopsis thaliana*, Brachypodium, Tobacco, Rice, Barley, Soybean, Wheat, Tomato, and Zea mays (Zhang et al., 2013). By targeting a certain nucleotide sequence and introducing a double-stranded break with an overhang, investigators are capable of modifying genes. Therefore, TALENs operate as a variety of extensible restriction enzyme (Bhushan et al., 2016).

A Cas9 protein that can cleave the genome at particular sites is aggregated with a guide RNA in the CRISPR/Cas9 GE technique. The CRISPR/Cas9 system has quickly gained popularity as the most practical method for modifying the genomes of plants and animals owing to its ease of use and effectiveness (Liu et al., 2022). Growing amounts of effort are being put into developing new germplasm resources and modifying the traits of numerous plants, including major crops (Zetsche et al., 2015). It is possible to lower the amylose content of rice grains and enhance their nutritional value and flavor by utilizing the CRISPR/Cas9 system (Ma et al., 2015b). Plants have shown diverse GE capabilities through the use of CRISPR/Cas9 and its derived system. such as ribonucleoprotein in plant cells, sgRNA synthesis and design, gene knock-in, knockout, and expression activation through editing, transformation, and screening of plants with altered genes. Additionally, editing several genes at once has aided in pathway-level research (Liu et al., 2021).

Many important agricultural traits are determined by indels or random point mutations at specific locations in coding or promoter regions, which limits the use of SSNs for crop improvement. In order to introduce specific point mutations and new gene functions, or to integrate them in a way that brings foreign genes to the desired safe harbor in a defined way, genome modification through SSNs-mediated HDR for GT, enabling either gene replacement or knock-in/out, offers an unprecedented ability to facilitate plant breeding (Miki et al., 2021; Sun et al., 2016).

A profound understanding of the different repair pathways is a prerequirement for targeted gene modifications. The DNA DSBs repair takes place by nonhomologous end joining or homology-dependent repair (HDR) mechanism through small insertions and deletions. It could be applied to knock out genes or to rearrange linkage groups specifically (Shen and Li, 2022). The basic requirement for HDR is the presence of a homologous template, therefore, HDR mechanism can be used for targeted integration of ectopic DNA sequences into the plant genome. The advancement of various types of artificial site-specific nucleases allows to introduce targeted DSBs in the desired plant genome (Schmidt et al., 2019).

MNs can alter their recognition sequence using protein engineering, enabling them to effectively and precisely replace, remove, or modify any desired sequence. MNs, which are produced by transgenic elements, can induce recombination through a process known as homing. Genes are encoded within an intron or intein, allowing for horizontal gene transfer. They are comprised of DNA-binding and DNA-cleavage domains. Repair of the lesion at the recognition site of MNs involves HR following the formation of DSBs (Iqbal et al., 2020).

3.7 MODERN BREEDING TECHNIQUES FOR TRAIT IMPROVEMENTS IN CROP PLANTS: CASE STUDIES

Elite plant varieties were chosen through hybridization and a continuous screening procedure in classical breeding. By crossing crops with specific agronomic traits with wild relatives and choosing the best genotypes, the basic goal of conventional plant breeding technique is to introduce genetic diversity (Lavarenne et al., 2018). Similarly, research on wild populations can offer a great path forward for agricultural improvement by fostering greater genetic diversity in desired traits for the spread of new crop varieties (Razzaq et al., 2021).

Numerous genetic and molecular approaches are used in modern plant breeding methods, which have enhanced agricultural improvement. Staple crops like wheat, rice, maize, sorghum, and others are frequently farmed using these contemporary methods. Modern biotechnological techniques are being developed and applied to increase the speed and efficiency of breeding programs. CRISPR-Cas9, high-throughput phenotyping, genomic selection, and marker-assisted selection are a few of these contemporary breeding techniques.

3.7.1 CASE STUDY 1—WHEAT

Synthetic wheat hexaploids cannot be used directly for wheat production because they are normally not adapted to the target growing region and express wild traits, such as poor threshing and extreme lodging (Figure 3.1). Therefore, the synthetic lines were crossed to an adapted wheat cultivar and then backcrossed once or twice to the adapted cultivar, followed by selection. This results in breeding lines that have mostly adapted genomes with segments of synthetic hexaploid chromosomes interspersed. These lines were then grown and evaluated like normal wheat in prospects of improvements in qualitative and quantitative traits due to the inserted synthetic hexaploid segments (Li et al., 2018; Ogbonnaya et al., 2013).

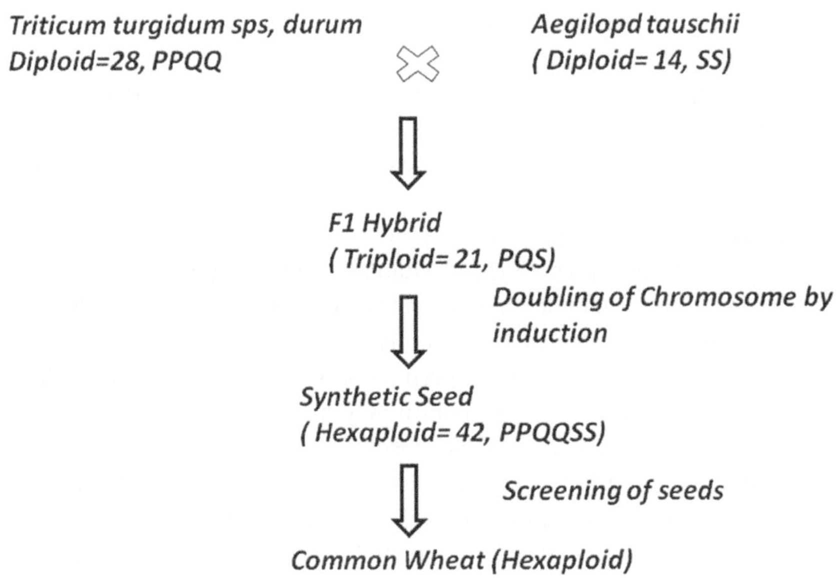

FIGURE 3.1 Strategy for the development of synthetic hexaploidy wheat.

Source: Adapted from Ogbonnaya et al. (2013)

3.7.2 CASE STUDY 2—MAIZE

In Gujarat, India, a collaborative effort between plant breeders and farmers resulted in the development of improved varieties of corn tailored for resource-poor farmers. By crossbreeding yellow and white endosperm corn varieties to achieve a desirable level of early maturity, the population was then refined through mass selection for traits identified by farmers. The successful breeding program yielded several high-performing varieties, with GDRM-187, officially approved as GM-6 for maize cultivation in hill areas of Gujarat, standing out for its 18% higher yield compared to locally grown maize. In fields with lower average yields, the advantage was even greater at 28%, with farmers reporting GDRM-187 to have superior grain quality compared to local varieties. This innovative breeding program was able to produce a variety that matured earlier than any conventionally bred corn, requiring fewer years to reach maturity. The higher yields from plant breeding programs are attributed to their cost-effectiveness and quicker realization of benefits for farmers compared to conventional breeding methods (Fenzi and Couix, 2022; Witcombe et al., 2003).

3.7.3 CASE STUDY 3—RICE

The green revolution was sparked by the early 20th century development of modern agricultural techniques. However, there have been rumors of a decline in yield in some Asian large-scale rice farms, indicating that efforts to improve rice quality may be missing a genetic advantage in potential yield. By using these data, it was possible to investigate yield differences that resulted from either a lack of adaptation to changing environmental conditions or from genetic changes in the rice variety (IR8) brought about by repeated planting cycles. Continuous research was done on IR8 growth in both field studies and pot experiments. Both seed sources produced IR8 yields that were 15% less than newly created varieties. These findings imply that IR8's low yield is caused by a failure to adapt to shifting environmental conditions. The study offers compelling evidence in favor of carrying out conservation breeding initiatives to preserve rice yield potential by enhancing resistance to biotic and abiotic stresses that are changing quickly (Peng et al., 2010; Tang et al., 2022).

Since rice was domesticated thousands of years ago, there has been a significant increase in selection pressure, which has resulted in a reduction in genetic variety. A variety of mutant databases have made it feasible to transition from structural to functional genomics, emphasizing the role that mutagenesis has played in this development. Both random and intentional mutations have been used to create successful gene changes. Techniques were used to find mutations in the rice genome and to understand the molecular regulation of DNA damage repair processes (Viana et al., 2019).

3.8 GE FOR CROP IMPROVEMENT: CHALLENGES

GE is a form of biological mutagenesis method that has helped to secure the global food supply. It is set to revolutionize crop breeding method, which does not have ethical and off-targeting issues. Although, some studies have been conducted across the globe and consumer perception. However, there are many technical challenges to overcome, one of them is public acceptance of the technology and governmental regulatory frameworks pose the greatest barriers to aid of this approach in agriculture. Furthermore, various plant species have successfully endured by genetic modification, and in some instances, USDA/APHIS has taken the regulatory status of the modified plants into consideration (Waltz, 2018).

GE serves as improvement of the crop genetics via two different ways: (1) through CRISPR-Cas tools that aim to study gene function in complex biological traits, and (2) through introduction of specific required genetic changes into a plant. Scientists remarkably use GE to better harness genetic diversity, to find out the useful candidate genes and genetic variants serving desirable characteristics. Wolter et al. (2019) carried out GE by introduction of one specific genetic change into elite plant varieties without simultaneously transferring other undesired genetic changes.

Genome-modified crops are thought to be typically just as safe as crops with the same DNA modifications generated through conventional means. These crops have DNA changes that can arise spontaneously in nature or because of mutation breeding techniques. In other words, a crop that has undergone GE with a particular mutation is just as safe as a crop that has undergone conventional breeding. Furthermore, crops with a particular mutation created by GE are seen as being less dangerous than crops with the same mutation created through mutational breeding. The second scenario would result in a crop with several extra random mutations, the effects of which are unknown and unpredictable. As a result, the traditional crop's safety is far less reliable than the crop that has genome edited.

The application of genetically modified crops will surely increase in the US if clear regulations are put in place. But according to a recent ruling by the Court of Justice of the European Union, plants created through GE are not exempt from regulation and instead need to be handled similarly to transgenic plant lines. The European Union contends that genetically modified plants may have detrimental effects on both the environment and human health because gene editing modifies genetic material in a way that is not natural. Genetically modified crops are still not subject to a clear and uniform regulatory framework in the majority of nations. Even if laws governing the utilization of genetically modified crops were to be passed, it would be very difficult or impossible to enforce them because most gene editing events are not distinguishable from "natural" mutations.

In order to accurately alter genome sequences, scientists can use a number of innovative techniques in GE. Research is being done to develop crop varieties with higher yields, strengthened stress tolerance, pest resistance, and increased nutritive quality. The strategies also allow scientists to change the patterns of gene regulation and expression in a predefined region and expedite novel insights into the functional genomics of desired crop varieties.

With a SSNs approach, a large range of tools is available for GE. Single- or double-strand DNA breaks are produced by the SSN mechanism, which also stimulates endogenous DNA repair mechanisms. Depending on the position of the particular sites (loxP, FLP, etc.) flanking a target site, such as Cre/loxP and Flp/FRT driven systems, can knockdown or knock-in genes in the genome of the desired crop variety. Mega-nucleases (homing endonucleases), ZFNs, TALENs) and the CRISPR/Cas nuclease system are the four major classes of SSNs established to snip genomic sequences.

GE may result in three different sorts of changes: Type I focuses on changing a few nucleotides, Type II entails substituting in a preexisting allele, and Type III permits the introduction of new genes at specific locations across the genome. Crops developed by gene editing might circumvent the strict regulatory processes often connected with the creation of GM crops since the majority of genome-editing techniques can leave behind evidence of DNA mutations visible in a small proportion of nucleotides. Because of this, many scientists think that plants modified using more exact gene editing methods will become more popular with the general people than the transgenic plants. Among the foregoing GE strategies, the type-2 mode is the most problematic for researchers and paradoxically the most valuable for plant breeding since it permits editing of existing alleles or new genes to be "knocked-in." The gene editing might be intended either to keep the function of existing alleles but were changed or lost its functioning during traditional breeding or redesigning the allele for producing unique features.

Scientists have been actively developing and refining techniques to increase crop quality and quantity for over a century. These techniques range from conventional breeding to contemporary biotechnological methods. GE technologies have a number of benefits over conventional breeding techniques. Even though crop properties have improved through traditional breeding methods, the new generation of varieties is extremely resource-intensive due to extended breeding cycles, excessive heterozygosity, lack of variable hybridization accuracies, and low frequency of beneficial mutations. Transgenic technology offers a wide range of applications in crop improvement because it can overcome the obstacles caused by species incompatibilities. However, there are a lot of challenges this method must overcome, especially those related to regulations, public acceptance, and expense of risk studies needed before commercialization. Using the latest gene editing techniques today is more accurate and more effective for enhancing features. Since they are not deemed to be transgenic, crops created by GE are anticipated to be more acceptable to consumers. Combining the above three technologies will make it easier to improve agricultural attributes. Wisely using GE methods will result in the production of more food, fodder, and fiber that is generated in a much more sustainable way, improving global food security.

3.9 FUTURE PROSPECTS AND CONCLUDING REMARKS

This chapter summarizes how GE technologies can support the global health and agricultural sectors. The clean gene technique uses GE to securely remove genes used during experiments and transformations, leaving plants free of any extra genetic waste. Plants engineered with antibiotic resistance genes or selectable markers can no longer pose a risk thanks to technology. In addition, GE technology offers a quicker method of creating genetically modified plants than plants created via breeding methods. This provide benefits to both the seed industry and commercial growers.

It is possible that a species of choice may not have protoplast-based systems easily available or even feasible. Additionally, tissue-cultured regenerated crop plants could be challenging or restricted to a few genotypes. Designing techniques that do not involve regeneration, like use of pollens or immature embryos can be made to germinate *in vitro*, could be beneficial under these circumstances. GE promises to play a crucial role in improving crop breeding and satisfying the rising global demand for food. Additionally, the demands of climate change necessitate significant innovation and flexibility in crop resilience and productivity (Zhang et al., 2018).

To further study the genetics of complex characteristics like yield, quality, and drought tolerance, breeders are actively striving to reinvent the potato as a diploid crop. A better crop breeding strategy may be possible with the use of GE associated with the establishment of inbred diploid lines, which would represent a significant change in the potential for genetic improvement (Nadakuduti et al., 2018).

Precision breeding is one of the new breakthroughs that have accompanied changes in plant breeding techniques over time. The phrase "new plant technology" was recently used to refer to all current advancements made in the field of biotechnology to increase crop improvements (Limera et al., 2017). The success of GE in India and throughout the world will be demonstrated by the inclusion of all developed pertinent tools and techniques employing genomics, GE, artificial intelligence, and deep learning, among other technologies.

It must be mentioned that for developing GE products, communication between farmers, plant breeders, and users must be continuously active. It is also important to note the kind of delivery mechanisms employed to repair mutants. Therefore, it is believed that the new crop improvement technologies might satisfy the unmet promise in expediting agricultural development efforts.

The advent of GE technologies, which are based on bacterial or engineered nucleases, has made it possible to target and alter the genome sequences of nearly all eukaryotic cells directly. GE is an invaluable tool for high-precision molecular breeding of crops. It has the potential to revolutionize agriculture in a world where there is an there is an urgent need to produce more food with less detrimental effects on the environment, it has the potential to completely transform agriculture. Apart from being accurate, GE can lower the cost of plant breeding and swiftly create new plant varieties that are more nutrient-efficient, stress-tolerant, nutritious, and yield-oriented. To sum up, GE and clean genetic technologies can help allay public and regulatory concerns about the acceptability of using genetically modified crops, thereby promoting their global adoption.

KEYWORDS

- **clustered regularly interspersed short palindromic repeats/ Cas9**
- **double strand breaks**
- **gene targeting**
- **homology-directed repair**
- **sequence-specific nucleases**
- **transcription activator-like effector nucleases**
- **zinc-finger nucleases**

REFERENCES

Arnould, S.; Delenda, C.; Grizot, S.; Desseaux, C.; Pâques, F.; Silva, G.H.; Smith, J. *I-CreI Meganuclease and Its Engineered Derivatives: Applications from Cell Modification to Gene Therapy*. OUP Academic, 2011. Available at: https://academic.oup.com/peds/article/24/1-2/27/1470637 [Accessed December 31, 2022]

Bapela, T.; Shimelis, H.; Tsilo, T.J.; Mathew, I. *Genetic Improvement of Wheat for Drought Tolerance: Progress, Challenges and Opportunities*. MDPI, 2022. Available at: https://www.mdpi.com/2223-7747/11/10/1331 [Accessed December 31, 2022]

Bhushan, K.; Pratap, D.; Sharma, P.K. Transcription activator-like effector nucleases (TALENs): An efficient tool for plant genome editing. *Eng. Life Sci.* 2016, 16(4), 330–337.

Bonawitz, N.D.; Ainley, W.M.; Itaya, A.; Chennareddy, S.R.; Cicak, T.; Effinger, K.; Jiang, K.; Mall, T.K..; Marri, P.R.; Samuel, J.P.; Sardesai, N.; Simpson, M.; Folkerts, O.; Sarria, R.; Webb, S.R.; Gonzalez, D.O.; Simmonds, D.H.; Pareddy, D.R. Zinc finger nuclease-mediated targeting of multiple transgenes to an endogenous soybean genomic locus via non-homologous end joining. *Plant Biotechnol. J.* 2008, 17(4), 750–761. https://doi.org/10.1111/pbi.13012

Bortesi, L.; Fischer, R. The CRISPR/Cas9 system for plant genome editing and beyond. *Biotechnol. Adv.* 2015, 33(1), 41–52. https://doi.org/10.1016/j.biotechadv.2014.12.006

Brooks, C.; Nekrasov, V.; Lippman, Z.B.; Van Eck, J. Efficient gene editing in tomato in the first generation using the clustered regularly interspaced short palindromic Repeats/

CRISPR-Associated9 system. *Plant Physiol.* 2014, 166(3), 1292–1297. https://doi.org/10.1104/pp.114.247577

Cao, J.; Xiao, Q.; Yan, Q. The multiplexed CRISPR targeting platforms. *Drug Discov. Today Technol.* 2018, 28, 53–61. https://doi.org/10.1016/j.ddtec.2018.01.001

Carroll, D. Genome engineering with zinc-finger nucleases. *Genetics* 2011, 188(4), 773–782. https://doi.org/10.1534/genetics.111.131433

Chen, K.; Gao, C. Targeted genome modification technologies and their applications in crop improvements. *Plant Cell Rep.* 2013, 33(4), 575–583. https://doi.org/10.1007/s00299-013-1539-6

Davies, J.P.; Kumar, S.; Sastry-Dent, L. Use of zinc-finger nucleases for crop improvement. *Prog. Mol. Biol. Transl. Sci.* 2017, 149, 47–63. https://doi.org/10.1016/bs.pmbts.2017.03.006

Fenzi, M.; Couix, N. Growing maize landraces in industrialized countries: From the search for seeds to the emergence of new practices and values. *Int. J. Agri. Sustain.* 2022, 20(3), 327–345.

He, J.; Zhang, K.; Tang, M.; Zhou, W.; Chen, L.; Chen, Z.; Li, M. CRISPR-based genome editing technology and its applications in oil crops. *Oil Crop Sci.* 2021, 6(3), 105–113. https://doi.org/10.1016/j.ocsci.2021.07.002

Hidema, J.; Teranishi, M.; Iwamatsu, Y.; Hirouchi, T.; Ueda, T.; Sato, T.; Burr, B.; Sutherland, B.M.; Yamamoto, K.; Kumagai, T. Spontaneously occurring mutations in the cyclobutane pyrimidine dimer photolyase gene cause different sensitivities to ultraviolet-B in rice. *Plant J.* 2005, 43(13), 57–67. https://doi.org/10.1111/j.1365-313x.2005.02428.x

Hossain, A.; Rahman, M.M.E.; Ali, S.; Islam Tanjina; Syed, M.A.; Syed, T.; Zafar, S.A.; Behera, L.; Skalicky, M.; Brestic, M.; Islam Tofazzal. CRISPR-Cas9-mediated genome editing technology for abiotic stress tolerance in crop plant. *Plant Perspect. Glob. Clim. Chang.* 2022, 331–354. https://doi.org/10.1016/b978-0-323-85665-2.00008-x

Hua, K.; Zhang, J.; Botella, J.R.; Ma, C.; Kong, F.; Liu, B.; Zhu, J.K. Perspectives on the application of genome-editing technologies in crop breeding. *Mol. Plant.* 2019, 12(8), 1047–1059.

Iqbal, Z.; Iqbal, M.S.; Ahmad, A.; Memon, A.G.; Ansari, M.I. New prospects on the horizon: Genome editing to engineer plants for desirable traits. *Curr. Plant Biol.* 2020, 24, 100171.

Jansing, J.; Schiermeyer, A.; Schillberg, S.; Fischer, R.; Bortesi, L. Genome editing in agriculture: Technical and practical considerations. *Int. J. Mol. Sci.* 2019, 20(12), 2888. https://doi.org/10.3390/ijms20122888

Kaur, H.; Pandey, D.K.; Goutam, U.; Kumar, V. CRISPR/Cas9-mediated genome editing is revolutionizing the improvement of horticultural crops: Recent advances and future prospects. *Sci. Horticult.* 2021, 289, 110476. https://doi.org/10.1016/j.scienta.2021.110476

Komor, A.C.; Badran, A.H.; Liu, D.R. CRISPR-based technologies for the manipulation of eukaryotic genomes. *Cell* 2017, 168(1–2), 20–36.

Kumar, K.; Gambhir, G.; Dass, A.; Tripathi, A.K.; Singh, A.; Jha, A.K.; Yadava, P.; Choudhary, M.; Rakshit, S. Genetically modified crops: Current status and future prospects. *Planta* 2020, 251(4), 91. https://doi.org/10.1007/s00425-020-03372-8

Lavarenne, J.; Guyomarc'h, S.; Sallaud, C.; Gantet, P.; Lucas, M. The spring of systems biology-driven breeding. *Trends Plant Sci.* 2018, 23(8), 706–720.

Li, A.; Liu, D.; Yang, W.; Kishii, M.; Mao, L. Synthetic hexaploid wheat: Yesterday, today, and tomorrow. *Engineering* 2018, 4(4), 552–558.

Liu, H.; Chen, W.; Li, Y.; Sun, L.; Chai, Y.; Chen, H.; Huang, C. CRISPR/Cas9 technology and its utility for crop improvement. *Int. J. Mol. Sci.* 2022, 23(18), 10442–10458.

Liu, Q.; Yang, F.; Zhang, J.; Liu, H.; Rahman, S.; Islam, S.; She, M. Application of CRISPR/Cas9 in crop quality improvement. *Int. J. Mol. Sci.* 2021, 22(8), 4206–4220.

Li, S.; Zhang, C.; Li, J.; Yan, L.; Wang, N.; Xia, L. Present and future prospects for wheat improvement through genome editing and advanced technologies. *Plant Commun.* 2021, 2(4), 100211. https://doi.org/10.1016/j.xplc.2021.100211

Li, T.; Liu, B.; Chen, C.Y.; Yang, B. TALEN-mediated homologous recombination produces site-directed DNA base change and herbicide-resistant rice. *J. Genet. Genom.* 2016, 43(5), 297–305. https://doi.org/10.1016/j.jgg.2016.03.005

Limera, C.; Sabbadini, S.; Sweet, J.B.; Mezzetti, B. New biotechnological tools for the genetic improvement of major woody fruit species. *Front. Plant Sci.* 2017, 8, 1418–1424.

Ma, L.; Zhu, F.; Li, Z.; Zhang, J.; Li, X.; Dong, J.; Wang, T. TALEN-based mutagenesis of lipoxygenase LOX3 enhances the storage tolerance of rice (*Oryza sativa*) seeds. *PLoS One* 2015a, 10(12), e0143877. https://doi.org/10.1371/journal.pone.0143877

Ma, X.; Zhang, Q.; Zhu, Q.; Liu, W.; Chen, Y.; Qiu, R.; Liu, Y. A robust CRISPR/Cas9 system for convenient, high-efficiency multiplex genome editing in monocot and dicot plants. *Mol. Plant* 2015b, 8(8), 1274–1284.

Miki, D.; Wang, R.; Li, J.; Kong, D.; Zhang, L.; Zhu, J.K. Gene targeting facilitated by engineered sequence-specific nucleases: Potential applications for crop improvement. *Plant Cell Physiol.* 2021, 62(5), 752–765. https://doi.org/10.1093/pcp/pcab034

Nadakuduti, S.S.; Buell, C.R.; Voytas, D.F.; Starker, C.G.; Douches, D.S. Genome editing for crop improvement—Applications in clonally propagated polyploids with a focus on potato (*Solanum tuberosum* L.). *Front. Plant Sci.* 2018, 9, 1607–1622.

Nadolska-Orczyk, A.; Gasparis, S.; Orczyk, W. The determinants of grain texture in cereals. *J. Appl. Genet.* 2009, 50(3), 185–197. https://doi.org/10.1007/bf03195672

Negi, C.; Vasistha, N.K.; Singh, D.; Vyas, P.; Dhaliwal, H.S. Application of CRISPR-mediated gene editing for crop improvement. *Mol. Biotechnol.* 2022, 64(11), 1198–1217.

Novak, S. Plant biotechnology applications of zinc finger technology. In: Kumar, S.; Barone, P.; Smith, M.; (Eds.), *Transgenic Plants. Methods in Molecular Biology*. Humana Press: New York, USA, 2019, pp. 295–310, Vol. 1864.

Ogbonnaya, FC.; Abdalla, O.; Mujeeb-Kazi, A.; Kazi, AG.; Xu, SS.; Gosman, N.; Lagudah, ES.; Bonnett, D.; Sorrells, ME.; Tsujimoto, H. Synthetic hexaploids: Harnessing species of the primary gene pool for wheat improvement. *Plant Breed. Rev.* 2013, 37, 35–122.

Paques, F.; Duchateau, P. Meganucleases and DNA double-strand break-induced recombination: Perspectives for gene therapy. *Curr. Gene Ther.* 2007, 7(23), 49–66. https://doi.org/10.2174/156652307779940216

Peng, S.; Huang, J.; Cassman, K.G.; Laza, RC.; Visperas, RM.; Khush, GS. The importance of maintenance breeding: A case study of the first miracle rice variety-IR8. *Field Crops Res.* 2010, 119(2–3), 342–347.

Razzaq, A.; Kaur, P.; Akhter, N.; Wani, S.H.; Saleem, F. Next-generation breeding strategies for climate-ready crops. *Front. Plant Sci.* 2021, 12–28.

Samanta, M.K.; Dey, A.; Gayen, S. (CRISPR/Cas9: An advanced tool for editing plant genomes. *Transgenic Res.* 2016, 25(5), 561–573. https://doi.org/10.1007/s11248-016-9953-5

Schmidt, C.; Pacher, M.; Puchta, H. DNA break repair in plants and its application for genome engineering. *Methods Mol. Biol.* 2019, 1864, 237–266.

Shen, H.; Li, Z. DNA Double-strand break repairs and their application in plant DNA integration. *Gene* 2022, 13(2), 322–338.

Sun, Y.; Li, J.; Xia, L. Precise genome modification via sequence-specific nucleases-mediated gene targeting for crop improvement. *Front. Plant Sci.* 2016, 7, 19–28.

Sedeek, K.E.M.; Mahas, A.; Mahfouz, M. Plant genome engineering for targeted improvement of crop traits. *Front. Plant Sci.* 2019, 10. https://doi.org/10.3389/fpls.2019.00114

Talakayala, A.; Ankanagari, S.; Garladinne, M. CRISPR-Cas genome editing system: A versatile tool for developing disease resistant crops. *Plant Stress* 2022, 3, 100056. https://doi.org/10.1016/j.stress.2022.100056

Tang, W.; Lin, J.; Wang, Y.; An, H.; Chen, H.; Pan, G.; Zhang, Y. Selection and validation of 48 KASP markers for variety identification and breeding guidance in conventional and hybrid rice (*Oryza sativa* L.). *Rice* 2022, 15(1), 1–11.

Tilman, D.; Balzer, C.; Hill, J.; Befort, B.L. Global food demand and the sustainable intensification of agriculture. *Proc. Natl. Acad. Sci.* 2011, 108(50), 20260–20264. https://doi.org/10.1073/pnas.1116437108

Ul Haq, S.I.; Zheng, D.; Feng, N.; Jiang, X.; Qiao, F.; He, J.-S.; Qiu, Q.-S. Progresses of CRISPR/Cas9 genome editing in forage crops. *J. Plant Physiol.* 2022, 279, 153860. https://doi.org/10.1016/j.jplph.2022.153860

Viana, V.E.; Pegoraro, C.; Busanello, C.; Costa de Oliveira, A. Mutagenesis in rice: The basis for breeding a new super plant. *Front. Plant Sci.* 2019, 10, 13–26.

Veillet, F.; Perrot, L.; Chauvin, L.; Kermarrec, M.P.; Guyon-Debast, A.; Chauvin, J.E.; Nogué, F.; Mazier, M. Transgene-free genome editing in tomato and potato plants using agrobacterium-mediated delivery of a CRISPR/Cas9 cytidine base editor. *Int. J. Mol. Sci.* 2019, 20(2), 402. https://doi.org/10.3390/ijms20020402

Voytas, D.F. Plant genome engineering with sequence-specific nucleases. *Ann. Rev. Plant Biol.* 2013, 64(1), 327–350. https://doi.org/10.1146/annurev-arplant-042811-105552

Waltz, E. With a free pass, CRISPR-edited plants reach market in record time. *Nat. Biotechnol.* 2018, 36(1), 6–7. https://doi.org/10.1038/nbt0118-6b

Witcombe, J.R.; Joshi, A.; Goyal, S.N. Participatory plant breeding in maize: A case study from Gujarat, India. *Euphytica* 2003, 130(3), 413–422.

Wolter, F.; Schindele, P.; Puchta, H. Plant breeding at the speed of light: The power of CRISPR/Cas to generate directed genetic diversity at multiple sites. *BMC Plant Biol.* 2019, 19(1), 176. https://doi.org/10.1186/s12870-019-1775-1

Xue, M.; Wang, J.; Jiang, L.; Wang, M.; Wolfe, S.; Pawlowski, W.P.; Wang, Y.; He, Y. The number of meiotic double-strand breaks influences crossover distribution in arabidopsis. *Plant Cell* 2018, 30(10), 2628–2638. https://doi.org/10.1105/tpc.18.00531

Youssefian, S.; Kirby, E.J.M.; Gale, M.D. Pleiotropic effects of the GA-insensitive Rht dwarfing genes in wheat. 1. Effects on development of the ear, stem and leaves. *Field Crops Res.* 1992, 28(3), 179–190. https://doi.org/10.1016/0378-4290(92)90039-c.

Zetsche, B.; Gootenberg, J.S.; Abudayyeh, OO.; Slaymaker, I. M.; Makarova, K.S.; Essletzbichler, P.; Zhang, F. Cpf1 is a single RNA-guided endonuclease of a class 2 CRISPR-Cas system. *Cell* 2015, 163(3), 759–771.

Zhang, F.; Wen, Y.; Guo, X. CRISPR/Cas9 for genome editing: Progress, implications and challenges. *Hum. Mol. Genet.* 2014, 23(R1), R40–R46. https://doi.org/10.1093/hmg/ddu125

Zhang, H.; Zhang, J.; Lang, Z.; Botella, J.R.; Zhu, J.K. Genome editing—Principles and applications for functional genomics research and crop improvement. *Crit. Rev. Plant Sci.* 2017, 36(4), 291–309. https://doi.org/10.1080/07352689.2017.1402989

Zhang, Y.; Massel, K.; Godwin, I. D.; Gao, C. Applications and potential of genome editing in crop improvement. *Genome Biol.* 2018, 19(1), 1–11.

Zhang, Y.; Zhang, F.; Li, X.; Baller, J.A.; Qi, Y.; Starker, C.G.; Voytas, D.F. Transcription activator-like effector nucleases enable efficient plant genome engineering. *Plant Physiol.* 2013, 161(1), 20–27.

CHAPTER 4

Tissue Culture Techniques and Its Role in Crop Improvement

VATSALA TOMAR,[1,*] SURABHI SINGHAL,[1] and NAVNEET SHARMA[2]

[1]*Department of Botany, School of Life Science and Technology, IIMT University, Meerut, Uttar Pradesh, India*

[2]*Department of Biotechnology, School of Life Science and Technology, IIMT University, Meerut, Uttar Pradesh, India*

*Corresponding author

ABSTRACT

Plant tissue culture includes a variety of *in vitro* techniques, methods, and strategies owing to advent of plant biotechnology. Tissue culture has been used to generate genetic mutations. This can upgrade the quality of crops by increasing the health of and increasing their numbers. When a desirable germplasm is available to plant breeders, several tissue culture protocols can be designed for most plant species. However, many crops, especially cereals and woody plants, require further optimization. Tissue culture techniques combined with molecular techniques have been successfully used to incorporate specific traits by gene transfer. *In vitro* culture techniques make use of various materials such as protoplasts, anthers, microspores, and ovules. Embryos have been used to create new genetic variations in breeding lines, often through the production of haploids. Cell cultures have also generated somatic and gametophytic clonal variants that may enhance harvesting. The culture of single cells and meristems can be effectively used to eradicate pathogens from plants. Tissue culture

Technological Advancements in Sustainable Crop Improvement and Production.
Devendra Kumar, Alok Kumar Singh, and Nisha Malik (Eds.)

DOI: 10.1201/9781779643155-4

has dramatically improved the yields of established cultivars by providing a large-scale micropropagation of plants for the commercial, ornamental, and agricultural purposes.

4.1 INTRODUCTION

Plant tissue culture is defined as culturing of plant seeds, organs, explants, tissues, cells, or protoplasts on a chemically designed, synthetic nutrient media under sterile and controlled conditions of light, temperature, and humidity.

Plant tissue culture is the process of growing and multiplying plant cells, tissues, and organs on specified solid or liquid media in an aseptic and controlled environment (Gaikwad et al., 2017). It started as an experiment to illustrate the totipotency notion, which is described as a cell's genetic capacity to become a full multicellular creature, and the cell doctrine, which holds that all living entities are composed of cells, which are the basic units of structure and reproduction (Loyola-Vargas and Ochoa-Alejo, 2018).

Each plant cell contains a totipotency component that will be divided into a whole plant as part of the plant portion's plant process.

According to Hussain et al. (2012), tissue culture is the *in vitro* aseptic growing of cells, tissues, organs, or the complete plant under controlled dietary and environmental conditions, mostly to produce plant clones. The actual clones that are produced are used to organize the selected genotype.

Controlled environments provide the best conditions for the development and repetition of culture. These parameters include the right amount of nutrients, the right pH environment, the right temperature, and the right gas and liquid environment. Partial vegetation culture refers to the growth or preservation of plant cells, tissues, organs, or plants *in vitro* (Tissue and Terminology, 2003). Aseptic culture of plant cells, tissues, organs, and their components is known as plant tissue culture.

In order to clarify some of the fundamental issues in developmental biology, Gottlieb Haberlandt conducted studies on the culture of single cells in the year 1902, which led to the theoretical underpinnings for plant tissue culture. The hypothesis of cell totipotency, which states that any cell possesses all necessary information and has the power to build a complete plant, was developed in response to the finding that single, detached cells of lower plant forms might continue to grow. But only cells capable of mitotic division—meristematic cells in embryos, stem/root tips, cambium, and undifferentiated parenchymal cells—are covered by the hypothesis of totipotency.

Plant tissue tradition has substantially impacted each agriculture and industry. It has made widespread contributions to the development of agricultural sciences nowadays and today; those strategies represent an essential device in cutting-edge agriculture. Apart from its use as a device of research, plant tissue tradition strategies have grown to be of most important business significance withinside the place of micropropagation, sickness elimination, crop development, and manufacturing of secondary metabolites. Plant tissue tradition strategies additionally play an essential function in advent, conservation, and usage of genetic variability for crop development.

The effect of tissue tradition strategies like anther tradition, meristem tradition, mobileular tradition, somaclonal version, and somatic hybridization may be located in advent of latest genetic version by and large through haploid manufacturing, germplasm conservation, micropropagation, molecular farming for manufacturing of secondary metabolites inclusive of meals flavors and fit to be eaten colors, dyes, perfumes, tablets, and eradication of plant pathogens consequently substantially enhancing the yield, specially of vegetatively propagated and perennial crops. With the provision of crop unique protocols, plant tissue tradition strategies could have ever lasting effect on crop development.

It usually takes 10–20 years for select lab supplies to reach the market. Plant breeding: this technology will have a greater impact than ever on crop improvement approaching the new millennium.

Since it offers enormous potential for crop improvement programs like selecting disease, insect, or stress resistant plants, regeneration of the novel hybrid technology, saving the embryos from wide crosses through embryo culture, producing haploid and di-haploid offspring in a short amount of time, etc., plant tissue culture has been given the highest priority in plant breeding work.

4.2 TYPES OF TISSUE CULTURE TECHNIQUES

Plant tissue culture applications include: plant tissue culture has recently become indispensable in both agriculture and industry, providing the plants required to meet the continuously rising global demand. As a result, the following list of unique applications of plant tissue culture, particularly in the field of plant breeding for crop improvement.

4.2.1 EMBRYO CULTURE

Embryo culture was one of the first methods of tissue culture employed in plant breeding. It involves separating the embryo from the seed and raising it in a lab until the plant is prepared to be planted and raised until maturity. Although early-stage embryos (globular and heart-shaped phases) and even unfertilized ovules have been tested, embryo cultivation is typically done with late-stage embryos (Tilton and Russell, 1984). Late-stage embryos can typically be grown on simple nutrient media, but when more immature embryos are cultured, the hormone and protein requirements become increasingly specialized, and success becomes less common.

Embryo culture, an *in vitro* method, is known as an embryo rescue method since it can prevent hybrids from degenerating. Embryo rescue is a common term for the *in vitro* condition used to improve crops. In order to effectively produce the plants, Hanning first grew an isolated mature embryo from a hybrid between Cochlearia and Raphanus in 1904. On a salt and sugar medium, crucifer embryos that were fully developed were employed. Due to the miscarriage of the majority of embryo crosses, interspecific hybridization is the specific approach for embryo cultivation utilized in plant breeding. Immature embryos have been shown to be the most popular explants for wheat transformation. However, obtaining immature embryos necessitates unique procedures, and its application is eventually limited. Somatic embryo production and callus growth are important alternatives to mature embryos that can be exploited in wheat transformation. According to research by Turhan and Baser (2004), the production of a callus in wheat from mature embryos is beneficial for improving wheat crops using both genetic engineering and contemporary plant tissue culture techniques.

In the majority of situations, substantial hybridization initiatives including sexual crosses between unrelated plants require embryo cultivation to recover plants (embryos). In many cases, the disintegration of the endosperm, which feeds the growing embryo, causes incompatibility. By "rescuing" an embryo and cultivating it on an appropriate medium, a plant can frequently be produced to maturity (Collins and Grosser, 1984).

The primary reason for generating wide crosses is to transmit desired traits (such disease resistance or stress tolerance) from distantly related species to cultivated varieties. Embryo culture has been used to successfully cross various plant species, including cotton, tomato, barley, rice, cabbage, melons, beans, and jute.

4.2.2 MICROPROPAGATION

It is common to think of micropropagation as a development of more traditional plant propagation techniques. Its objective is to use clonal multiplication to quickly increase superior genotypes of pest- and disease-free plants. Buds from a desired plant are planted on an appropriate medium under certain growing conditions to support the development of axillary shoots. Repeated subculture of the buds and shoots results in a huge number of plants with identical genetic characteristics to the original (Hussey, 1983).

One of the most significant applications of micropropagation is the bulk production of superior plants. Traditional propagation is frequently a drawn-out process that may be impeded by illness and insect worries. Once a year, micropropagation has the capacity to produce hundreds, if not millions, of plants, but the number of plants that can be handled is frequently constrained. Before using conventional methods of multiplication, fast expansion is frequently a helpful first boost for thc crcation of enormous populations (Damiano et al., 1983). Additionally, germplasm can be stored through micropropagation to keep a healthy supply (Wilkins and Dodds, 1983).

4.2.3 MERISTEM CULTURE

Apical/axillary shoot meristems from explants are cultured during meristem culture for the growth of the entire plantlet. Meristematic tissues are categorized according to where they are found as follows: (1) apical meristems, which are found at the plant's root and shoot tips. This area's division aids in the development of the root and shoot; (2) the intercalary meristem, which is situated in the intercalary location, lengthens the internodal period; and (3) lateral meristems are situated at the lateral side of stems and roots and are engaged in thickening the plant. Such meristem is seen in monocots and pines. The size of the explant is crucial to achieving the culture's specific objective. The process of producing virus-free plants using meristem tips between 0.2 and 0.5 mm is known as meristem-tip culture. This tissue culture approach has uses in the elimination of viruses, especially in crop species that are vegetatively propagated and have a high risk of transmitting systemic viral infections during the propagation process, mass multiplication, preservation of germplasm, and genetic alterations.

4.2.4 ANTHER CULTURE

The haploid production by anther culture or pollen culture, which was first established by Guha and Maheshwari (1964, 1966) in Datura, is a crucial component of plant tissue culture. The embryogenesis of *in vitro* cultivated anthers, also known as androgenesis, which involves a succession of cell division and differentiation, promotes the generation of haploid plantlets. The quick recovery of homozygous lines from heterozygous source materials is made possible by the pollen grain's ability to produce haploids. Through the diploidization of regenerated haploid plants, anther culture enables fixation of homozygosity, leading to the creation of inbred lines that are helpful for hybrid production in cross-pollinated crops (and as pure lines in self-pollinated crops). The absence of the masking effects of recessive alleles makes it easier to detect mutations and recover rare combinations in haploids produced by anther culture.

When compared to conventional breeding, anther culture has the advantage of producing haploid plants with a higher frequency, which reduces both time and expense. Anther culture, on the other hand, has drawbacks such as low frequency of haploid production, lack of technique for haploid production in many crops, difficulty in identifying spontaneously doubled haploids and diploids originating from somatic tissues, and the significant time and labor required to produce large numbers of plants needed for evaluation purposes.

4.2.5 SOMACLONAL VARIATIONS

It is crucial to maintain the genetic integrity of plants when they are being multiplied *in vitro*, however, following micropropagation or plant regeneration from calli or somatic embryos, so-called somaclonal variations have been noted. The observed variation in the regenerated plants may be heritable or epigenetic. In micropropagated plants, somaclonal variation has been seen at several morphological, cytological, cytochemical, biochemical, and molecular levels. Somaclonal changes brought on by tissue culture are comparable to those brought on by physical and chemical mutagens. Such variances could be kept to a minimum by shortening the *in vitro* culture period and skipping the callus phase. Somaclonal variants have significant advantages over other genetic modification techniques in that they are less expensive, do not necessitate containment procedures, and can be used as an alternative tool to introduce variation into breeding programs and produce commercial varieties with new traits like pathogen resistance, chilling tolerance, drought tolerance, salinity tolerance, herbicide tolerance, and submergence tolerance.

However, because somaclonal variations are unpredictable and unreproducible, it is impossible to foresee how they will turn out. Somaclonal variation would also be a constraint in commercial-scale micropropagation, when the goal is to preserve consistency.

4.2.6 SOMATIC EMBRYOGENESIS

Somatic embryogenesis is the process by which nonsexual (somatic) cells develop into structures resembling embryos. Despite the perception that somatic embryogenesis is a challenging process, there are a number of benefits to recovering plants from cells utilizing somatic embryogenesis as opposed to micropropagation. Cells grown in suspension are widely used to create somatic embryos, which enables batch culture processes to be scaled up with little additional handling expense. Some plants, such carrots, tobacco, potatoes, celery, and tomatoes, reproduce incredibly quickly. Because of this, the embryos of these plants are encapsulated and used as artificial seeds (Ng, 1986).

Callus synthesis is fairly easy for many plant species, but regeneration from disorganized cells, such callus and cell suspensions, is frequently more challenging.

Furthermore, there are certain crucial problems when employing somatic embryogenesis to quickly clone a vital plant. Numerous studies have demonstrated that genetic variety in regenerated plants can be produced by callus and cell suspension regeneration. This variance in tissue-grown plants is referred to as a somaclonal variation (Larkin and Scowcroft, 1981).

4.2.7 SOMATIC HYBRIDIZATION

Protoplast fusion, also known as somatic hybridization, is a method of manipulating cellular genome technology that entails fusing plant protoplasts from somatic cells of two different species that cannot be crossed naturally and then regenerating hybrid plants from the merged protoplasts. Somatic hybridization shortens the breeding cycle and makes it possible to develop crop plants through unusual methods. It also allows for the hybridization of plants at any stage, even those in their infancy. This tissue culture method is particularly beneficial for improving heterozygotic crops with various ploidy levels, such as potatoes. Somatic hybridization can also be used to transmit cytoplasmically encoded features that are significant in agriculture, such as cytoplasmic male sterility, antibiotic resistance, and herbicide tolerance. Even between easily crossable species, hybrids can be created between absolutely sterile clones, and mitochondria from one species can unite with chloroplasts from another species without doing so sexually. However, the capacity to regenerate the fused product and interspecific hybrids' sterility, which limits their employment in breeding programs, limit the development of distinct somatic hybrid plants. Another restriction of somatic hybridization is the removal of chromosomes from the hybrid cell in specific wide crossovers.

4.3 ADVANTAGES OF TISSUE CULTURE TECHNIQUES

Concerns concerning the extinction of species, genetic variety loss, habitat degradation, and loss of plant populations have emerged as a result of the rising market demand for goods made from natural matrices (Roberto and Francesca, 2011). An innovative propagation technique called tissue culture has had a profoundly favorable effect on the agricultural sector. Micropropagation enables the bulk replication of outstanding clones by enabling the construction of vast numbers of plants from minute fragments of the mother plant. It only needs a few minutes of your time during the assembly to grow plants. One ex-plant might quickly multiply into thousands of plants in a single year, depending on the species being produced. It is important to eliminate diseases from planting materials.

The purchase, upkeep, and widespread proliferation of specific disease-free plants is another area where plant component culture thrives.

The concept of using tissue culture as a variety system and indexing pest-free plants are pretty similar. As tissue culture explants, plant tissues that are known to be disease-free (viral, bacterial, or fungal) are physically chosen. To prevent or get rid of disease, which can ruin plants, tissue culture might be a viable technique.

Tissue culture for plant enhancement By using tissue culture, it is feasible to develop agricultural crops that are superior to those that would otherwise be impossible to develop using conventional plant breeding techniques. Production of plants that are true to type: Many faithful sort plants can be propagated in a small space and at any time of year. For instance, it will be feasible to develop two to four lakhs of tissue-grown plants from a single shrub or rose, as opposed to the 10–15 plants that may be multiplied now using existing methods.

Furthermore, the newest method of plant propagation takes at least 6–8 months for many species, whereas tissue culture can produce good planting material in about 2–4 months.

Enhanced Branching and Flowering, Greater Vigor, and Better Production: Due to the potential for disease eradication, tissue culture plants may exhibit increased branching and flowering, greater vigor, and better yield. Effective way to save energy and space: The plan saves the farmer's energy and space. For instance, in a traditional method, plants are grown in an open farm that needs about 25,000 m of land. In a tissue culture laboratory, however, the same number of plants only need 10 m of room.

Flexible approach: Childcare facilities usually become more adaptable. Due to the almost zero capital investment in the mother plant, it will be simpler to adapt to changing circumstances. Furthermore, a substantially more efficient production schedule is conceivable due to the improved plant uniformity and consequent availability within the bulk at any time. Tissue culture is commonly employed in the generation of novel types.

4.4 DISADVANTAGES OF TISSUE CULTURE TECHNIQUES

With a labor cost of up to 70%, it is prohibitively expensive. A monoculture develops after micropropagation, which results in a lack of general disease resistance because all progeny plants could be vulnerable to the same ailments. A contaminated plant sample can result in the birth of infected progeny. If the stock plants are carefully examined and screened to prevent developing plants that are afflicted with viruses or fungi, this is uncommon. Due to a lack of understanding about the ideal growing medium or the plants' production of secondary metabolic chemicals that stunt or kill the explant, not all plants can be successfully tissue grown (Sharma et al., 2015). Losses brought on by infections; potential for off-types/spontaneous mutations (Consultant, 2019).

Somaclonal variation: Any variation that occurs during multiplication may be undetected, and many tree species, such as mango, may not react to *in vitro* growth because of species/genotype recalcitrance (Biotechnology, 2007).

4.5 SUMMARY AND CONCLUSION

Plant tissue culture is now a well-recognized technology that has significantly impacted the growth and propagation of agricultural crops. The use of *in vitro* reproduction has the potential to be an effective tool in both basic and applied biology. Fundamental research is increasingly using the *in vitro* method as a paradigm for in-depth studies at the cellular and molecular level. The technology of plant tissue culture has been widely used to enhance both important agricultural crops and natural species that are in danger of extinction. Mutation and somatic hybridization are employed in a variety of ways with this special technology to promote crop species diversity.

With its many objectives, sustainable development seeks to utilize plants as economically as possible while ensuring a secure future for future generations. The management of remaining natural and semi-natural lands, ecological restoration, and, more broadly, the restoration of natural capital, which includes ecological restoration, ecological, and economic rehabilitation of production systems, will all be necessary to ensure that plant resources and wild habitats are preserved. In addition to ensuring that this biodiversity is available to support current and future use, safeguarding its *in situ* and *ex situ* components will also help to ensure that it is accessible for management and restoration projects.

The introduction of novel *in vitro* techniques that use inexpensive tissue culture and *ex vitro* rooting and adaptation to lower the value of production per plant presents difficulties for producers.

Tissue culture is one of the most important components of applied biotechnology. The population of the world will keep expanding in the coming decades, necessitating more living space and farmland. Another thing to think about is the effects of global warming. We must give our kids and grandkids a pleasant, healthy, and hunger-free environment in light of this. For this, plant part cultivation is the only option.

KEYWORDS

- **micropropagation**
- **gene transfer**
- **crop improvement**
- **genetic variations**
- **haploid production**

REFERENCES

Biotechnology, P. *Introduction to Plant Biotechnology*, 2007, pp. 1–42.

Collins, G.B.; Grosser, J.W. Culture of embryos. In: Vasil, I.K.; (Ed.), *Cell Culture and Somatic Cell Genetics of Plants*. Academic Press: Orlando, 1984, pp. 241–257, Vol. 1.

Consultant, M.H.H. *Plant Tissue Culture 2019 Medicinal Cannabis Conference My Back Ground*, 2019 March, pp. 1–25.

Gaikwad, A.V.; Singh, S.K.; Gilhotra, R. Plant tissue culture-A review. *SGVU J. Pharmacy. Res. Edu.* 2017, 2(1), 217–220.

Guha, S.; Maheshwari, S.C. In vitro production of embryos from anthers of Datura. *Nature* 1964, 204(4969), 497–498. https://doi.org/10.1038/204497a0

Guha, S.; Maheshwari, S.C. Cell division and differentiation of embryos in the cultured anthers of Datura. *Nature* 1966, 212(5069), 97–98.

Turhan, H.; Baser, I. Callus induction from mature embryo of winter wheat (Triticum aestivum L.). *Asian J. Plant S.* 2004, 3.1, 17–19.

Hussain, A.; Qarshi, I.A.; Nazir, H.; Ullah, I. Plant tissue culture: Current status and opportunities. In: *Recent Advances in Plant in Vitro Culture*, 2012, pp. 1–28.

Hussey, G. In vitro propagation of horticultural and agricultural crops. In: Mantell, S.H.; Smith, H.; (Eds.), *Plant Biotechnology*. Cambridge University Press: Cambridge, 1983, pp. 111–138.

Larkin, P.J.; Scowcroft, W.R. Somaclonal variation—A novel source of variability from cell culture for plant improvement. *Theoret. Appl. Genetics* 1981, 60, 197–214.

Loyola-Vargas, V.M.; Ochoa-Alejo, N. An introduction to plant tissue culture: Advances and perspectives. In: *Plant Cell Culture Protocols*, 2018, pp. 3–13.

Ng, T.J. Use of tissue culture for micropropagation of vegetable crops. In: Zimmerman, R.H.; Griesbach, R.J.; Hammerschlag, F.A.; Lawson, R.H.; (Eds.), *Tissue Culture as a Plant Production System for Horticultural Crops*. Martinus Nijhoff: The Netherlands, 1986, pp. 259–270.

Roberto, T.; Francesca, M. Sustainable sourcing of natural food ingredients by plant cell cultures. *Agro. Food Ind. Hi Tech*. 2011, 22, 26–28.

Sharma, G.K.; Jagetiya, S.; Dashora, R. *General Techniques of Plant Tissue Culture*, Lulu Press Inc.: Raleigh, NC, United States, 2015.

Tilton, V.R.; Russell, H. Applications in in vitro pollination/fertilization technology. *Biocience*. 1984, 34, 239–242.

Plant Tissue Culture Terminology. Phyto Technology Laboratories, Inc., 2003, pp. 0–2.

Wilkins, C.P.; Dodds, J.H. The application of tissue culture to plant genetic conservation. *Sci. Prog. Oxford* 1983, 68, 259–284.

CHAPTER 5

Application of Genome Editing for Sustainable Development of Agriculture: A Perspective View

SOUMYAYAN ROY[1], MAINAK BARMAN[1], ARPITA DAS,[1*] and SANJEEV GUPTA[2]

[1]Department of Genetics and Plant Breeding, Bidhan Chandra Krishi Viswavidyalaya, Mohanpur, Nadia, West Bengal, India

[2]Crop Science Division, ICAR, Krishi Bhawan, New Delhi, India

*Corresponding author

ABSTRACT

Genome editing has revolutionized plant science by offering precise manipulation of gene expression to improve plant architecture, agronomic performance, and quality. Unlike traditional mutation breeding, genome editing, particularly through engineered nucleases, has shown great promise in enhancing diverse crop plants. Recent technological advancements have provided scientists with innovative tools to generate new variability in desired traits, supporting sustainable and boosted crop production. This chapter focuses on the CRISPR/Cas9 system, which has transformed agricultural science over the past decade. The CRISPR-edited plant products are often classified as non-GMOs, bypassing stringent regulatory processes, thus opening new opportunities globally and in India. This chapter updates the basic steps and cutting-edge tools used in plant genome editing, emphasizing CRISPR/Cas9. It explores the emerging applications of genome editing in plant science, highlighting future prospects for enhancing crop productivity and resilience. By leveraging

Technological Advancements in Sustainable Crop Improvement and Production.
Devendra Kumar, Alok Kumar Singh, and Nisha Malik (Eds.)

DOI: 10.1201/9781779643155-5

genome editing, scientists can address pressing agricultural challenges, contributing to global food and nutritional security. This technology holds significant potential for improving crop traits and ensuring sustainable agricultural practices. As genome editing continues to evolve, it is poised to play a crucial role in shaping the future of agriculture, offering solutions to enhance crop performance and meet growing global demands.

5.1 INTRODUCTION

The rapidly growing scenario of the global population is anticipated to touch 9.8 billion by 2050. To meet the needs of the growing population, it is essential to produce twice as much food as we currently harvest (UNDESA, 2017). Another major challenge for agriculture now is the rising temperature, which is predicted to increase by 1.1–5.4°C by the end of this century (Tollefson, 2020). These alarming predictions clearly indicate that crops will struggle with heat stress throughout their growing period, along with more frequent drought conditions (Rustgi et al., 2021). These vicissitudes might result in instability and insecurity of nutritional aspects of crops along with declined crop productivity (Molotoks et al., 2021). Breeders are gruelingly trying to find innovative paths to encounter this possible mandate for sustainable food grain production despite changing climatic scenarios and accompanying alterations of the pest dynamics (Maja and Ayano, 2021). Diminishing resources are the next layer of complexity in the way of improving agriculture in a sustainable manner. To meet these scarcities, improved cultivars of diverse crops have been developed mostly by adopting plant breeding tactics by conventional means entailing genetic crossing followed by selection for the trait(s) sought, which largely emphasized on primary gene pool of the crop (Kaiser et al., 2020). However, current developments nowadays display that plant breeding by molecular biology approaches comprises genomic as well as biotechnological methods, enabling introgression or editing of favorable genetic alterations from a more diverse gene pool with more accuracy and rapidity in the crop genome.

Consequently, conventional crop breeding strategies are belligerently being complemented by methods involving molecular approaches with the aim of attaining anticipated outcomes in a comparatively lesser time frame (Hasan et al., 2021). Use of cutting-edge biotechnologies for advancing yield and nutritional excellence is facilitating to preserve agriculture of modern era. Genome editing in crops permits us to make alterations of sequence at a definite locus and accordingly allows us to precisely modulate attributes of interest. The simple operating nature, cost-effectiveness, and efficiency of genome editing technologies have aroused the eagerness of researchers and have rapidly developed in the last few years (Figure 5.1). This chapter deals with the mechanisms of genome-editing technologies in this current era, along with their applications toward food security, and also puts light on the regulatory mechanisms for the release of genome-edited crops.

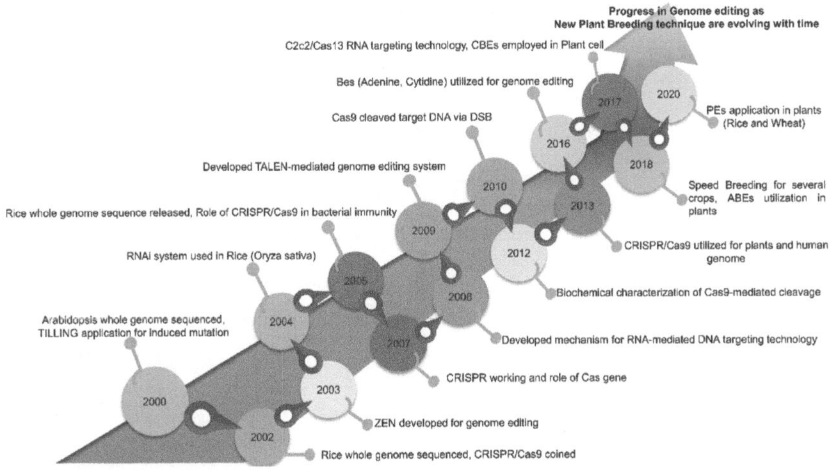

FIGURE 5.1 Last 20 years timeline of genome editing technology.

5.2 HOW GENOME EDITING WORKS?

Genome editing mainly utilizes two techniques for modifying the genome at specific positions. The major one involves DNA repair post-breakage.

5.2.1 DOUBLE-STRAND BREAK-BASED GENOME EDITING

DNA damage by several external factors, like radiation and internal agents of the cell, namely, free radicals and metabolic by-products, ensues naturally. Such damages can lead to several breaks in the DNA strands of an organism. Among these forms of DNA damage, compared to single-strand break, a double-strand break (DSB) is more fatal, and the cell repairs it immediately before DNA replication.

DNA repair is mediated by two pathways in eukaryotes: (1) nonhomologous end joining (NHEJ) and (2) homology-directed repair (HDR) (Figure 5.2). Repair processes of the cell hold immense importance in genome editing as the process of repairing one or more DNA DSB by an engineered endonuclease is the central methodology that governs the genome-editing methodologies (Jansing et al., 2019).

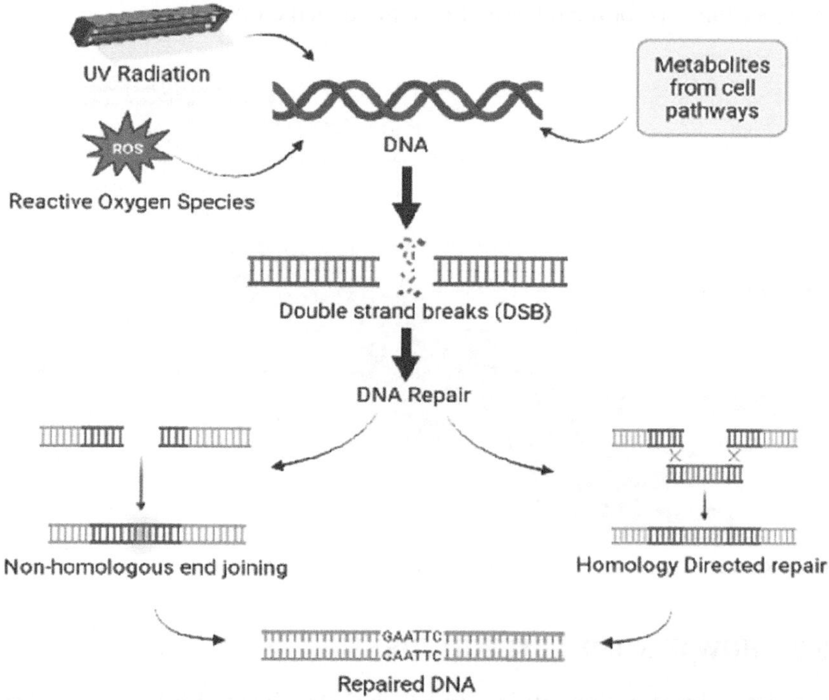

FIGURE 5.2 DNA damage for the formation of DSBs and DNA repair through NHEJ and HDR.

5.2.2 DNA REPAIR FOLLOWING DOUBLE-STRAND BREAKS

Knowledge of DNA repair processes is necessary for understanding the engineered endonucleases used to manipulate the repair process for genome editing.

DNA repair, followed by the formation of DSBs, is separated by their use or nonuse of homology for repair. NHEJ does not use homology for repair, so it is quick but imprecise, and is carried out in cells utilizing a classical method wherein there is no use of homology and a micro-homology-based alternative in NHEJ.

In classical NHEJ, broken DNA ends are bound together by multiple proteins like Ku70, Ku80, and so on, and ligase joins the ends together, forming Insertion and Deletion of nucleotides. Alternatively, in micro-homology NHEJ, the 5′ end of the broken region is cleaved until there is homology with the overhanging 3′ end. The complementary sequences bind, and the nonhomologous DNA flaps are excised, leading to deletions in the DNA preceding the homologous sections and forming frameshift mutations. Overall, NHEJ can result in indel formation and nonfunctional genes due to the formation of premature stop codons (Malzahn et al., 2017). The comparison between the NHEJ and HDR is represented in Figure 5.3.

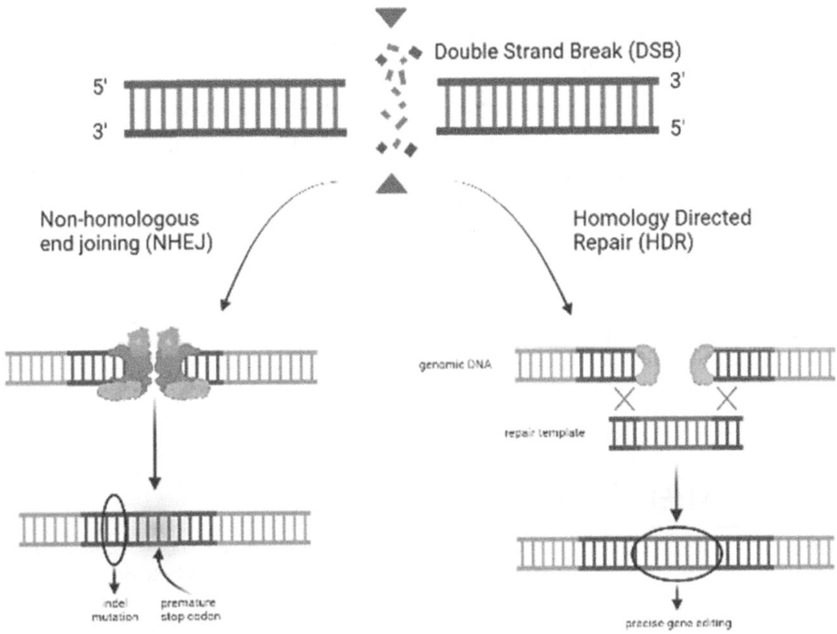

FIGURE 5.3 DNA repair of DSBs using NHEJ and HDR: a comparison.

The HDR-based DNA repair system utilizes template DNA's homology to repair the DSBs. Synthesis-dependent strand annealing pathway mostly mediates the HRD-based DSB repair in plants. In case of a DSB, the 3′ overhang is extended, the 5′ end forms a D-loop with the extended region, and the gaps are filled using homologous DNA as the template. Finally, the second 3′ end is visualized to be reannealed with the 3′ end without any crossover. Thus, HDR is precise, resulting in an accurately repaired DNA (Bonawitz et al., 2019).

5.2.3 UTILIZATION OF DNA REPAIRING MECHANISMS FOR GENOME EDITING

The DNA repair methods can be utilized for genome editing as these methods can be used for precise modification of preselected sequences if DSBs can be inserted in the sequences.

In case of NHEJ, if the DSB is within the coding sequence, the indel created by NHEJ can enable to generate knockout, plants which can be utilized in gene function studies as well as to exclude undesirable traits in organisms. In contrast, HDR uses an externally introduced donor DNA template having target site homology for implementing the required sequence in the HDR region. This modification can help to engineer the endogenous genes precisely and accurately add sequences or genes at the predetermined loci (Chen et al., 2019) (Figure 5.4).

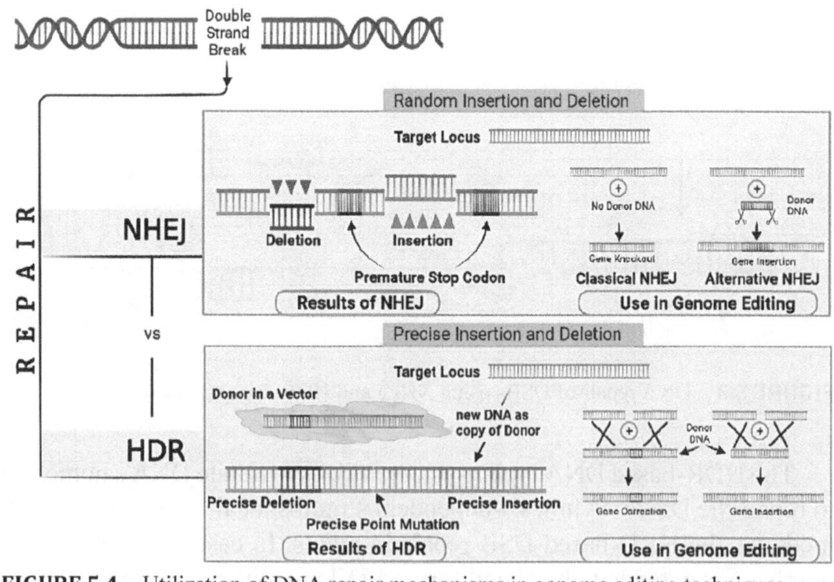

FIGURE 5.4 Utilization of DNA repair mechanisms in genome editing techniques.

5.2.4 IMPLEMENTING DSBS PRECISELY IN SEQUENCES

Consideration of the addition of DSBs in the required region is achieved using programmable nucleases. Currently, engineered nucleases comprise meganucleases, transcription activator-like effector nucleases (TALENs), zinc finger nucleases (ZFNs), and Cas9 (Table 5.1). The terminal one exists in the CRISPR/Cas9 system (Chen et al., 2019).

5.2.1.1 CLASSIFICATION OF CRISPR/CAS SYSTEMS

Marraffini (2015) differentiated three central CRISPR/Cas systems. Out of which, type II is used in genome editing applications. Table 5.2 shows a comparison of the three systems based on the significant parameters.

5.3 GENOME EDITING THROUGH MODIFYING THE BASES

Specific base changes through base-editing systems provide an alternative to the DSB-based genome editing strategy. The base-editing techniques offer an efficient and universal approach for single-base editing and provide a solution to the unpredictability of NHEJ and the low efficiency of HDR. The base editing techniques include a cytosine base editor (CBE) and an Adenine base editor system.

5.3.1 CYTOSINE BASE EDITOR SYSTEMS

The CBE system consists of nicking Cas9 (nCas9) bearing the D10A mutation for deactivating the nuclease domain RuvC, a uracil glycosylase inhibitor, and a cytidine deaminase. The cytosine in the DNA is converted to uracil by cytidine deaminase, which gets converted to thymine during DNA replication. But as there is a high chance of uracil excision before DNA replication, the glycosylase inhibitor prevents the excision. It ensures base excision repair activity for the conversion of cytosine to thymine (Chenet al., 2019) (Figure 5.10).

5.3.2 ADENINE BASE EDITOR (ABE) SYSTEMS

The ABE system developed by Gaudelli et al. in 2017 can mediate Adenine (A) to Guanine (G) conversion with high purity and precision in a wide range of targets. The ABE converts Adenine to inosine. The final iteration of the

TABLE 5.1 Details of Different Sequence-Specific Nucleases (SSNs) Applied for Editing of Genomes

Sequence-Specific Nucleases	Mode of Working	Diagrammatic Representation	Drawbacks	References
Meganucleases	Meganucleases identify 12 bp long DNA sequences for site-specific endonuclease activity. Modified meganucleases can target endogenous genes rather than a single sequence, as only a small number of amino acids attach to nucleotides and altering them produces the desired result (Figure 5.5).	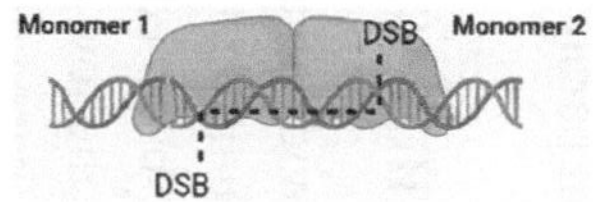 **FIGURE 5.5** Meganuclease creating a DSB.	Meganucleases retain their nuclease activity with the needed DNA-binding properties, making them difficult to use in genome editing experiments.	Grizot et al. (2010)
Zinc finger nucleases (ZFNs)	Zinc finger nuclease acts as a dimer; each monomer has three to four domains binding Zinc Finger DNA. Individually zinc finger module identifies a nucleotide triplet and the Zinc Finger domains are attached to a catalytic domain of the restriction endonuclease FokI (Type IIS) which is sequence-independent. To cleave the DNA, the nuclease then forms a dimer (Figure 5.6).	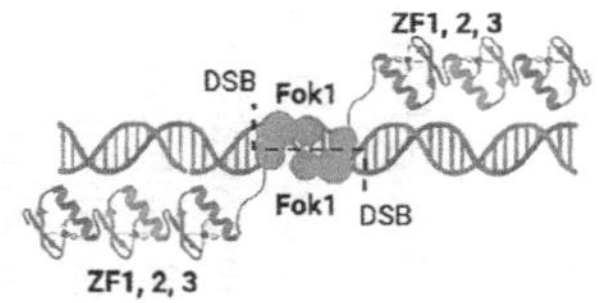 **FIGURE 5.6** ZFN creating a DSB using Fok1 endonuclease.	ZFNs require identified matching sites, so short regulatory regions and small noncoding RNA genes are hard to target. Modular ZFNs have poor activity and high toxicity.	Szczepek et al. (2007), Miller et al. (2007), and Doyon et al. (2011)
Transcription activator-like effector nucleases (TALENs)	Transcription activator-like effector (TALE) consists of 33–35 amino acid domains identifying a single nucleotide. Engineered TALE domains consist of multiple TALE domains that recognize 15–30 nucleotides. This is bound to a Fok1 endonuclease and works as a dimer to cleave the DNA at the specific position (Figure 5.7). Due to the individual specificity of TALE domains, this is much more specific than ZFNs.	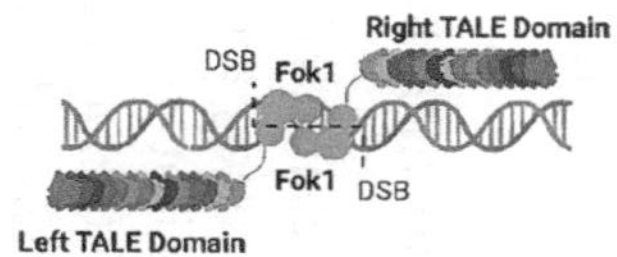 **FIGURE 5.7** TALEN creating a DSB using Fok1 endonuclease.	The TALE domain has multiple repetitive sequences known as repeat variable di-residues. This makes the TALE construction via PCR difficult and can be constructed via multiple assembly methods using Golden Gate cloning.	Moscou and Bogdanove, 2009; Boch et al. (2009)

TABLE 5.1 *(Continued)*

Sequence-Specific Nucleases	Mode of Working	Diagrammatic Representation	Drawbacks	References
CRISPR/Cas9	Cas9 system is a monomeric nuclease forming a complex of ribonucleoprotein (RNP) with a chimeric guide RNA (gRNA) to cleave the DNA at the desired location. CRISPR are multiple short, direct repeats containing loci producing CRISPR RNA that couples with tracrRNA to form the guide RNA (gRNA). The guide RNA provides a 20 nucleotide-specific sequences complementarity for identification by Cas9 (Figure 5.8).	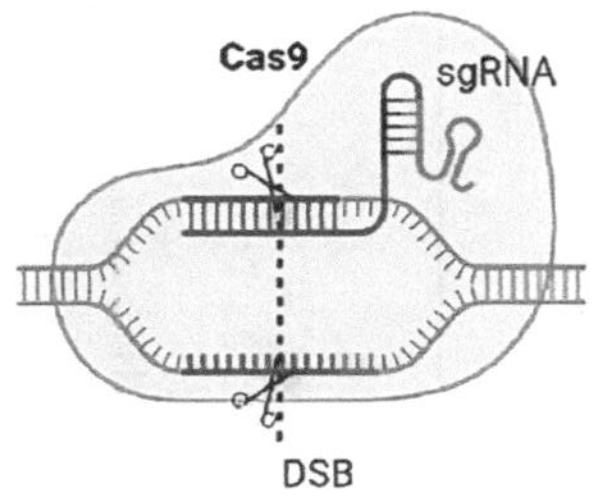 **FIGURE 5.8** Cas9 creating a DSB using gRNA to identify the sequence.	The CRISPR/Cas9 system generates several off-target mutations if not implemented correctly, and for proper usage, whole genome sequencing is a necessity in most cases.	Shan et al. (2013), Nekrasov et al. (2013), and Li et al. (2013)

TABLE 5.2 Comparison Between Three CRISPR/Cas Systems

Parameters	Type I System	Type II System	Type III System
Complex composition	The crRNA-cascade complex comprises of Cas6e and Cas8 proteins.	Only a single Cas9 gene is present	Cas10 complex is present along with precursor CRISPR RNA (crRNA).
Process	Cas8 recognizes the protospacer adjacent motif (PAM) and Cas6e cleaves precursor crRNA.	tracrRNA mediates the association of Cas9 through sequence complementarity, and the PAM binding motif of Cas9 recognizes the PAM	Precursor crRNA is cleaved by Cas6 (not part of the complex) and loaded on the Cas10 complex
Cleaving of DNA	Cas3 cleaves the DNA	Cas9 cleaves the double strand	Mature Cas10 complex cleaves the DNA
PAM position	PAM is present upstream of the target sequence	PAM is present immediately downstream of the target sequence	PAM not required

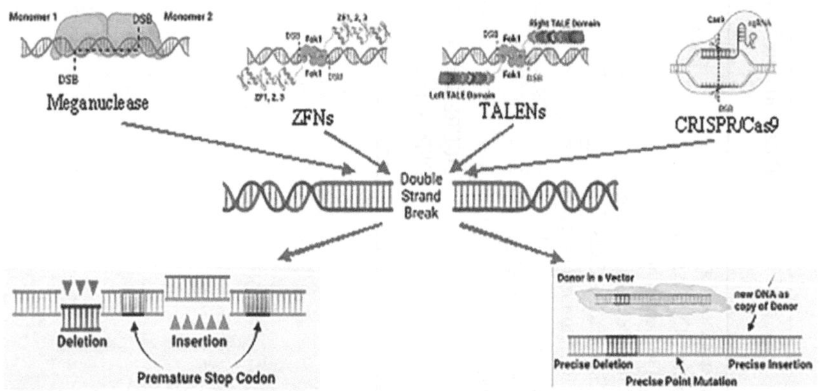

FIGURE 5.9 Strategies involved in DNA repair-based genome editing techniques.

ABE developed through protein engineering and directed evolution includes a tRNA adenosine deaminase extracted from *Escherichia coli* combined with nCas9 for precise modification of adenine (Chen et al., 2019; Figure 5.10).

5.3.3 DUAL-BASE EDITING

Dual-base editor simultaneously edits adenine and cytosine bases using a single-guide RNA (sgRNA19). The unit comprises a nCas9 (D10A) with two deminases: cytidine deaminase (APOBEC3A) and adenosine deaminase

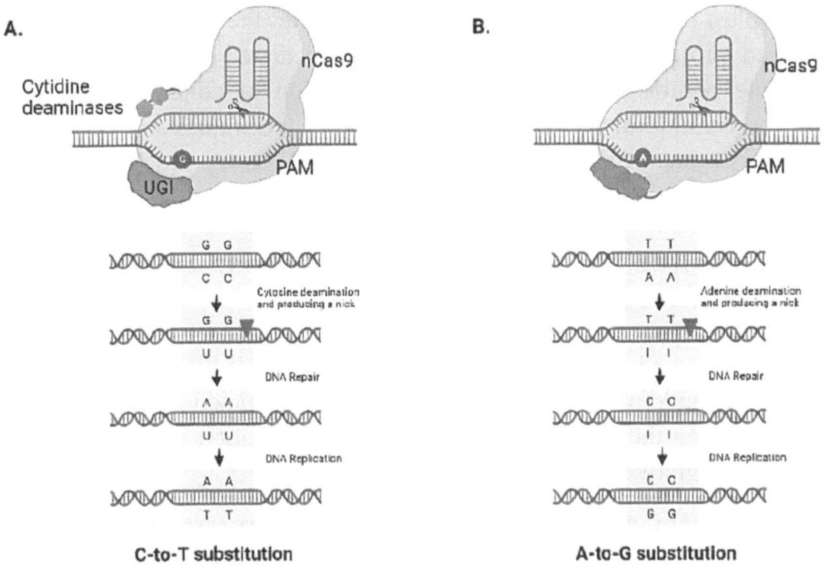

FIGURE 5.10 Base editing mechanisms: (A) C-to-T substitution and (B) A-to-G substitution.

(ecTadA–ecTadA*), and a fusion of UGI. This component is denoted as "saturated targeted endogenous mutagenesis editor" (Zhu et al., 2020).

5.3.4 ADVANTAGES OF BASE EDITING OVER DNA REPAIR-BASED METHODS

The advantages that base editing provides over DSB-based methods are as follows:

1. Base editing produces nonsense mutations comparatively easier than indels in DSB-based methods.
2. The whole-gene base or multiplex editing will not lead to large-scale inversions or deletions.
3. The undesired product generation is far less than DSB-based methods, as the process is more efficient.

5.4 DELIVERY OF GENOME EDITING SYSTEMS

The delivery system for editing plant cells is carried out through particle bombardment, *Agrobacterium*-mediated DNA transformation, and protoplast fusion. The protoplast fusion mediates the transient expression-related projects, whereas *Agrobacterium*-mediated transformation and particle bombardment mediate the production of edited plants. Some newer techniques, such as nanoparticle-based techniques and virus-based delivery systems, are also used (Chen et al., 2019).

5.5 EXPRESSION ANALYSIS OF GENOME-EDITED PLANTS FOR CONFIRMATION

Cas protein and sgRNA related to the desired CRISPR/Cas system may integrate randomly in the plant genome or be degraded in the cytoplasm. So, checking their expression for further confirmation is necessary. It is done through two major ways:

1. **Genome Editing Followed by Stable Expression of CRISPR/CAS DNA**: This method is based on the traditional DNA transformation principle, where the CRISPR/Cas system is transferred through *Agrobacterium*-mediated transfer or particle bombardment and marked through a transmarker gene for expression after the needed genome editing is performed. But the transgene can be integrated at unknown sites where they might show expression and increase the off-target chances, limiting the commercial application on a large scale (Chen et al., 2019).
2. **Genome Editing Followed by Transient Expression of CRISPR/CAS DNA**: In this method, the use of a transgene is eliminated; thus, the requirement for herbicide or antibiotic screening is removed. This prevents the need for foreign DNA incorporation in the plant genome. Moreover, it also reduces the time for tissue culture by a month. Importantly, the frequency of mutation is nearly similar to the conventional method. So, it is used in base editing applications also (Chen et al, 2019).

Another method is being developed by Zhang et al. (2016), where DNA is not transferred; rather RNA is delivered through Cas9/sgRNA, a ribonucleoprotein (RNPs) complex in plants. This reduces off-target effects to a great extent.

5.6 APPLICATIONS OF GENOME EDITING TECHNOLOGY TOWARDS FOOD SECURITY

5.6.1 ABIOTIC STRESS TOLERANCE

Global climate-changing scenario depicts irregularity in rainfall, more frequent periods of dryness, an increase in sea-water level touching new land areas, and overall, a changed climate (Pendergrass et al., 2020). New varieties and cultivation practices are required to withstand these changes. Not only is the yield of concern to feed the increasing global population, but the quality of economic harvest is also going to deteriorate (Kumar, 2016; Zhao et al., 2017). In this situation, it is necessary to advance plants that can adapt well changing climate. There are two major approaches at the molecular level to improve the abiotic stress tolerance mechanism in crops: (1) to boost the expression or regulation of genes mediating the tolerance, and/or (2) to make modifications in the sensitivity genes like the reactive oxygen species production-controlling transcription factors (Zafar et al., 2020). If we can describe some linked genes along with their role in how they act in stress, then modification of all differentially regulated genes under stress may not be required to achieve a certain stress tolerance.

If we consider the perilous abiotic stresses that affect economic harvest, drought is counted as one of the most severe in the list, and it is a complex trait. Numerous physiological properties and pathways contributing to plants' responses linked with increased drought tolerance have already been identified. Furthermore, stability in yield performance is a must to be assured. One such achievement is the hybridization of knock-out lines with intensified tolerance to drought with edited *OsSRL1*, *OsSRL2*, and wild-type rice genotypes (Liao et al., 2019). The factors regulating abiotic stress response pathways negatively are not well documented; hence, reports on genome-edited studies are insufficient on abiotic stress tolerance. Lim et al. (2022) have reported a very recent work where Nam-Chon Paek et al. targeted the *WRKY5* gene in rice using gene editing to generate loss-of-function mutation, where gene editing-based *OsWRKY5* inactivation not only led to decreased water loss and increased drought tolerance but also improved the yield of the crop. Opportunity is there to alter the cis-regulatory element(s) sequence adopting new genomic techniques (NGT) to change some genes' regulation engaged in drought stress tolerance (Nakashima and Yamaguchi-Shinozaki, 2013; Planas-Vishwakarma et al., 2017; Planas-Riverola et al., 2019). Use of multiplex CRISPR techniques to concurrently edit genes of sensitivity and tolerance can be another choice (Zafar et al., 2020). An effective negative regulator of abscisic acid (ABA)-mediated stress signaling is receptor for activated C kinase 1 (*RACK1*), which is a protein from the WD40-repeat family or beta-transducing repeats. Subsequent RNA interference (RNAi)-mediated knockdown of a *RACK1* homolog pointedly improved salt and drought tolerance in soybean crops. The same happened in the case of rice, implying *RACK1* as a decent target for genome editing (Zhang et al., 2013). Crops like rapeseed and wheat with gene-edited mutants of *ENHANCED RESPONSE TO ABA1* (*ERA1*) exhibited tolerance to drought as well as seed dormancy. These mutants displayed higher susceptibility to pathogen attack but enhanced heat tolerance capability (Wu et al., 2017). Down-regulation of the genes, namely, *BnERA1* and *BnFTA* (farnesyltransferase), induced by drought in *Brassica napus* were reported to defend yield under aquatic scarcity throughout blooming. Enhanced sensitivity to drought was reported in the case of ABA hypersensitive mutants of *Arabidopsis*, such as cyp85a2, *abo1*, and *abh1*, although not much information is accessible regarding latent pleiotropic effects or roles in further plant species (Sathee et al., 2022). In a very recent time, Ganie et al. (2021) have compiled positive results from CRISPR-mediated editing of genes in rice linked with stress signaling, hormonal homeostasis, and carbohydrate metabolism to mitigate abiotic stress bottleneck.

5.6.2 BIOTIC STRESS TOLERANCE

It is now well-documented to consider gene editing technologies can alter host–plants interaction. Current tools to engineer the immunity of plants to deliberate resistance against pathogens, more specifically viruses affecting plant cells, can be found in detail from Ali and Mahfouz (2022). In the majority, Cas9 systems are employed to inhibit viral DNA replication, and Cas13 systems help to fight against the RNA viruses. CRISPR/Cas system has another vital facet of being a potential diagnostic tool to detect plant viruses in the field (Ali and Mahfouz, 2022). A well-considered transcription factor credited for a mechanism of quantitative resistance is *OsERF922,* which controls genes of plant defense in rice negatively. The knockdown or knockout provides resistance to fungal infections and boosts salt stress tolerance (Liu et al., 2012; Wang et al., 2016a). Genome-editing technologies have been utilized to identify and target the conforming homologs for establishing enduring resistance to powdery mildew in barley and some other crops like grape, apple, etc. (Wang et al., 2014; Nekrasov et al., 2017). CRISPR/Cas9-based concurrent knockout of three MLO loci was achieved in bread wheat (Zhang et al., 2017). Researchers also targeted other S gene loci that confer surface protein in wheat, watermelon, tomato, and rice for developing resistance by elimination of functional gene locus' transcription (Wang et al., 2016; Paula de et al., 2016; Li et al., 2020; Zhang et al., 2020; Santillán Martínez et al., 2020). One successful and noteworthy attempt was pathogen-specific gene exploitation by altering the conserved promoter parts for bacterial blight-causing *Xanthomonas oryzae* pv. oryzae (Li et al., 2012; Blanvillain-Baufumé et al., 2017). Thus, it is now possible to avoid unexpected yield loss using genome editing technologies while maintaining the normal cellular functions of gene products.

5.6.3 NUTRITIONAL IMPROVEMENT AND BIOFORTIFICATION

Principally in developing nations, along with other economic and social denials, inadequate nutrition has been recognized as a major cause of extensive malnutrition. Biofortification is a biological approach with sustainability and cost-effectiveness to supplement consumable food crops with nutrients to enrich the nutritional supply to consumers (Singh et al., 2022). Lower nutrient uptake and their accrual in edible plant parts, along with the existence of absorption-affecting inhibitors in high echelons, are the major attributes for micronutrient deficiencies. Genome-editing technologies have immense capability to manipulate these pathways for increasing the vitamin and micronutrient levels in plant bodies, and their effective exploitation can augment the nutritional content of crops. Nutrient bioavailability is reduced by anti-nutritional factors like amylase inhibitors, saponins, glucosinolates, phytic acid, etc. To weaken these antinutritional factors, genome editing can be a potent weapon. Role of *HvITPK1* (*Hordeum vulgare Inositol-Tetrakisphosphate 1-Kinase*) in stress signaling and inositol hexakisphosphate synthesis was explicated in *Hordeum vulgare,* generating homozygous mutants for the gene *HvITPK1* by CRISPR/Cas9, where the insertion mutants expressed enhanced salinity tolerance with a linked reduction in grain phosphorus

(P) content (Vlčko and Ohnoutkova, 2020). Gene editing of *Inositol Pentakisphosphate 2-Kinase 1* (*TaIPK1)* using CRISPR/Cas9 efficiently lowered phytate content and thus improved grain zinc and iron contents in wheat (Ibrahim et al., 2022). Genome editing tools have been confirmed to be successful in enhancing carotenoid range in globally significant cereal rice and one of the most popular edible crops, banana. Mutation creation by Kaur et al. (2018) in phytoene desaturase (RAS-PDS) of banana cv. *Rasthali* by genome editing through CRISPR/Cas9 is an example of the efficiency of banana genome editing. Cavendish banana cultivar (cv.) Grand Naine enriched with β-carotene-was further developed by them in 2020, where β-carotene content was enhanced by sixfold (Dong et al., 2020). In rice, quality enrichment has been the prime goal. Targeted editing of the enzymes linked with starch branching formed lines with a mutation having lesser amylose content (Sun et al., 2017). Genome editing by CRISPR/Cas9 was used to target the first exon of *Oryza sativa Inositol 1,3,4-trisphosphate 5/6-kinase* (*OsITPK6*) in rice, where the mutant lines displayed noteworthy lower content of phytic acid and increased inorganic P (Jiang et al., 2019). In *Citrullus lanatus*, knockout mutations were created in the Phytoene desaturase gene using the sgRNA-guided type II system of the same genome editing tool (Tian et al., 2017). In mushrooms, tools of genome editing have been taken into use to minimize browning genes in white truffles, leading to their life span extension, thus offering an advantage from the business point of view (Waltz, 2018). Likewise, in cut apples, a mutation in the polyphenol oxidase enzyme-encoding gene resulted in less browning (Nishitani et al., 2016). Development of corn with low phytic acid in addition to higher accessible phosphorus content and acrylamide-free potatoes are some notable achievements by researchers (Liang et al., 2014; Halterman et al., 2016). One of the greatest achievements of genome-editing technology in recent times was the targeted alteration of *Glycene max omega-6 desaturase* (*GmFAD2*) gene in soybean, resulting in superior oleic acid accumulation by decreasing the α-linolenic as well as linoleic acids (Al Amin et al., 2019).

5.6.4 DE NOVO DOMESTICATION

De novo domestication of novel crop species from their wild ancestors is one of the utmost striking feats of current-era gene editing technologies

(Zsogon et al., 2018). Earlier studies on genomics and genetic aspects have enabled us to identify genes crucial for the domestication of numerous species of crops. Gene-editing technologies and their progress have allowed us to modify multiple genes simultaneously. Recent advancements in the domestication of crops adopting CRISPR-based gene editing methodologies are well-documented by Agustin Zsogon and colleagues (Curtin et al., 2022). One burning resource using these technologies was reported for tomato introgression breeding, which was a protoplast-based, foreign DNA-free system of CRISPR–Cas and also generated *Solanum peruvianum* mutants for a total of 110 genes entailed in biogenesis of siRNAs and resistance to disease (Hsu et al., 2022). Their report tells us about the techniques of protoplast regeneration in addition to gene editing techniques that are free of transgene(s), will allow domestication of *S. peruvianum* and also lead to a significant upsurge in polyploidization of tomato. Gene editing makes pathway divergence and multiple traits' heaping very competent too. Some Brassinosteroid mutant combinations by CRISPR/Cas9-based gene editing multiplexing reported by Hongning Tong and colleagues showed reformed patterns of development and provided resources for Brassinosteroid research by which numerous agronomically imperative characters are affected along with hints to improve the architecture of rice (Liu et al., 2022).

5.7 GLOBAL REGULATORY LEGISLATION SCENARIO

The intergovernmental organization for Economic and Co-operative Development conference in June 2018 in Paris held a conversation on *Regulatory aspects* of genome editing. Six different countries participated there on the frameworks of policy relating to genome editing in their corresponding nations and addressed the regulatory position of agricultural methodologies related to the same in their particular realms and emphasized the consequential deliberations concerning the safety of plant breeding approaches in addition to prevailing agricultural product-related regulations (Friedrichs et al., 2019). Figure 5.11 shows the findings.

FIGURE 5.11 Current global legislation scenario of genome-edited crops.

5.8 REGULATORY SCENARIO IN INDIA

India has invited public comments for updating their resolutions regarding upcoming genome editing policies. The country announced a draft concerning genome-edited organisms: "Regulatory Framework and Guidelines for Risk Assessment" in the year 2020. It recommends a tier-up tactic for grouping items depending upon hazard, like Site-Directed Nuclease (SDN)-1 resting use of DNA sequence template; SDN-II using a DNA sequence template pointed as low risk having edits on one or a few base-pairs, and SDN-III site-directed insertion of gene/large DNA sequence using a DNA sequence template as higher risk with insertion of large or foreign DNA. Recently, the Government of India, through legislation, allowed the exemption to gene-edited plants in the SDN1 and SDN2 classes, which are free from exogeneous introduced DNA.

5.9 FUTURE PROSPECTS

Genome editing can help in genetic improvement in plants through various methods other than the methods already implemented. Some of them are as follows:

1. Reducing the time of breeding programs by a huge margin as the need for multiple generations to achieve a particular combination is not necessary.
2. Weed control can also be attempted through genome editing by implementing resistance genes for herbicides in the crop plants.
3. High-throughput mutant library can be developed for the identification of mutations and their effects.
4. Self-incompatibility can also be removed from incompatible plants to facilitate crossing in previously noncompatible plants.
5. Domestication of wild plants can be accelerated through genome editing methods.
6. Transfer of the nodulation property of pulses to cereals can mitigate the excessive nitrogen application in plants.
7. Plants can also be modified to produce various secondary metabolites to ward off pests and also for higher production of useful secondary metabolites.

Several other uses can be postulated for the novel method of genome editing, and further research will bring forward other methods that will be healthier for the environment and also be healthier for human consumption to mitigate the global concern related to the cultivation of genome-edited crops (Ricroch, 2019).

5.10 CONCLUSION

Creation of potently aimed, sequence-described, genome-wide genetic variability in plants using genome editing has significantly advanced plant science and crop breeding as a whole. It mediates precision plant advancement through knockin, point mutations, replacement, gene deletion, fine-tuning of gene regulation, in addition to further alterations at any locus of the plant's genes. It is also helpful for creating high-throughput mutant libraries and breeding antivirals. Among the methods, CRISPR/Cas systems' ease of use, adaptability, and durability make genome editing accessible for a large part of the scientific community, making huge strides in development. Table 5.3 shows some success stories of genome editing driven by CRISPR/Cas tools. It will be possible to build advanced crops with significantly better attributes by combining the ideas of system and synthetic biology, incorporating developments in functional genomics,

TABLE 5.3 Implementation of CRISPR/Cas System-Based Genome Editing Technologies in Agriculture

Plant	Perspective	Sequence(s) Targeted	Strategy	References
Implementation for Biofortification, Yield, and Post-Harvest Management				
Biofortification				
Papaver somniferum	Benzylisoquinoline alkaloids (BIAs) biosynthesis	3′-hydroxyl-*N* methylcoclaurine 4′ OMT2 (4′-*O*-methyltransferase isoform 2) gene	Cas9/gRNA-mediated gene knockout	Alagoz et al. (2016)
Solanum tuberosum	Amylopectin starch-rich potato	*GBSS* (*granule-bound starch synthase*) gene regions for amylose synthesis	–	Andersson et al. (2016)
Salvia miltiorrhiza	Tanshinone biosynthesis-related diterpene synthase gene knockout	*SmCPS1* (*Carbamoyl phosphate synthetase 1*) gene for diterpene synthase	–	Li et al. (2017a)
Oryza sativa	Enriched amylose content	Starch branching enzymes (SBEI and SBEIIb) for starch branching	–	Sun et al. (2017)
Nicotiana tabacum	Biotherapeutic protein production	Β (1,2)-xylosyltransferase (XylT) and α (1,3)-	–	Hanania et al. (2017)
		fucosyltransferase (FucT) gene for glycan biosynthesis	CRISPR–Cas9-mediated multiplex genome editing	Mercx et al. (2017)
Camelina sativa	Improvement of fatty acid composition	*Fatty acid desaturase 2* (*FAD2*) gene for linoleic acid formation from oleic acid	–	Morineau et al. (2016) and Jiang et al. (2017)
	Biosynthesis of seed oil	*CsDGAT1* (*Diacylglycerol acyltransferases 1*) or *CsPDAT1* (*Phospholipid:diacylglycerol acyltransferase1*) homoeologous genes for triacylglycerol (TAG) synthesis	–	Aznar-Moreno and Durrett (2017)
Hordeum vulgare *cv.* "Golden Promise"	Modification of n-glycans in grains	N-glycans biosynthesis associated ENGase gene	–	Kapusi et al. (2017)

TABLE 5.3 *(Continued)*

Plant	Perspective	Sequence(s) Targeted	Strategy	References
		Yield		
Zea mays	Improve the high-frequency targeted mutagenesis	*Argonaute 18* (*ZmAgo18a* and *ZmAgo18b*), dihydroflavonol 4-reductase for sporogenesis or *anthocyaninless* genes (*a1* and *a4*) for anthocyanin biosynthesis	*Agrobacterium*-mediated transformation	Char et al. (2017)
	Linkage drag reduction during breeding	*IG1* (*liguleless-1*) gene for upright architecture of maize leaves	Cas9/gRNA-mediated gene knockout	Li et al. (2017b)
Oryza sativa	Early maturing rice cultivar breeding	*Heading date* (*Hd2, Hd4 and Hd5)* genes for flowering suppression in *early heading date-1* (*Ehd1*)-dependent photoperiodic flowering pathway	–	Li et al. (2017c)
	Stomatal development modification	Positive regulator *Epidermal Patterning Factor like-9* (*EPFL9*) gene for stomatal development pathway	Gene knockout with CRISPR–Cas9/ Cpf1 system	Yin et al. (2017)
	Regulation of seed development	Maternally Expressed Genes (MEG)s for nutrient metabolism and Paternally Expressed Genes (PEGs) for endosperm development	Cas9/gRNA-mediated gene knockout	Yuan et al. (2017)
	Grain yield performance	QTLs related to Grain Yield: Grain number 1a (Gn1a) and Grain size3 (GS3)	–	Shen et al. (2016)
	Rice formation and development	*Mitogen activated protein kinases* (*MPK1 and MPK6*) genes	–	Minkenberg et al. (2017)
	Starch synthesis pathway modification	ADP-glucose pyrophosphorylase (*OsAGPL4*) large plastic subunit	–	Lee et al. (2016)

TABLE 5.3 *(Continued)*

Plant	Perspective	Sequence(s) Targeted	Strategy	References
	Marker free transgenic line development	GUS (beta-glucuronidase) marker gene	–	Srivastava, Underwood and Zhao (2017)
	Dense upright panicles, larger grain size, higher grain quantity development	γ-subunit of G protein (DEP1), Cytokinin dehydrogenase2 (Gn1a), squamosa promoter binding protein (IPA1) and γ-subunit of G protein (GS3)	–	Li et al. (2016c)
	Pollen tube growth and integrity	Ruptured pollen tube (RUPO), Plant-specific receptor-like kinase CrRLK1LS subfamily of rice members	–	Liu et al. (2016)
	Photo-sensitive genic male sterile japonica rice lines	*CSA* (*Carbon starved anther*) gene related to pho-sensitive male sterile lines	–	Li et al. (2016b)
	Caryopsis development	Sugar transporter *OsSWEET11* gene	–	Ma et al. (2017)
	Grain weight increment	Negatively regulating genes for thousand-grain weight (*TGW6*) and Grain width (*GW2* and *GW5*)	CRISPR—Cas9-mediated multiplex genome editing	Xu et al. (2016)
	Floral meristem determinacy and maintenance	*Meristem maintenance Floral organ number 2* (*FON2*) gene and stamen specification *OsMADS3* gene	Cas9/gRNA-mediated gene knockout	Yasui et al. (2017)
Dendrobium officinale	Biosynthesis of lignocellulose	Lignocellulose biosynthesis pathway genes C4H, C3H, CCR, 4CL and IRX.	–	Kui et al. (2017)
Solanum lycopersicum	Parthenocarpic tomato	Parthenocarpic *SlIAA9* gene	–	Ueta et al. (2017)
Nicotiana sp.	Auxiliary bud growth regulation	Auxin biosynthesis NtPIN4 gene	–	Xie et al. (2017)

TABLE 5.3 *(Continued)*

Plant	Perspective	Sequence(s) Targeted	Strategy	References
Brassica oleracea and *Hordeum vulgare*	Seed dormancy and pod shattering prevention	Grain dormancy regulator gene *HvPM19* and *BolC.GA4.a* gene for valve development at the pod margin	–	Lawrenson et al. (2015)
Taraxacum kok-saghyz	Hairy root rubber biosynthesis	Inulin biosynthesis TK 1-FFT (*fructan:fructan 1-fructosyltransferase*) gene	–	Iaffaldano et al. (2016)
		Post-Harvest Management		
Solanum lycopersicum	Fruit Ripening Inhibition	Removal of three locus of RIN gene for ripening inhibition through MADS-box transcription factor	CRISPR-Cas9-mediated editing of multiplex genome	Ito et al. (2015)
		Abiotic Stress Tolerance		
		Drought		
Zea mays	Drought mitigation through enhanced grain yield under stress	Ethylene negative regulator *ARGOS8* gene	Donor template-based insertion or replacement	Shi et al. (2017)
		Salinity		
Oryza sativa	Tolerance to salinity	*OsRAV2* (*ETHYLENE RESPONSE DNA BINDING FACTOR 2*) gene GT-1 element for brassinosteroid synthesis, flowering time and leaf senescence regulation	Cas9/Grna-mediated gene knockout	Duan et al. (2016)
		Tolerance to Herbicide Application		
Linum usitatissimum	Tolerance to Glyphosate	5′-Enolpyruvylshikimate-3-phosphate synthase (EPSPS) in shikimate pathway for formation of secondary metabolites	Gene replacement	Sauer et al. (2016)

TABLE 5.3 *(Continued)*

Plant	Perspective	Sequence(s) Targeted	Strategy	References
Arabidopsis thaliana	Glufosinate resistance combined with small trichomes	Glufosinate resistance BAR gene (Phosphinothricin Acetyltransferase) and trichome formation *GLABROUS1* (*GL1*) gene	Cas9/gRNA-mediated gene knockout	Hahn et al. (2017)
	Drought, salt and cold tolerance	Gene for small sugar molecules for regulation of stress related processes UDP-glycosyltransferases *UGT79B2* and *UGT79B3*	–	Li et al. (2017d)
Oryza sativa	Glyphosate tolerance	5′-Enolpyruvylshikimate-3-phosphate synthase (EPSPS) in shikimate pathway for formation of secondary metabolites	Gene insertion and replacement by donor template	Li et al. (2016c)
	Imazamox resistance	C287 gene mutation	CRISPR-Cas9-mediated editing of multiplex genome	Shimatani et al. (2017)
	Herbicide tolerance	Amino acid biosynthesis gene Acetolactate synthase (ALS)	Gene replacement by donor template	Sun et al. (2016)
Lotus japonicus	Increase nitrogen fixation through intracellular accommodation of nitrogen-fixing bacteria and increaseBioavailability of soil organic nitrogen	Single and multiple *symbiotic nitrogen fixation* genes: *simbiosis receptor-like kinase* (*SYMRK*), leghemoglobin loci (LjLb1, LjLb2, LjLb3)	Cas9/Grna-mediated gene knockout	Wang et al. (2016b)
Solanum tuberosum	Increased Tolerance to ALS-inhibiting herbicides	Amino acid biosynthesis gene *Acetolactate synthase 1* (*ALS1*)	-do-	Butler et al. (2016)

TABLE 5.3 *(Continued)*

Plant	Perspective	Sequence(s) Targeted	Strategy	References
		Biotic Stress Resistance		
		Bacterial Diseases		
Oryza sativa	*Xanthomonas oryzae pv. oryzae* (Bacterial blight) tolerance	Sucrose transporter gene *OsSWEET13* for Tal receptor gene in pathogen	Cas9/gRNA-mediated gene knockout	Zhou et al. (2015)
Citrus sinensis Osbeck	Canker resistance	*Citrus sinensis lateral organ boundary 1* (*CsLOB1*) promoter	–	Peng et al. (2017)
Citrus paradisi	Canker tolerance	PthA4 effector binding elements (EBEs) in the Type I *CsLOB1* promoter (EBE_{PthA4}-CsLOBP) of the *CsLOB1* gene	–	Jia et al. (2016)
		Fungal Diseases		
Triticum aestivum	*Blumeria graminis f. sp. Tritici* (Powdery mildew) resistance	*TaMLO-A1* allele	Cas9/gRNA-mediated gene knockout	Wang et al. (2014)
Solanum lycopersicum	Powdery mildew resistance	Powdery mildew susceptibility *SlMlo* gene	–	Nekrasov et al. (2017)
Oryza sativa	*Magnaporthe oryzae* (Blast) tolerance	*OsERF922* gene responsible for Ethylene responsive factor ERF transcription	Single and multiplex Cas9/gRNA-mediated gene knockout	Wang et al. (2016a)
		Viral Diseases		
Nicotiana benthamiana	Virus tolerance	Virus specific antiviral role of *Argonaute RISC Catalytic Component 2* (*AGO2*) gene	Cas9/gRNA-mediated gene knockout	Ludman et al. (2017)

TABLE 5.3 *(Continued)*

Plant	Perspective	Sequence(s) Targeted	Strategy	References
	Resistance to Tomato yellow leaf curl virus (TYLCV)	TYLCV noncoding and coding sequences	–	Ali et al. (2015)
Arabidopsis thaliana	Resistance to Potyvirus (TuMV)	Eukaryotic translation initiation factor (eIF) member: eIF(iso) 4E involved in recessive resistance against potyvirus	–	Pyott et al. (2016)
Cucumis sativus	Tolerance to Zucchini yellow mosaic virus, Papaya ring spot mosaic virus-W and Ipomovirus immunity	Host factor for multiple viruses: eIF4E	–	Chandrasekaran et al. (2016)
Nicotiana benthamiana and *Arabidopsis thaliana*\	Tolerance to Beet severe curly top virus (BSCTV)	Transient expression assays in BSCTV genome noncoding or coding sequences to select two sites to induce transgenic lines due to their importance in viral replication	–	Ji et al. (2015)

genome-editing machineries, next-generation sequencing, and many other relevant tools, if combined with genome editing for the betterment of agriculture, and can lead toward food security. Finally, it is convenient to state that this chapter encompasses all the relevant and progressing mechanisms of genome editing technologies in this current era, along with their applications toward food security, and also puts light on the global legislation scenario of genome-edited crops.

KEYWORDS

- **application**
- **CRISPR/Cas9**
- **genome editing**
- **legislation**
- **plant**
- **strategies**

REFERENCES

Al Amin, N.; Ahmad, N.; Wu, N.; Pu, X.; Ma, T.; Du, Y. CRISPR-Cas9 mediated targeted disruption of FAD2-2 microsomal omega-6 desaturase in soybean (Glycinemax. L). *BMC Biotechnol.* 2019, 19, 9–10.

Alagoz, Y.; Gurkok, T.; Zhang, B.; Unver, T. Manipulating the biosynthesis of bioactive compound alkaloids for next-generation metabolic engineering in opium poppy using CRISPR-Cas 9 genome editing technology. *Sci. Rep.* 2016, 6, 30910 https://doi.org/10.1038/srep30910

Ali, Z.; Abulfaraj, A.; Idris, A.; Ali, S.; Tashkandi, M.; Mahfouz, M.M. CRISPR/Cas9-mediated viral interference in plants. *Genome Biol.* 2015, 16, 238 https://doi.org/10.1186/s13059-015-0799-6

Ali, Z.; Mahfouz, M. CRISPR/Cas systems versus plant viruses: Engineering plant immunity and beyond. *Plant Physiol.* 2022, 186, 1770–1785.

Andersson, M.; Turesson, H.; Nicolia, A.; Fält, A.-S.; Samuelsson, M.; Hofvander, P. Efficient targeted multiallelic mutagenesis in tetraploid potato (*Solanum tuberosum*) by transient CRISPR-Cas9 expression in protoplasts. *Plant Cell Rep.* 2016, 36, 117–128 https://doi.org/10.1007/s00299-016-2062-3

Aznar-Moreno, J.A.; Durrett, T.P. Simultaneous targeting of multiple gene homeologs to alter seed oil production in *Camelina sativa*. *Plant Cell Physiol.* 2017, 58, 1260–1267 https://doi.org/10.1093/pcp/pcx058

Blanvillain-Baufumé, S.; Reschke, M.; Solé, M.; Auguy, F.; Doucoure, H.; Szurek, B.; Meynard, D.; Portefaix, M.; Cunnac, S.; Guiderdoni, E. Targeted promoter editing for rice resistance to Xanthomonas oryzae pv. oryzae reveals differential activities for SWEET14-inducing TAL effectors. *Plant Biotechnol. J.* 2017, 15, 306–317.

Boch, J.; Scholze, H.; Schornack, S.; Landgraf, A.; Hahn, S.; Kay, S.; Lahaye, T.; Nickstadt, A.; Bonas U. Breaking the code of DNA binding specificity of TAL-type III effectors. *Science* 2009, 326(5959):1509–12.

Bonawitz, N.D.; Ainley, W.M.; Itaya, A.; Chennareddy, S.R.; Cicak, T.; Effinger, K.; Jiang, K.; Mall, T.K.; Marri, P.R.; Samuel, J.P.; Sardesai, N.; Simpson, M.; Folkerts, O.; Sarria, R.; Webb, S.R.; Gonzalez, D.O.; Simmonds, D.H.; Pareddy, D.R.; Pareddy, D.R. Zinc finger nuclease-mediated targeting of multiple transgenes to an endogenous soybean genomic locus via non-homologous end joining. *Plant Biotechnol. J.* 2019, 17(4), 750–761.

Butler, N.M.; Baltes, N.J.; Voytas, D.F.; Douches, D.S. Geminivirus-mediated genome editing in potato (*Solanum tuberosum* L.) using sequence-specific nucleases. *Front. Plant Sci.* 2016, 7, 1045 https://doi.org/10.3389/fpls.2016.01045

Chandrasekaran, J.; Brumin, M.; Wolf, D.; Leibman, D.; Klap, C.; Pearlsman, M. et al. Development of broad virus resistance in non-transgenic cucumber using CRISPR/Cas9 technology. *Mol. Plant Pathol.* 2016, 17, 1140–1153 https://doi.org/10.1111/mpp.12375

Char, S.N.; Neelakandan, A.K.; Nahampun, H.; Frame, B.; Main, M.; Spalding, M.H. et al. An *Agrobacterium*-delivered CRISPR/Cas9 system for high-frequency targeted mutagenesis in maize. *Plant Biotechnol. J.* 2017, 15, 257–268 https://doi.org/10.1111/pbi.12611

Chen, K.; Wang, Y.; Zhang, R.; Zhang, H.; Gao, C. CRISPR/Cas genome editing and precision plant breeding in agriculture. *Ann. Rev. Plant Biol.* 2019, 70, 667–697.

Curtin, S.; Qi, Y.; Peres, L.; Fernie, A.; Zso¨go¨n, A. Pathways to de novo domestication of crop wild relatives. *Plant Physiol.* 2022, 188, 1746–1756.

Dong, O.X.; Yu, S.; Jain, R.; Zhang, N.; Duong, P.Q.; Butler, C.; et al. Marker-free carotenoid-enriched rice generated through targeted gene insertion using CRISPR-Cas9. *Nat. Commun.* 2020, 11, 1178.

Doyon, Y.; Vo, T.D.; Mendel, M.C.; Greenberg, S.G.; Wang, J.B.; Xia, D.F.; Miller, J.C. et al. Enhancing zinc-finger-nuclease activity with improved obligate heterodimeric architectures. *Nat. Methods* 2011, 8, 74–79.

Duan, Y.-B.; Li, J.; Qin, R.-Y.; Xu, R.-F.; Li, H.; Yang, Y.-C. et al. Identification of a regulatory element responsible for salt induction of rice OsRAV2 through ex situ and in situ promoter analysis. *Plant Mol. Biol.* 2016, 90, 49–62 https://doi.org/10.1007/s11103-015-0393-z

Friedrichs, S.; Takasu, Y.; Kearns, P.; Dagallier, B.; Oshima, R.; Schofield, J.; et al. Meeting report of the OECD Con-ference on "genome editing: Applications in agriculture—Implications for health, environment and regulation. *Transgenic Res.* 2019, 28, 419–463, http://dx.doi.org/10.1007/s11248-019-00154-1

Ganie, S.A.; Wani, S.H.; Henry, R.; Hensel, G. Improving rice salt tolerance by precision breeding in a new era. *Curr. Opin. Plant Biol.* 2021, 60, 101996.

Grizot, S.; Epinat, J.C.; Thomas, S.; Duclert, A.; Rolland, S.; Pa^ques, F.; Duchateau, P. Generation of redesigned homing endonucleases comprising DNA-binding domains derived from two different scaffolds. *Nucleic Acids Res.* 2010, 38(6), 2006–2018. https://doi.org/10.1093/nar/gkp1171.

Hahn, F.; Mantegazza, O.; Greiner, A.; Hegemann, P.; Eisenhut, M.; Weber, A.P.M. An efficient visual screen for CRISPR/Cas9 activity in *Arabidopsis thaliana. Front. Plant Sci.* 2017, 8, 39 https://doi.org/10.3389/fpls.2017.00039

Halterman, D.; Guenthner, J.; Collinge, S.; Butler, N.; Douches, D. Biotech potatoes in the 21st century: 20 years since the first biotech potato. *Am. J. Potato Res.* 2016, 93, 1–20.

Hanania, U.; Ariel, T.; Tekoah, Y.; Fux, L.; Sheva, M.; Gubbay, Y. et al. Establishment of a tobacco BY2 cell line devoid of plant-specific xylose and fucose as a platform for the production of biotherapeutic proteins. *Plant Biotechnol. J.* 2017, 15, 1120–1129 https://doi.org/10.1111/pbi.12702

Hasan, N.; Choudhary, S.; Naaz, N.; Sharma, N.; Laskar, R.A. Recent advancements in molecular marker-assisted selection and applications in plant breeding programmes. *J. Genet. Eng. Biotechnol.* 2021, 19(1), 1–26.

Hsu, C.; Yuan, Y.; Zheng, P.; Wu, F.; Cheng, Q.; Wu, Y.; Lin, S.; Yue, J.; Cheng, Y.; Lin, S.; et al. DNA-free CRISPR-Cas9 gene editing of wild tetraploid tomatoes Solanum peruvianum using protoplast regeneration. *Plant Physiol.* 2022, 188, 1917–1930.

Iaffaldano, B.; Zhang, Y.; Cornish, K. CRISPR/cas9 genome editing of rubber producing dandelion *Taraxacum kok-saghyz* using *Agrobacterium rhizogenes* without selection. *Ind. Crops Prod.* 2016, 89, 356–362 https://doi.org/10.1016/j.indcrop.2016.05.029

Ibrahim, S.; Saleem, B.; Rehman, N.; Zafar, S.A.; Naeem, M.K.; Khan, M.R. CRISPR/Cas9 mediated disruption of inositol pentakisphosphate 2-kinase 1 (TaIPK1) reduces phytic acid and improves iron and zinc accumulation in wheat grains. *J. Adv. Res.* 2022, 37, 33–41.

Ito, Y.; Nishizawa-Yokoi, A.; Endo, M.; Mikami, M.; Toki, S. CRISPR/Cas9-mediated mutagenesis of the RIN locus that regulates tomato fruit ripening. *Biochem. Biophys. Res. Commun.* 2015, 467, 76–82 https://doi.org/10.1016/j.bbrc.2015.09.117

Jansing, J.; Schiermeyer, A.; Schillberg, S.; Fischer, R.; Bortesi, L. Genome editing in agriculture: Technical and practical considerations. *Int. J. Mol. Sci.* 2019, 20(12), 2888.

Ji, X.; Zhang, H.; Zhang, Y.; Wang, Y.; Gao, C. Establishing a CRISPR–Cas-like immune system conferring DNA virus resistance in plants. *Nat. Plants* 2015, 1, 15144 https://doi.org/10.1038/nplants.2015.144

Jia, H.; Orbovic, V.; Jones, J.B.; Wang, N. Modification of the PthA4 effector binding elements in type I CsLOB1 promoter using Cas9/sgRNA to produce transgenic Duncan grapefruit alleviating XccΔpthA4:dCsLOB1.3 infection. *Plant Biotechnol. J.* 2016, 14, 1291–1301 https://doi.org/10.1111/pbi.12495

Jiang, M.; Liu, Y.; Liu, Y.; Tan, Y.; Huang, J.; Shu, Q. Mutation of inositol 1,3,4-trisphosphate 5/6-kinase6 impairs plant growth and phytic acid synthesis in rice. *Plants.* 2019, 8, 114.

Jiang, W.Z.; Henry, I.M.; Lynagh, P.G.; Comai, L.; Cahoon, E.B.; Weeks, D.P. Significant enhancement of fatty acid composition in seeds of the allohexaploid, *Camelina sativa*, using CRISPR/Cas9 gene editing. *Plant Biotechnol. J.* 2017, 15, 648–657 https://doi.org/10.1111/pbi.12663

Kaiser, N.; Douches, D.; Dhingra, A.; Glenn, K.C.; Herzig, P.R.; Stowe, E.C.; Swarup, S. The role of conventional plant breeding in ensuring safe levels of naturally occurring toxins in food crops. *Trends Food Sci. Technol.* 2020, 100, 51–66.

Kapusi, E.; Corcuera-Gómez, M.; Melnik, S.; Stoger, E. Heritable genomic fragment deletions and small indels in the putative ENGase gene induced by CRISPR/Cas9 in barley. *Front. Plant Sci.* 2017, 8, 540. https://doi.org/10.3389/fpls.2017.00540

Kaur, N.; Alok, A.; Shivani, N.; Kaur, N.; Pandey, P.; Awasthi, P.; et al. CRISPR/Cas9-mediated efficient editing in Phytoene Desaturase (PDS) demonstrates precise manipulation in banana Cv. rasthali genome. *Funct. Integr. Genomics.* 2018, 18, 89–99.

Kui, L.; Chen, H.; Zhang, W.; He, S.; Xiong, Z.; Zhang, Y. et al. Building a genetic manipulation tool box for orchid biology: Identification of constitutive promoters and application of CRISPR/Cas9 in the orchid, *Dendrobium officinale. Front. Plant Sci.* 2017, 7, 2036 https://doi.org/10.3389/fpls.2016.02036

Kumar, M. Impact of climate change on crop yield and role of model for achieving food security. *Environ. Monit. Assess.* 2016, 188, 465.

Lawrenson, T.; Shorinola, O.; Stacey, N.; Li, C.; Østergaard, L.; Patron, N. et al. Induction of targeted, heritable mutations in barley and *Brassica oleracea* using RNA-guided Cas9 nuclease. *Genome Biol.* 2015, 16, 258 https://doi.org/10.1186/s13059-015-0826-7

Lee, S.-K.; Eom, J.-S.; Hwang, S.-K.; Shin, D.; An, G.; Okita, T.W. et al. Plastidic phosphoglucomutase and ADP-glucose pyrophosphorylase mutants impair starch synthesis in rice pollen grains and cause male sterility. *J. Exp. Bot.* 2016, 67, 5557–5569 https://doi.org/10.1093/jxb/erw324

Li JF, Norville JE, Aach, J.; McCormack, M.; Zhang, D.; et al. Multiplex and homologous recombination-mediated genome editing in *Arabidopsis* and *Nicotiana benthamiana* using guide RNA and Cas9. *Nat. Biotechnol.* 2013, 31:688–691.

Li, B.; Cui, G.; Shen, G.; Zhan, Z.; Huang, L.; Chen, J. et al. Targeted mutagenesis in the medicinal plant *Salvia miltiorrhiza. Sci. Rep.* 2017a, 7, 43320. https://doi.org/10.1038/srep43320

Li, C.; Liu, C.; Qi, X.; Wu, Y.; Fei, X.; Mao, L. et al. RNA-guided Cas9 as an in vivo desired-target mutator in maize. *Plant Biotechnol. J.* 2017b, 15(12), 1566–1576, https://doi.org/10.1111/pbi.12739

Li, J.; Meng, X.; Zong, Y.; Chen, K.; Zhang, H.; Liu, J. et al. Gene replacements and insertions in rice by intron targeting using CRISPR–Cas9. *Nat. Plants* 2016c, 2, 16139 https://doi.org/10.1038/nplants.2016.139

Li, M.; Li, X.; Zhou, Z.; Wu, P.; Fang, M.; Pan, X. et al. Reassessment of the four yield-related genes Gn1a, DEP1, GS3, and IPA1 in rice using a CRISPR/Cas9 system. *Front. Plant Sci.* 2016a, 7, 377 https://doi.org/10.3389/fpls.2016.00377

Li, P.; Li, Y.-J.; Zhang, F.-J.; Zhang, G.-Z.; Jiang, X.-Y.; Yu, H.-M. et al. The Arabidopsis UDP-glycosyltransferases UGT79B2 and UGT79B3, contribute to cold, salt and drought stress tolerance via modulating anthocyanin accumulation. *Plant J.* 2017d, 89, 85–103 https://doi.org/10.1111/tpj.13324

Li, Q.; Zhang, D.; Chen, M.; Liang, W.; Wei, J.; Qi, Y. et al. Development of japonica photo-sensitive genic male sterile rice lines by editing carbon starved anther using CRISPR/Cas9. *J. Genet. Genomics* 2016b, 43, 415–419 https://doi.org/10.1016/j.jgg.2016.04.011

Li, R.; Wang, L.; Li, Y.; Zhao, R.; Zhang, Y.; Sheng, J.; Ma, P.; Shen, L. Knockout of SlNPR1 enhances tomato plants resistance against Botrytis cinerea by modulating ROS homeostasis and JA/ET signaling pathways. *Physiol. Plant.* 2020, 170, 569–579.

Li, T.; Liu, B.; Spalding, M.H.; Weeks, D.P.; Yang, B. High-efficiency TALEN-based gene editing produces disease-resistant rice. *Nat. Biotechnol.* 2012, 30, 390–392.

Li, X.; Zhou, W.; Ren, Y.; Tian, X.; Lv, T.; Wang, Z. et al. High-efficiency breeding of early-maturing rice cultivars via CRISPR/Cas9-mediated genome editing. *J. Genet. Genomics* 2017c, 44, 175–178 https://doi.org/10.1016/j.jgg.2017.02.001

Liang, Z.; Zhang, K.; Chen, K.; Gao, C. Targeted Mutagenesis in Zea mays Using TALENs and the CRISPR/Cas system. *J. Genet. Genomics*. 2014, 41, 63–68.

Liao, S.; Qin, X.; Luo, L.; Han, Y.; Wang, X.; Usman, B.; Nawaz, G.; Zhao, N.; Liu, Y.; Li, R. CRISPR/Cas9-induced mutagenesis of semi-rolled Leaf1,2 confers curled leaf phenotype and drought tolerance by influencing protein expression patterns and ROS scavenging in rice (Oryza sativa L.). *Agronomy* 2019, 9, 728.

Lim, C.; Kang, K.; Shim, Y.; Yoo, S.; Paek, N. Inactivating transcription factor OsWRKY5 enhances drought tolerance through abscisic acid signaling pathways. *Plant Physiol.* 2022, 188, 1900–1916.

Liu, D.; Chen, X.; Liu, J.; Ye, J.; Guo, Z. The rice ERF transcription factor OsERF922 negatively regulates resistance to *Magnaporthe oryzae* and salt tolerance. *J. Exp. Bot.* 2012, 63, 3899–3912.

Liu, D.; Yu, Z.; Zhang, G.; Yin, W.; Li, L.; Niu, M.; Meng, W.; Zhang, X.; Dong, N.; Liu, J.; et al. Diversification of plant agronomic traits by genome editing of brassinosteroid signaling family genes in rice. *Plant Physiol.* 2022, 187, 2563–2576.

Liu, L.; Zheng, C.; Kuang, B.; Wei, L.; Yan, L.; Wang, T. Receptor-like kinase RUPO interacts with potassium transporters to regulate pollen tube growth and integrity in rice. *PLoS Genet.* 2016, 12, e1006085 https://doi.org/10.1371/journal.pgen.1006085

Ludman, M.; Burgyán, J.; Fátyol, K. CRISPR/Cas9 mediated inactivation of argonaute 2 reveals its differential involvement in antiviral responses. *Sci. Rep.* 2017, 7, 1010 https://doi.org/10.1038/s41598-017-01050-6

Ma, L.; Zhang, D.; Miao, Q.; Yang, J.; Xuan, Y.; Hu, Y. Essential role of sugar transporter OsSWEET11 during the early stage of rice grain filling. *Plant Cell Physiol.* 2017, 58, 863–873 https://doi.org/10.1093/pcp/pcx040

Maja, M.M.; Ayano, S.F. The impact of population growth on natural resources and farmers' capacity to adapt to climate change in low-income countries. *Earth Syst. Environ.* 2021, 5(2), 271–283.

Malzahn, A.; Lowder, L.; Qi, Y. Plant genome editing with TALEN and CRISPR. *Cell Biosci.* 2017, 7(1), 1–18.

Marraffini, L. CRISPR-Cas immunity in prokaryotes. *Nature*. 2015, 526, 55–61. https://doi.org/10.1038/nature15386

Mercx, S.; Smargiasso, N.; Chaumont, F.; De Pauw, E.; Boutry, M.; Navarre, C. Inactivation of the β(1,2)-xylosyltransferase and the α (1,3)-fucosyltransferase genes in nicotiana tabacum BY-2 cells by a multiplex CRISPR/Cas9 strategy results in glycoproteins without plant-specific glycans. *Front. Plant Sci.* 2017, 8, 403 https://doi.org/10.3389/fpls.2017.00403

Miller, J.C.; Holmes, M.C.; Wang, J.B.; Guschin, D.Y.; Lee, Y.L.; Rupniewski, I.; Beausejour, C.M. et al. An improved zinc-finger nuclease architecture for highly specific genome editing. *Nat. Biotechnol.* 2007, 25, 778–785.

Minkenberg, B.; Xie, K.; Yang, Y. Discovery of rice essential genes by characterizing a CRISPR-edited mutation of closely related rice MAP kinase genes. *Plant J.* 2017, 89, 636–648 https://doi.org/10.1111/tpj.13399

Molotoks, A.; Smith, P.; Dawson, T.P. Impacts of land use, population, and climate change on global food security. *Food Energy Secur.* 2021, 10(1), e261.

Morineau, C.; Bellec, Y.; Tellier, F.; Gissot, L.; Kelemen, Z.; Nogué, F. et al. Selective gene dosage by CRISPR-Cas9 genome editing in hexaploidy *Camelina sativa. Plant Biotechnol.* 2016, 15, 729–739 https://doi.org/10.1111/pbi.12671

Moscou MJ, Bogdanove AJ. A simple cipher governs DNA recognition by TAL effectors. *Science* 2009, 326(5959):1501.

Nakashima, K.; Yamaguchi-Shinozaki, K. ABA signaling in stress-response and seed development. *Plant Cell Rep.* 2013, 32, 959–970.

Nekrasov, V.; Staskawicz, B.; Weigel, D.; Jones JD, Kamoun S. Targeted mutagenesis in the model plant *Nicotiana benthamiana* using Cas9 RNA-guided endonuclease. *Nat. Biotechnol.* 2013, 31, 691–693

Nekrasov, V.; Wang, C.; Win, J.; Lanz, C.; Weigel, D.; Kamoun, S. Rapid generation of a transgene-free powdery mildew resistant tomato by genome deletion. *Sci. Rep.* 2017, 7, 482 https://doi.org/10.1038/s41598-017-00578-x

Nishitani, C.; Hirai, N.; Komori, S.; Wada, M.; Okada, K.; Osakabe, K.; et al. Efficient genome editing in apple using a CRISPR/Cas9 system. *Sci. Rep.* 2016, 6, 31481–31488.

Paula de, D.; Brail, Q.; Dahlbeck, D.; Staskawicz, B. CRISPR-Cas9 mediated mutagenesis of a DMR6 ortholog in tomato confers broad-spectrum disease resistance. *bioRxiv.* 2016, 064824.

Pendergrass, A.G.; Meehl, G.A.; Pulwarty, R.; Hobbins, M.; Hoell, A.; AghaKouchak, A.; Bonfils, C.J.W.; Gallant, A.J.E.; Hoerling, M.; Hoffmann, D. Flash droughts present a new challenge for subseasonal-to-seasonal prediction. *Nat. Clim. Chang.* 2020, 10, 191–199.

Peng, A.; Chen, S.; Lei, T.; Xu, L.; He, Y.; Wu, L. et al. Engineering canker-resistant plants through CRISPR/Cas9-targeted editing of the susceptibility gene CsLOB1 promoter in citrus. *Plant Biotechnol. J.* 2017, 15(12), 1509–1519, https://doi.org/10.1111/pbi.12733

Planas-Riverola, A.; Gupta, A.; Betego´ n-Putze, I.; Bosch, N.; Ibanes, M.; Cano-Delgado, A.I. Brassinosteroid signalling in plant development and adaptation to stress. *Development* 2019, 146, dev151894.

Pyott, D.E.; Sheehan, E.; Molnar, A. Engineering of CRISPR/Cas9-mediated potyvirus resistance in transgene-free Arabidopsis plants. *Mol. Plant Pathol.* 2016, 17, 1276–1288 https://doi.org/10.1111/mpp.12417

Ricroch, A. Global developments of genome editing in agriculture. In: *Transgenic Research.* Springer International Publishing, 2019, pp. 45–52, Vol. 28.

Rustgi, S.; Kakati, J.P.; Jones, Z.T.; Zoong, Z.S.; Narayanan, S. Heat tolerance as a function of membrane lipid remodeling in the major US oilseed crops (soybean and peanut). *J. Plant Biochem. Biotechnol.* 2021, 30, 652–667.

Santillán Martínez, M.I.; Bracuto, V.; Koseoglou, E.; Appiano, M.; Jacobsen, E.; Visser, R.G.F.; Wolters, A.M.A.; Bai, Y. CRISPR/Cas9-targeted mutagenesis of the tomato susceptibility gene PMR4 for resistance against powdery mildew. *BMC Plant Biol.* 2020, 20, 284.

Sathee, L.; Jagadhesan, B.; Pandesha, P.H.; Barman, D.; Adavi, B.S.; Nagar, S.; Krishna, G.K.; Tripathi, S.; Jha, S.K.; Chinnusamy, V. Genome editing targets for improving nutrient use efficiency and nutrient stress adaptation. *Front. Genet* 2022, 13, 900897.

Sauer, N.J.; Narváez-Vásquez, J.; Mozoruk, J.; Miller, R.B.; Warburg, Z.J.; Woodward, M.J. et al. Oligonucleotide-mediated genome editing provides precision and function to engineered nucleases and antibiotics in plants. *Plant Physiol.* 2016, 170, 1917–1928 https://doi.org/10.1104/pp.15.01696

Shan, Q.; Wang, Y.; Li, J.; Zhang, Y.; Chen, K.; et al. Targeted genome modification of crop plants using a CRISPR-Cas system. *Nat. Biotechnol.* 2013, 31, 686–688.

Shen, L.; Wang, C.; Fu, Y.; Wang, J.; Liu, Q.; Zhang, X. et al. QTL editing confers opposing yield performance in different rice varieties. *J. Integr. Plant Biol.* 2016, 60(2), 89–93, https://doi.org/10.1111/jipb.12501

Shi, J.; Gao, H.; Wang, H.; Lafitte, H.R.; Archibald, R.L.; Yang, M. et al. ARGOS8 variants generated by CRISPR-Cas9 improve maize grain yield under field drought stress conditions. *Plant Biotechnol. J.* 2017, 15, 207–216 https://doi.org/10.1111/pbi.12603

Shimatani, Z.; Kashojiya, S.; Takayama, M.; Terada, R.; Arazoe, T.; Ishii, H. et al. Targeted base editing in rice and tomato using a CRISPR-Cas9 cytidine deaminase fusion. *Nat. Biotechnol.* 2017, 35, 441–443 https://doi.org/10.1038/nbt.3833

Singh, S.K.; Barman, M.; Sil, A. et al. Nutrient biofortification in wheat: Opportunities and challenges. *Cer. Res. Commun.* 2022, 51, 15–28, https://doi.org/10.1007/s42976-022-00288-9.

Srivastava, V.; Underwood, J.L.; Zhao, S. Dual-targeting by CRISPR/Cas9 for precise excision of transgenes from rice genome. *Plant Cell Tissue Organ Culture* 2017, 129, 153–160 https://doi.org/10.1007/s11240-016-1166-3

Sun, Y.; Jiao, G.; Liu, Z.; Zhang, X.; Li, J.; Guo, X. et al. Generation of high-amylose rice through CRISPR/Cas9-mediated targeted mutagenesis of starch branching enzymes. *Front. Plant Sci.* 2017, 8, 298. https://doi.org/10.3389/fpls.2017.00298

Sun, Y.; Zhang, X.; Wu, C.; He, Y.; Ma, Y.; Hou, H. et al. Engineering herbicide-resistant rice plants through CRISPR/Cas9-mediated homologous recombination of acetolactate synthase. *Mol. Plant* 2016, 9, 628–631 https://doi.org/10.1016/j.molp.2016.01.001

Szczepek, M.; Brondani, V.; Buchel, J.; Serrano, L.; Segal, D.J.; Cathomen, T. Structure-based redesign of the dimerization interface reduces the toxicity of zinc-finger nucleases. *Nat. Biotechnol.* 2007, 25, 786–793.

Tian, S.; Jiang, L.; Gao, Q.; Zhang, J.; Zong, M.; Zhang, H.; et al. Efficient CRISPR/Cas9-based gene knockout in watermelon. *Plant Cell. Rep.* 2017, 36, 399–406.

Tollefson, J. How hot will earth get by 2100? *Nature* 2020, 580, 444–446.

Ueta, R.; Abe, C.; Watanabe, T.; Sugano, S.S.; Ishihara, R.; Ezura, H. et al. Rapid breeding of parthenocarpic tomato plants using CRISPR/Cas9. *Sci. Rep.* 2017, 7, 507 https://doi.org/10.1038/s41598-017-00501-4

UNDESA (United Nations Department of Economic and Social Affairs, Population Division) *World Population Prospects: The 2017 Revision.* United Nations: New York, 2017, Available at: https://www.un.org/development/desa/publications/world-population-prospects-the-2017-revision.html.

Vishwakarma, K.; Upadhyay, N.; Kumar, N.; Yadav, G.; Singh, J.; Mishra, R.K.; Kumar, V.; Verma, R.; Upadhyay, R.G.; Pandey, M.; et al. Abscisic acid signaling and abiotic stress tolerance in plants: A review on current knowledge and future prospects. *Front. Plant Sci.* 2017, 8, 161.

Vlčko, T.; Ohnoutkova, L. Allelic variants of CRISPR/Cas9 induced mutation in an inositol trisphosphate 5/6 kinase gene manifest different phenotypes in barley. *Plants* 2020, 9, 195.

Waltz, E. With a free pass, CRISPR-edited plants reach market in record time. *Nat. Biotechnol.* 2018, 36, 6–7.

Wang, F.; Wang, C.; Liu, P.; Lei, C.; Hao, W.; Gao, Y. et al. Enhanced rice blast resistance by CRISPR/Cas9-targeted mutagenesis of the ERF transcription factor gene OsERF922. *PLoS One* 2016a, 11, e0154027 https://doi.org/10.1371/journal.pone.0154027

Wang, L.; Wang, L.; Tan, Q.; Fan, Q.; Zhu, H.; Hong, Z. et al. Efficient inactivation of symbiotic nitrogen fixation related genes in *Lotus japonicus* using CRISPR-Cas9. *Front Plant. Sci.* 2016b, 7, 1333 https://doi.org/10.3389/fpls.2016.01333

Wang, Y.; Cheng, X.; Shan, Q.; Zhang, Y.; Liu, J.; Gao, C. et al. Simultaneous editing of three homoeoalleles in hexaploid bread wheat confers heritable resistance to powdery mildew. *Nat. Biotechnol.* 2014, 32, 947–951 https://doi.org/10.1038/nbt.2969

Wu, Y.; Yang, W.; Wei, J.; Yoon, H.; An, G. Transcription factor *OsDOF18* controls ammonium uptake by inducing ammonium transporters in rice roots. *Mol. Cells*. 2017, 40, 178–185.

Xie, X.; Qin, G.; Si, P.; Luo, Z.; Gao, J.; Chen, X. et al. Analysis of Nicotiana tabacum PIN genes identifies NtPIN4 as a key regulator of axillary bud growth. *Physiol. Plant*. 2017, 160, 222–239 https://doi.org/10.1111/ppl.12547

Xu, R.; Yang, Y.; Qin, R.; Li, H.; Qiu, C.; Li, L. et al. Rapid improvement of grain weight via highly efficient CRISPR/Cas9-mediated multiplex genome editing in rice. *J. Genet. Genomics* 2016, 43, 529–532 https://doi.org/10.1016/j.jgg.2016.07.003

Yasui, Y.; Tanaka, W.; Sakamoto, T.; Kurata, T.; Hirano, H.-Y. Genetic enhancer analysis reveals that FLORAL ORGAN NUMBER2 and OsMADS3 co-operatively regulate maintenance and determinacy of the flower meristem in rice. *Plant Cell Physiol.* 2017, 58, 893–903 https://doi.org/10.1093/pcp/pcx038

Yin, X.; Biswal, A.K.; Dionora, J.; Perdigon, K.M.; Balahadia, C.P.; Mazumdar, S. et al. CRISPR-Cas9 and CRISPR-Cpf1 mediated targeting of a stomatal developmental gene EPFL9 in rice. *Plant Cell Rep.* 2017, 36, 745–757 https://doi.org/10.1007/s00299-017-2118-z

Yuan, J.; Chen, S.; Jiao, W.; Wang, L.; Wang, L.; Ye, W. et al. Both maternally and paternally imprinted genes regulate seed development in rice. *New Phytol.* 2017, 216, 373–387 https://doi.org/10.1111/nph.14510

Zafar, S.A.; Zaidi, S.S.E.A.; Gaba, Y.; Singla-Pareek, S.L.; Dhankher, O.P.; Li, X.; Mansoor, S.; Pareek, A.; Foyer, C. Engineering abiotic stress tolerance via CRISPR/Cas-mediated genome editing. *J. Exp. Bot.* 2020, 71, 470–479.

Zhang, Y.; Liang, Z.; Zong, Y.; Wang, Y.; Liu, J.; et al. Efficient and transgene-free genome editing in wheat through transient expression of CRISPR/Cas9 DNA or RNA. *Nat. Commun.* 2016, 7, 12617

Zhang, D.; Wang, Q.; Zhu, T.; Cao, J.; Zhang, X.; Wang, J.; et al. RACK1 promotes the proliferation of THP1 acute myeloid leukemia cells. *Mol. Cell. Biochem.* 2013, 384, 197–202.

Zhang, M.; Liu, Q.; Yang, X.; Xu, J.; Liu, G.; Yao, X.; Ren, R.; Xu, J.; Lou, L. CRISPR/Cas9-mediated mutagenesis of Clpsk1 in watermelon to confer resistance to Fusarium oxysporum f.sp. niveum. *Plant Cell Rep.* 2020, 39, 589–595.

Zhang, Y.; Bai, Y.; Wu, G.; Zou, S.; Chen, Y.; Gao, C.; Tang, D. Simultaneous modification of three homoeologs of TaEDR1 by genome editing enhances powdery mildew resistance in wheat. *Plant J.* 2017, 91, 714–724.

Zhao, C.; Liu, B.; Piao, S.; Wang, X.; Lobell, D.B.; Huang, Y.; Huang, M.; Yao, Y.; Bassu, S.; Ciais, P.; et al. Temperature increase reduces global yields of major crops in four independent estimates. *Proc. Natl. Acad. Sci. USA*. 2017, 114, 9326–9331.

Zhou, J.; Peng, Z.; Long, J.; Sosso, D.; Liu, B.; Eom, J.-S. et al. Gene targeting by the TAL effector PthXo2 reveals cryptic resistance gene for bacterial blight of rice. *Plant J.* 2015, 82, 632–643 https://doi.org/10.1111/tpj.12838

Zhu, H.; Li, C.; Gao, C. Applications of CRISPR–Cas in agriculture and plant biotechnology. *Nat. Rev. Mol. Cell Biol.* 2020, 21(11), 661–677.

Zsogon, A.; Cermak, T.; Naves, E.R.; Notini, M.M.; Edel, K.H.; Weinl, S.; Freschi, L.; Voytas, D.F.; Kudla, J.; Peres, L.E.P. De novo domestication of wild tomato using genome editing. *Nat. Biotechnol.* 2018, 36, 1211–1216.

CHAPTER 6

Application of Nanotechnology for Sustainable Development of Agriculture

MOHAMMAD IRFAN ALI[1*], ANKUR BHARDWAJ[2], and SURENDRA PRAKASH GUPTA[3]

[1]*School of Allied Sciences, JECRC University, Jaipur, Rajasthan, India*

[2]*Department of Biotechnology, Shri Vaishnav Institute of Sciences, Shri Vaishnav Vidyapeeth Vishwavidyalaya, Indore, Madhya Pradesh, India*

[3]*Department of Life Science, Shri Vaishnav Vidhyapeeth Viswavidhayalaya, Indore, Madhya Pradesh, India*

*Corresponding author

ABSTRACT

Nanotechnology is a promising tool for revolutionizing agriculture by enhancing productivity and sustainability. It offers efficient solutions such as nanofertilizers, nanopesticides, and nanoherbicides, which reduce chemical usage and minimize environmental contamination. Nanotechnology enhances soil characteristics, improves water management, and facilitates pathogen detection, thereby contributing to improved food safety and quality. By using nanoscale nutrients and growth-stimulating materials, agriculture can reduce inputs while increasing yields, thereby advancing sustainability and food security. The application of nanomaterials, including nanofertilizers and pesticides, helps control diseases and pests without polluting water and soil. Nanobiosensors are developed for monitoring soil quality, supporting sustainable agriculture. Additionally, nanotechnology enhances industrial food processing by improving

Technological Advancements in Sustainable Crop Improvement and Production.
Devendra Kumar, Alok Kumar Singh, and Nisha Malik (Eds.)

DOI: 10.1201/9781779643155-6

nutritional value, safety, and sensory properties while reducing postharvest losses. This chapter explores recent applications of nanotechnology in agriculture, focusing on challenges related to sustainability, climate change, and food security. By leveraging nanotechnology, agriculture can become more efficient and environmentally friendly, addressing global food security needs while mitigating environmental impacts. The integration of nanotechnology in agriculture holds significant potential for transforming the sector into a more sustainable and resilient system, ensuring better crop management and resource utilization. This transformation supports long-term agricultural sustainability.

6.1 INTRODUCTION

Agriculture has progressed in parallel with human evolution since ancient times. Traditional agriculture requires the frequent application of chemical fertilizers, as well as time-honored farming methods that can greatly enhance crop growth, yield, productivity, and nutritional value, while also causing various negative consequences such as reducing soil quality, weakening resistance to pests and diseases, and harming the environment. Hence, agriculture as a whole is currently facing many significant challenges, including losses in production due to pests, global climate change, and the depletion of natural resources (Lowry et al., 2019).

The world population is expected to rise to 9 to 10 billion by 2050, indicating that food production will need to increase by 25% to 70% compared to the current state (Scott et al., 2018). Therefore, new approaches and technologies need to be focused on agriculture to ensure the sustainable development of agriculture (Fraceto et al., 2016). The current challenge is in creating sustainable agricultural practices to increase crop production quickly. Nanotechnology has also been applied in agriculture to regulate the release of pest-control substances and to monitor soil health and nutrition. This is part of the growing field of sustainable agriculture, which aims to decrease energy consumption and waste production. Farmers are utilizing nanotechnology to effectively manage water, fertilizer, soil quality, and fertility, leading to improvements in crop quality and yield (Fellet et al., 2021).

The effective use of natural resources, including nutrients, water, and chemicals in agricultural areas by using nanotechnology is environment-benign and cost-effective (Prasad et al., 2017). Nanomaterials have the potential to assist farmers in identifying crop pests by detecting viruses and microbial pathogens, as well as weeds, and accurately determining the levels of stress, such as drought, salinity, floods, and soil nutrient deficiencies. They can also help improve water and fertilizer efficiency, thereby aiding in the reduction of environmental pollution (Jhanzab et al., 2015; Pandey et al., 2016; Wang and Xie, 2018). Many nanomaterials have been developed for uses in agricultural practices, including new innovative solutions for soil remediation, water remediation, nanofertilizers, and nanopesticides, used to reduce the toxic effects while increasing the production and quality of agricultural production (do Espirito et al., 2021).

The use of nanopesticides and nanofertilizers contributed to enhancing pest control, increasing plant nutrition, and developing environmentally benign methods for sustainable agriculture (Acharya et al., 2020). Nanomaterials, due to specific physical–chemical properties, size, zeta potential, and concentration, and have key roles in uptake and translocation in plants that affect the biological responses and act as a stimulant that improves seed metabolism, seedling, and plant growth by acting in signaling pathways (do Espirito et al., 2021).

Zeolites and nanoclays nanomaterials improve the water or liquid agrochemicals retention in the soil for their sustained release. The release of nanofertilizers is an attractive alternative aiding in the reduction of nitrogen losses via leaching, emissions, and extended soil absorption (Duhan et al., 2017). Traditional fertilizers typically require large quantities for application due to their inefficiency in uptake. The main challenges for phosphorus and nitrogen-based fertilizers are low efficiency of nutrient uptake and rapid transformation into chemical forms that cannot be used by plants. Ultimately, this results in the release of harmful greenhouse gases and eutrophication (Raliya et al., 2018). Nanotechnological advancements are also being applied to enhance plants and genetically transform them (Torney et al., 2007). Nanosilver is effective against different fungal pathogens (Ouda, 2014). Nanoencapsulation of insecticides helps plant absorption of insecticides efficiently against insect pests (Suman et al., 2010), weeds (Kumar et al., 2017). Nanoemulsions are effective against disease and insect pests (Dimetry and Hussein, 2016).

Hence, the use of nanotechnology is considered to be one of the new technologies that can help maintain crop production systems in order to meet the rising food needs of the growing global population. In this section, the most recent advancements, potential exploitation, challenges, and constraints of nanotechnology for the sustainable development of agriculture are discussed.

6.2 NANOTECHNOLOGY FOR SUSTAINABLE AGRICULTURAL DEVELOPMENT

The goal of sustainable agriculture development is to fulfill the need for human food and fiber, maintain environmental quality, and the efficient use of natural resources and nonrenewable resources, on-farm resources, and the economic sustainability of farming activities improves the well-being of farmers and society as a whole without endangering the ability of future generations to fulfill their requirements. The implementation of sustainable agricultural practices can be enhanced through the innovative use of nanotechnology (Duhan et al., 2017). Nanomaterials, with their high surface area to volume ratio and strong penetration capability of nanofertilizers, offer a promising substitute for chemical fertilizers. This can help in reducing environmental risks, preventing eutrophication, and increasing crop productivity by promoting seed germination, photosynthesis, nitrogen metabolism, protein, lipid, and carbohydrate synthesis, and stress tolerance (Jiang et al., 2021; Rautela et al., 2021). Nanomaterials utilized for the improvement and sustenance of soil and crop are either of organic (chitosan, clay, polyacrylic acid or zeolite, lipids, polymers, and CNTs), inorganic (Fe, SiO_2, AgO, MgO, ZnO, aluminium oxide, TiO_2, cerium oxide, gold nanorods, ZnCdSe, ZnS core-shell, Mn/ZnSe quantum dots and InP/ZnS core-shell or both hybrid (nanodiamonds, encapsulated carbon nanotubes, graphene) (Duhan et al., 2017; Ghormade et al., 2011; Morales-Díaz et al., 2017; Singh et al., 2021) that are used for the nanobased agricultural tools like nanofertilizer to maintain soil dynamic nature as well as to improve agriculture productivity and nanopesticides protect the crops from biotic stress. These collective efforts have made an unimpeachable impact on the sustainable agriculture development and agricultural research (Chhipa, 2017).

6.2.1 NANOFERTILIZERS

Nanofertilizers are fertilizers loaded with nutrients and wrapped or surrounded by nanomaterial to facilitate nanoscale dimension to enhance the soil fertility, controlled release, bioavailability efficiently and minimize nutrient loss by leaching, run-off. The rapid breakdown and evaporation improve the nutrient content and soil fertility, leading to increased crop yields in a more environmentally friendly manner compared to traditional fertilizers (Chhipa, 2017; Rautela et al., 2021). Nanofertilizers are categorized as either macronutrient or micronutrient nanofertilizers depending on the nutrients needed by plants (Khan et al., 2021).

Nanofertilizers contain nutrients that are attached to nanoabsorbents and can be used alone or together to release nutrients at a slower pace compared to traditional fertilizers. This helps reduce nitrogen leaching and runoff loss (Liu and Lal, 2015; Tan et al., 2017). Nanofertilizers can be prepared in three different ways according to the plant's nutritional needs for growth and development: (1) nanoscale coating fertilizers, (2) nanoscale additive fertilizers, and (3) nanoporous materials (Al-Juthery et al., 2021).

Both intrinsic and extrinsic factors are highly dependent on the absorption, accumulation, transportation, and the route of exposure of nanofertilizers. The primary internal factors that influence the effectiveness of nanofertilizers are the types of coatings on the surface and the size of the particles, while external factors such as soil pH, soil texture, and organic matter significantly impact the efficacy of nanofertilizers (Ma et al., 2018).

Different types of nanofertilizers containing nitrogen, potassium, phosphorus, zinc, iron, copper, manganese, molybdenum, and carbon nanotubes have displayed impressive effectiveness (Liu and Lal, 2014). Nanofertilizer based on calcium, urea, and phosphorus hydroxyapatite nanoparticles gradually releases calcium, phosphorus, and nitrogen and showed higher yield in soybean production in comparison to conventional fertilizer (Ahmed et al., 2021; Liu and Lal, 2015). Kottegoda et al. (2017) developed biocompatible urea functionalized hydroxyapatite nanoparticles used for controlled release of nitrogen. Zinc oxide (ZnO) nanofertilizer improved the germination, protein contents, sugar contents antioxidant activity in *Brassica oleracea* and *Lycopersicon esculentum* (Singh et al., 2013). Nanoiron fertilizer improved the plant growth, total protein contents, and pigment composition (Askary et al., 2016). The growth of Borago officinalis plants and the yield of essential oils were enhanced by the use of nanourea and nanoiron fertilizers in combination with traditional fertilizers (Mahmoodi et al., 2017).

Lahiani et al. (2013) reported that the use of multiwalled carbon nanotubes (MWCNTs) enhanced seed germination and plant growth in *Hordeum vulgare*, *Glycine max*, and *Zea mays*. Abdel-Aziz et al. (2016) revealed the efficiency of chitosan nanoparticle encapsulating N, P, and K in *Triticum aestivum* grown in sandy soil and observed improvement in the harvest index, mobilization index crop index. McGehee et al. (2017) found that adding MWCNT to hydroponically grown tomatoes can increase fruit yield.

Various concentrations of silver nanoparticles have been found to increase the rate of seed germination in *Zea mays*, *Citrullus lanatus*, and *Cucurbita pepo*. A study conducted by Almutairi and Alharbi in 2015 demonstrated this effect. *Lactuca sativa* seed germination can also be improved by the application of TiO_2 nanoparticles using single-capillary electrospraying, as indicated by research conducted by Wu et al. (2014). Biocompatible silver nanoparticles have been shown to significantly enhance seed germination, α-amylase activity, and upregulation of aquaporin genes in *Oryza sativa* through nanopriming, according to a study by Mahakham et al. (2017).

According to research conducted on Zn-deficient soil, the incorporation of ZnO NPs with other fertilizers significantly improved the productivity of Hordeum vulgare and Pennisetum Americanum. This finding was reported by Kale and Gawade in 2016, as well as by Moghaddasi et al. (2017). Studies have shown that nanocomposite fertilizers have a positive impact on rhizosphere bacteria by promoting the production of secondary metabolites. This leads to enhanced plant growth through increased colonization of the root surface. These results were observed in research conducted by Panichikkal et al. (2019). Researchers have utilized zinc–NPs to enhance the conversion of native P into plant-available P, resulting in an 11% increase in plant growth, yield, biomass, proteins, photosynthetic pigments, and nutritional content in legumes, cereals, and cotton plants. Enzymes such as phosphatase and phytase, which are essential for plant growth, require Zn as a cofactor. This was demonstrated in studies by Raliya et al. (2013b) and Venkatachalam et al. (2017). The use of rubber-type nanomaterials as a source of zinc has been shown to improve shoot and fruit yield compared to the use of commercial Zn-sulfate fertilizer in crops such as sunflower, sugarcane, potato, rice, maize, and wheat (Mattiello et al., 2015; Monreal et al., 2016).

Carbon nanoparticles and MWCNTs at concentrations of up to 200 mg/kg significantly increased the levels of urease, phosphatase, and dehydrogenase enzymes in the soil. Additionally, these nanoparticles also improved the soil's capacity to retain essential nutrients like ammonium, nitrate, and phosphorus (Xin et al., 2020; Zhao et al., 2021).

The effects of Zn, Cu, Mn, and iron oxide NPs on the germination of *Lactuca sativa* seeds were evaluated at low concentrations (i.e., less than 50 mg/L), and the study found that the effects of CuO NPs and Zn NPs were similar, and that the growth of lettuce seedlings was improved by MnO and FeO NPs (Lü et al., 2016). Fungal-mediated synthesis of P NPs by using tricalcium phosphate as a precursor salt has shown significant efficiency (Tarafdar et al., 2012). P nanofertilizer synthesized by using ultrasonic dispersion of raw phosphorite as precursor showed promising efficacy (Sharonova et al., 2015).

Silica nanoparticles enhance the growth of seedlings and trigger the activation of antioxidant enzymes to withstand biotic and abiotic pressures like high temperatures and irregular precipitation (Lin et al., 2004; Neethirajan et al., 2009). Si nanoparticles boost potassium absorption and decrease sodium uptake in the presence of salt stress (Rizwan et al., 2015).

A combination of biofertilizers and nanomaterials, that is nanobiofertilizers, is effective as nanomaterials interact with microorganisms. Although three categories of microorganisms, (1) plant growth-promoting rhizobacteria (PGPR), (2) nitrogen-fixing rhizobacteria, and (3) arbuscular mycorrhizal fungi, along with nanomaterials, may be used as nanobiofertilizers. Numerous studies have documented a positive impact, showcasing the interaction between gold nanoparticles and plant growth-promoting rhizobacteria (Nongbet et al., 2022; Thirugnanasambandan, 2018).

Marchiol et al. (2019) prepared hydroxyapatite ($Ca_{10}(PO_4)_6(OH)_2$) NPs nanofertilizers and found that increased levels of nanofertilizers were discovered to have no impact on the germination process. However, they did lead to a significant improvement in root elongation in *Solanum lycopersicum*.

In a study conducted by Francis et al. (2022), it was demonstrated that green-synthesized CuO and ZnO nanoparticles proved to be effective nanofertilizers in improving the agronomic characteristics of Amaranthus hybridus when applied through foliar application and hydroponics. Murgueitio-Herrera et al. (2022) prepared the nanofertilizer of ZnO_MnO-NPs and FeO_ZnO-NPs by using Andean blueberry extract. Prepared nanofertilizer showed improvement in plant growth in *Brassica oleracea* and *Lupinus mutabilis*. Algae as green biomass can be transformed into nanoscale particles having specific shape, size, and structure with unique surface properties to formulate modern-day agronanofertilizer with more efficient and drastically reduce our dependencies on synthetic chemical fertilizer (Mahapatra et al., 2022).

6.2.2 NANOPESTICIDES

The commonly used nanopesticides include nanofungicides, nanomolluscicides, nanoherbicides, nanonematicides and nanoparticulated growth regulators in the form of nanosuspensions, nanoemulsions, nanospheres, nanomicelles and nanocapsules that enhance the efficacy and environment and reduce the cost, properties of formulation such as dispersibility in water, stability in chemicals, and ability to adhere to targets, permeability and controlled release environmentally benign and eco-friendly and showed great potential in sustainable development of agriculture (Antonoglou et al., 2018; Pereira et al., 2017; Prasad et al., 2019; Zhao et al., 2017). The surfactants and organic ligands are the most frequently used types of nanoparticles in agriculture (Du et al., 2011; Elek et al., 2010). Silica, silver, carbon, and aluminum silicate are the key components that can be utilized in the creation of nanopesticides (Nuruzzaman et al., 2016).

Kumar et al. (2016) formulated a nanoform of fungicide improved antifungal effectiveness against *Macrophomina phaseolina* using a combination of 25% trifloxystrobin and 50% tubuconazole. Antonoglou et al. (2018) prepared inexpensive Cu–Zn bimetallic nanoparticles with effective antifungal activity against *Saccharomyces cerevisiae*.

A polymeric nanoformulation herbicide is less toxic than the free herbicides, consisting of poly(epsilon-caprolactone) and chitosan/tripoly-phosphate encapsulating atrazine/paraquat was prepared (Grillo et al., 2015). In 2015, Oliveira and colleagues created polymeric nanoparticles made of poly(epsilon-caprolactone) that contained atrazine. They assessed the effectiveness of these nanoparticles as postemergence herbicides on mustard greens (*Brassica juncea*) and found that they were more efficient when used at a lower rate.

In 2013, Guang and colleagues studied the toxicity of silver nanoparticles that were synthesized using green methods as a molluscicide on *Oncomelania hupensis* snails. They found that the synthesized molluscicide was toxic to the snails but had minimal toxicity to crucian seedlings. In 2017, Omobhude and team created polymeric nanoparticles made of polylactic acid that enclosed curcumin and nisin as molluscicidal, and the prepared nanoformulation was a potent molluscicide against *Biomphalaria. pfeifferi*.

Cromwell et al. (2014) assessed the effectiveness of silver nanoparticles as nematicidal against *Maloidogyne graminis* and *M. incognita* and found efficient against nematicidal. Abdellatif et al. (2016) devised a nanonematicide using green silver nanoparticles and found it effective against *M. javanica* infested in tomato.

According to Boehm et al. (2003), the penetration of ethiprole-loaded nanospheres made of polycaprolactone and polylactic acid was found to be superior to that of the reference formulation in terms of insecticidal activity.

The efficacy of controlled release of imidacloprid (1-(6-chloro-3-pyridinyl methyl)-*n*-nitro imidazolidin-2-ylideneamine) enclosed in polyethylene glycol was found to be effective against *Melanagromyza sojae Zehntmer* and *Bemisia tabaci Gennadius* (Sekhon, 2014). Utilization of lipid-based nanomaterials has shown promise in sustainable crop production systems as well (Naseri et al., 2015). Encapsulating two herbicides, atrazine and simazine, in solid lipid nanomaterials proved to be effective against *Raphanus raphanistrum* in maize without any negative impact on the crop (de Oliveira et al., 2015).

Nanomaterials improve the stability of active ingredients that easily undergo photolysis of active ingredients (Pang et al., 2020). Zhu et al. (2020) created nanoparticles containing Avermectin that were attached to the extended chain of poly(ethylene glycol)-carboxymethyl cellulose, which improved the stability of Avermectin significantly over the control. In 2018, Kaziem and colleagues enclosed Avermectin in hollow mesoporous silica and then attached cyclodextrin to the outermost layer of the particles to enhance the stability of Avermectin.

Nanomaterials can release active ingredients in a targeted and controlled manner at the most effective concentration. According to a study by Xiang et al. (2017), a magnetic nanocarrier was created using diatomite and Fe_3O_4. In this nanoformulation, both glyphosate herbicide and cypermethrin insecticide were enclosed and then coated with chitosan to regulate the release of the pesticide when in acidic conditions.

6.2.3 NANOBIOSENSORS

Crop production in a sustainable way could be enhanced by nutrient management, growth monitoring, pest management and disease assessment, detection of soil conditions, food production, and plant hormone detection by nanotechnology-based approaches that could be useful tools for the detection of the disease. Nanobiosensors, nanoimaging, nanoparticulated systems, and nanopore DNA sequencing tools have tremendously enhanced accuracy, persistence, and decreased the amount of time it takes to detect pathogens, and improved the high-throughput equipment used for pathogen detection (Khiyami et al., 2014; Yata et al., 2017). Nano-biosensors are an exceptional tool for determining the contaminants at a minimum concentration. Common nanosensor detection approaches include fluorescence resonance energy transfer, corona-phase molecular recognition, surface-enhanced Raman scattering, electrochemical nano-sensors, and piezoelectric nanosensors (Liu et al., 2021).

A nanobiosensor utilizing a graphene nanoribbon was utilized for detecting Mn impurities according to Enciu et al. (2014). Salinas et al. (2014) created a nanosensor that can detect argon using self-assembled monolayers on a gold substrate type (III) in a water-based solution. Gautam et al. (2014) worked on developing a DNA-based nanobiosensor embedded in a polymer coating for fertilizers that would only release the required amount of fertilizer based on the ionic signals from plant roots. A fluorescent ratiometric single-walled carbon nanotubes sensor developed for NO detection, which is based on a single-molecule detection technique (Giraldo et al., 2015). Nanosensors help monitor plant health at the molecular level. However, the stability, sensitivity, and reliability of nanosensors still need to be improved sufficiently for use in sustainable agricultural production (Liu et al., 2021).

6.2.4 SEED NANOPRIMING

Seed nanopriming is used to provide protection for seeds during storage, improve germination, enhance plant growth, and increase resistance under abiotic or biotic stress, which helps to reduce the use of pesticides and fertilizers (Malik et al., 2021). Seed nanopriming is able to activate different genes related to plant stress resistance during seed germination (Ye et al., 2020). Seed nanopriming is used for seed protection, biofortification to promote an increase in quality and production of food (Abbasi Khalaki et al., 2020).

In seed nanopriming, the nanosuspensions or nanoformulations used where the greatest fraction of nanoparticles is retained on the seed surface as a coating that acts as fungicides or bactericides and protects the seed against pathogens during storage and in the field (Gross et al., 2020). Nanomaterials, including metallic, polymeric nanoparticles, and biogenic metallic have shown potential application in seed nanopriming, resulting in the stimulation and changes in the defense system by increasing the antioxidant enzymes and providing more resistance against pests, biotic and abiotic stress conditions (do Espirito et al., 2021).

6.2.5 ROLE OF NANOPARTICLES IN GENETIC MODIFICATION OF CROPS

Nanoparticles act as a carrier to deliver the genetic materials into plant cells to obtain desirable traits in plants. Various types of nanoparticles, including mesoporous silica nanoparticles, gold, carbon nanotubes, and magnetic nanoparticles, have been utilized to deliver plasmid DNA, siRNA, and double-stranded RNA into protoplasts (Naik et al., 2022). Lv et al. (2020) have shown gene silencing and gene editing in plants with the use of magnetic nanoparticle-based pollen transformation. Similarly, gene editing has also been shown with small NPs-CRISPR/Cas9 vector complexes, and these complexes were microinjected into the plant parts like leaves (Duan et al., 2021). Carbon dots–siRNA complex has been utilized to silence the GFP in tomato and tobacco plants (Schwartz et al., 2020). Demirer et al. (2021) recently revealed genome editing in plants using the CRISPR/Cas9 system along with nanoparticles.

6.2.6 NANOTECHNOLOGY FOR GROWTH-STIMULATING ELEMENTS OF SUSTAINABLE AGRICULTURE

Carbon-based nanomaterials like carbon nanoparticles, carbon nanotubes, fullerols, and fullerenes are now recognized as significant regulators of plant growth. These materials have been found to enhance germination rates, chlorophyll levels, and protein content in plants. Kong et al. (2013) prepared a GSH-sensitive nanomaterial based on MSNs for the delivery of phytohormone, that is salicylic acid. Recently, foliar application of Chitosan nanoparticles significantly enhanced the growth, yield, and mineral content (Fe, Zn, Mn, P, Ca, and Mg) by inducing chitinase, beta-1,3 glucanase, protease inhibitors, chitosanase, peroxidase, and polyphenol oxidase in finger millet leaves (Sathiyabama and Manikandan, 2021). Salicylic acid nanoparticles could induce some resistant genes of sweet pepper against black mold disease (Abdel-Rahman et al., 2021).

6.2.7 NANOTECHNOLOGY FOR SOIL REMEDIATION

Certain organic and inorganic substances, like natural short-ordered aluminosilicate, titanium oxide surfaces, and humic acids, can be combined with Ni using MWCNTs to create an effective nanobioremediation tool for sustainable agriculture (Prasad et al., 2017; Raliya et al., 2013a).

Nanobioremediation is a cost-efficient approach of using plants and microbes for the breakdown of agri-contaminants and pollutants for improving soil quality and dipping pollution by breaking down, eradicate, retain, or reduce the pollutants (Ashraf et al., 2019). Polymeric nanoparticles, including nanocapsules or nanospheres, can be used in the elimination of persistent pesticides and long-chain hydrocarbons (Briffa et al., 2020).

FeO nanoparticles coated with polyvinylpyrrolidone have been employed for improving the bioremediation of the soil contaminated with Pb and Cd by *Halomonas* sp. (Alabresm et al., 2018). A biosorbent of Fe_3O_4 nanoparticles treated with *S. aureus*, with a surface encapsulated with phthalic acid employed for the removal of Cu, Ni, and Pb (Mahmoud et al., 2016). ZnO nanoparticles were employed for the removal of Cu, Cd, Cr, and Pb using heavy metal-resistant bacteria such as *B. cereus* and *L. macrolides* (Baragaño et al., 2020). However, the fate of nanoparticles should be carefully taken into account before the wide exploitation of nanobioremediation techniques (Rajput et al., 2022).

6.2.8 NANOTECHNOLOGY FOR CROP IMPROVEMENT

Effective enhancement of crops and productivity is a direct outcome of properly managing plant growth in various physiological and biological stress conditions. The implementation of nanotechnology could have a significant impact on sustaining and managing plant growth in the presence of both biotic and abiotic stressors. Utilizing nanometal oxides to combat soil pathogens and incorporating metallic nanomaterials like nanosilicon to improve water uptake efficiency in plants are regarded as highly auspicious methods within agricultural nanotechnology. The application of synthetic clay nanomineral and ZnO nanoparticles for eliminating arsenic appears to be beneficial for water purification systems. The use of silicon dioxide (SiO_2) nanoparticles has been found to enhance seed germination, plant growth, and yield in tomato and maize according to Siddiqui et al. (2015). In studies conducted by Pourkhaloee et al. (2011) and Zaytseva and Neumann (2016), it was observed that single and MWCNTs can also promote seed germination by causing perforation in the seed coat of *Salvia macrosiphon*, *Capsicum annuum*, and *Festuca arundinacea*. Therefore, the use of nanomaterials has the potential to significantly improve sustainable crop productivity.

6.2.9 NANOTECHNOLOGY FOR POSTHARVEST LOSS REDUCTION

Approximately 40% losses of harvested food occur processing and post-harvest stage (FAO, 2019). Nanotechnology can be employed for declining postharvest losses with better stability by designing functional packaging with the least amount of bioactives, mechanical properties, and improved gas with reduced effect on the qualities of food (Flores-López et al., 2016).

Nanocoatings with edible value employed on different foods that retain flavors, enzymes, colors, antioxidants, and browning resistant agents and enhance the shelf life of foods (Zambrano-Zaragoza et al., 2018).

Nanoparticles like Ag, Zn, Au, ZnO, TiO_2, SiO_2, and MgO usually employed in food processing, may also pose health risks. Such risks could be controlled in the future by using greener approaches, easy and cheaper protocols for the removal and degradation of existing nanomaterials (Neme et al., 2021).

6.2.10 ECOTOXICOLOGICAL ASSESSMENT OF THE NANOPARTICLES

Nanomaterials are released into soil and water during their use, therefore increasing concern about impacts on the environment on terrestrial and aquatic organisms, including human health. Nanoparticles, including carbon nanotubes, iron, titanium dioxide, and fullerenes nanoparticles, will aggregate to some degree (Fortner et al., 2005). Following release to the soil environment, nanoparticles are absorbed by soil particles and degraded by biotic and abiotic processes, and finally transported to water bodies through runoff, leaching, and drain flow (Tungittiplakorn et al., 2004). Several techniques have been utilized for the detection and characterization of Nanomaterials have included infrared, nuclear magnetic resonance spectrometry, mass spectrometry, ultraviolet, fluorescence spectrometry, inductively coupled plasma mass spectroscopy, X-ray spectroscopy and dynamic-light scattering, as well as imaging techniques including scanning-electron microscopy, transmission-electron microscopy and atomic-force microscopy. Many of these detection methods have been combined with separation methods, such as flow field fractionation, high-performance liquid chromatography, gel-permeation chromatography, and size-exclusion chromatography. There is now an imperative need to further develop and discover integrated approaches for environmental monitoring systems (Boxall et al., 2007).

6.2.11 NANOTECHNOLOGY IN FOOD PROCESSING

The European Commission and the United States Food and Drug Administration (US FDA) are the major legislation to regulate food nanotechnology. The US FDA has permitted the utilization of inorganic oxide chemicals, including MgO, TiO_2, and SiO_2, as anticaking agents, color additives, and flavor carriers, in various foods, including candies, white sauces, and puddings. The US FDA has also permitted the use of ZnO, copper oxide, and iron oxide as safe nutritional dietary supplements for animal feed (Khan et al., 2022).

Nanoscale enzymes have been used in food treatment methods to add dietary values, flavor, and aroma to processed food. In addition to helping in enhancing the enzymatic activities, and increases the shelf life of food. The nanoencapsulation has also been utilized for the improvement of food quality, flavor, and preservation to improve food products. Nanocapsules and nanoceramic pots are also used for changing absorption to diminish the time needed for cooking, and trans fatty acids that increase the absorption of nutrients in processed food. Nanoencapsulation not only accomplishes food interactions with active ingredients to prevent the exclusion of active agents but also prevents odors, resulting in shielding the food products from moisture, chemical, or biological damage during packaging and storage (Singh et al., 2021). Nanoparticles based on metallic oxides have also been employed as a coloring agent in food products such as SiO_2 nanomaterials are used for the development of aromas in food materials (Dekkers et al., 2011).

6.2.12 NANOTECHNOLOGY IN FOOD PACKAGING AND LABELING

Cellulosic paper-based products are commonly used in our everyday lives. These materials can serve as a means of transmitting various infectious diseases among humans (Angelakis et al., 2014). Applying a nanocoating to these materials results in antimicrobial cellulosic paper, making them safe for use and improving packaging to prolong the shelf life of fruits, vegetables, and other agricultural products. In a recent study, silver nanoparticles immobilized onto dopamine-modified cellulose paper displayed impressive antimicrobial properties against certain antibiotic-resistant fish pathogenic bacteria and fungal phytopathogens (Islam et al., 2018). Researchers have also developed daylight-driven rechargeable anti-bacterial and antiviral nanofibrous membranes by incorporating chemicals that can generate reactive oxygen species efficiently, thereby controlling bacterial and viral phytopathogens even in low-light or dark conditions (Si et al., 2022).

6.3 CONSTRAINTS IN THE USE OF NANOTECHNOLOGY IN AGRICULTURAL DEVELOPMENT

In some cases, nanoparticles can have unexpected negative impacts on sustainable agriculture, soil health, and the ecosystem (Zhang, 2003; Kelly, 2004). Over time, these nanoparticles can accumulate in soils and water (Gottschalk et al., 2009). Some nanoparticles may enter the food chain, posing a threat to human health and the ecosystem (Zhao et al., 2017). To mitigate these risks, research should focus on the safety and risks associated with nanoparticles, particularly their toxic effects and interactions with plants. This knowledge will help in the development of nanopesticides for sustainable agriculture, which can help in accelerating the degradation of toxic residues and reducing the residual effects of pesticides on soil and water by using biodegradable carriers and photocatalysts (Zhao et al., 2017).

Nanofertilizer shows low mobility and degradability that lead to deposition of the nanomaterials. Excessive amounts of this could result in the buildup of nanoparticles and toxicity at the nanoscale (Chhipa, 2017; Khan et al., 2018). Nanotechnology-based agrochemicals have to be associated with other nanotechnology approaches to meet the high demands and desired efficiency (Shao et al., 2022).

Understanding how nanomaterials work and their impact on nontarget organisms is a key area of focus in research and development. Additionally, there is a significant emphasis on utilizing nanotechnology in the development of nanopesticides, which remains a difficult task that could benefit from a more advanced approach in creating versatile nanomaterials.

6.4 FUTURE PERSPECTIVES

Nanomaterials can be utilized in creating nanofertilizers, nanopesticides, nanocarriers, nanosensors, nanopackaging, and nanochips to enhance crop productivity and resource efficiency by significantly reducing production costs and postharvest losses. By smartly delivering active ingredients, nanomaterials can help decrease the use of sprayed agrochemicals, reduce nutrient losses during fertilization, and improve yields through optimized water and nutrient management. The application of nanosensors in monitoring water, nutrients, and chemicals can further lower production costs and address environmental concerns. Additionally, nanomaterials can enhance soil water retention and support innovations in plant improvement and genomic transformation programs. However, improper and uncontrolled use of nanomaterials, like other synthetic agrochemicals, may have harmful effects on the environment. Therefore, a research-based and wise application of nanotechnology is essential for the sustainability of both the environment and crop production systems. Consequently, advanced techniques based on nanotechnology are being developed to enhance crop productivity in unique and innovative ways.

KEYWORDS

- **nanotechnology**
- **nanofertilizers**
- **nanopesticides**
- **sustainable agriculture**
- **nanomaterials**
- **nanobiosensors**

REFERENCES

Abbasi Khalaki, M.; Moameri, M.; Asgari Lajayer, B.; Astatkie, T. Influence of nano-priming on seed germination and plant growth of forage and medicinal plants. *Plant Growth Regul.* 2020, 93, 13–28.

Abdel-Aziz, H.M.M.; Hasaneen, M.N.A.; Omer, A.M. Nano chitosan-NPK fertilizer enhances the growth and productivity of wheat plants grown in sandy soil. *Span J Agric Res.* 2016, 14, e0902.

Abdellatif, K.F.; Abdelfattah, R.H.; El-Ansary, M.S.M. Green nanoparticles engineering on root-knot nematode infecting eggplants and their effect on plant DNA modification. *Iran J. Biotechnol.* 2016, 14, 250–259.

Abdel-Rahman, F.A.; Khafagi, E.Y.; Soliman, M.S.; Shoala, T.; Ahmed, Y. Preharvest application of salicylic acid induces some resistant genes of sweet pepper against black mold disease. *Eur. J. Plant Pathol.* 2021, 159, 755–768.

Acharya, P.; Jayaprakasha, G.K.; Crosby, K.M.; Jifon, J.L.; Patil, B.S. Nanoparticle-mediated seed priming improves germination, growth, yield, and quality of watermelons (*Citrullus lanatus*) at multi-locations in Texas. *Sci. Rep.* 2020, 10, 5037.

Ahmed, D.F.; Isawi, H.; Badway, N.A.; Elbayaa, A.A.; Shawky, H. Graphene oxide incorporated cellulose triacetate/cellulose acetate nanocomposite membranes for forward osmosis desalination. *Arab. J. Chem.* 2021, 14, 102995.

Alabresm, A.; Chen, Y.P.; Decho, A.W.; Lead, J. A novel method for the synergistic remediation of oil-water mixtures using nanoparticles and oil-degrading bacteria. *Sci. Total Environ.* 2018, 630, 1292–1297.

Al-Juthery, H.W.A.; Lahmod, N.R.; Al-Taee, R.A.H.G. Intelligent, nano-fertilizers: A new technology for improvement nutrient use efficiency (Article Review). *IOP Conf. Ser. Earth Environ. Sci.* 2021, 735, 012086.

Almutairi, Z.M.; Alharbi, A. Effect of silver nanoparticles on seed germination of crop plants. *Int. J. Nucl. Quantum Eng.* 2015, 9, 594–598.

Angelakis, E.; Azhar, E.I.; Bibi, F.; Yasir, M.; Al-Ghamdi, A.K.; Ashshi, A.M.; Elshemi, A.G.; Raoult, D. Paper money and coins as 772 potential vectors of transmissible disease. *Future Microbiol.* 2014, 9, 249–261.

Antonoglou, O.; Moustaka, J.; Adamakis, I.D.; Sperdouli, I,; Pantazaki, A.A.; Moustakas, M.; Dendrinou Samara, C. Nanobrass CuZn nanoparticles as foliar spray nonphytotoxic fungicides. *ACS Appl. Mater Interfaces* 2018, 10, 4450–4461.

Ashraf, S.; Siddiqa, A.; Shahida, S.; Qaisar, S. Titanium-based nanocomposite materials for arsenic removal from water: A review. *Heliyon* 2019, 5, e01577.

Askary, M.; Amirjani, M.R.; Saberi, T. Comparison of the effects of nano-iron fertilizer with iron-chelate on growth parameters and some biochemical properties of Catharanthus roseus. *J. Plant Nutr.* 2016, 40, 974–982.

Baragaño, D.; Forján, R.; Welte, L.; Gallego, J.L.R. Nanoremediation of As and metals polluted soils by means of graphene oxide nanoparticles. *Sci. Rep.* 2020, 10, 1896.

Boehm, A.L.; Martinon, I.; Zerrouk, R.; Rump, E.; Fessi, H. Nanoprecipitation technique for the encapsulation of agrochemical active ingredients. *J. Microencapsul.* 2003, 20, 433–441.

Boxall, A.B.; Tiede, K.; Chaudhry, Q. Engineered nanomaterials in soils and water: How do they behave and could they pose a risk to human health? *Nanomedicine* 2007, 2, 919–927.

Briffa, J.; Sinagra, E.; Blundell, R. Heavy metal pollution in the environment and their toxicological effects on humans. *Heliyon* 2020, 6, e04691.

Chhipa, H. Nanofertilizers and nanopesticides for agriculture. *Environ. Chem. Lett.* 2017, 15(1), 15–22.

Cromwell, W.A.; Yang, J.; Starr, J.L.; Jo, Y.K. Nematicidal effects of silver nanoparticles on root knot nematode in bermudagrass. *J. Nematol.* 2014, 46(3), 261–266.

de Oliveira, J.L.; Campos, E.V.R.; da Silva, C.M.G.; Pasquoto, T.; Lima, R.; Fraceto, L.F. Solid lipid nanoparticles co-loaded with simazine and atrazine: Preparation, characterization, and evaluation of herbicidal activity. *J. Agric. Food Chem.* 2015, 63(2), 422–432.

Dekkers, S.; Krystek, P.; Peters, R.J.; Lankveld, D.P.; Bokkers, B.G.; van Hoeven-Arentzen, P.H.; Bouwmeester, H.; Oomen, A.G. Presence and risks of nanosilica in food products. *Nanotoxicology* 2011, 5, 393–405.

Demirer, G.S.; Silva, T.N.; Jackson, C.T.; Thomas, J.B.; Ehrhardt, D.; Rhee, S.Y.; et al. Nanotechnology to advance CRISPR–Cas genetic engineering of plants. *Nat. Nanotechnol.* 2021, 16, 243–250. https://doi.org/10.1038/s41565-021-00854-y

Dimetry, N.Z.; Hussein, H.M. Role of nanotechnology in agriculture with special reference to pest control. *Int. J. Pharm. Technol. Res.* 2016, 9, 121–144.

do Espirito Santo Pereira, A.; Caixeta Oliveira, H.; Fernandes Fraceto, L.; Santaella, C. Nanotechnology potential in seed priming for sustainable agriculture. *Nanomaterials* 2021, 11(2), 267.

Du, W.; Sun, Y.; Ji, R.; Zhu, J.; Wu, J.; Guo, H. TiO2 and ZnO nanoparticles negatively affect wheat growth and soil enzyme activities in agricultural soil. *J. Environ. Monitor* 2011, 13(4), 822–828.

Duan, L.; Ouyang, K.; Xu, X.; Xu, L.; Wen, C.; Zhou, X.; et al. Nanoparticle delivery of CRISPR/ Cas9 for genome editing. *Front. Genet.* 2021, 12, 673286.

Duhan, J.S.; Kumar, R.; Kumar, N.; Kaur, P.; Nehra, K.; Duhan, S. Nanotechnology: The new perspective in precision agriculture. *Biotechnol. Rep.* 2017, 15, 11–23.

Elek, N.; Hoffman, R.; Raviv, U.; Resh, R.; Ishaaya, I.; Magdassi, S. Novaluron nanoparticles: For mation and potential use in controlling agricultural insect pests. *Coll Surfaces A: Physicochem. Eng. Aspects* 2010, 372(1–3), 66–72.

Enciu, D.; Toader, A.; Ursu, I. Magnetic field nanosensor based on Mn impurities. *Incas. Bulletin* 2014, 6(2), 51–60.

FAO, I. The state of food and agriculture 2019. In: *Moving Forward on Food Loss and Waste Reduction.* FAO: Rome, 2019, pp. 2–13.

Fellet, G.; Pilotto, L.; Marchiol, L.; Braidot, E. Tools for nano-enabled agriculture: Fertilizers based on calcium phosphate, silicon, and chitosan nanostructures. *Agronomy* 2021, 11, 1239.

Flores-López, M.L.; Cerqueira, M.A.; de Rodríguez, D.J.; Vicente, A.A. Perspectives on utilization of edible coatings and nano-laminate coatings for extension of postharvest storage of fruits and vegetables. *Food Eng. Rev.* 2016, 8(3), 292–305.

Fortner, J.D.; Lyon, D.Y.; Sayes, C.M.; et al. C-60 in water: Nanocrystal formation and microbial response. *Environ. Sci. Technol.* 2005, 39(11), 4307–4316.

Fraceto, L.F.; Grillo, R.; de Medeiros, G.A.; Scognamiglio, V.; Rea, G.; Bartolucci, C. Nanotechnology in agriculture: Which innovation potential does it have? *Front. Environ. Sci.* 2016, 4, 20.

Francis, D.V.; Sood, N.; Gokhale, T. Biogenic CuO and ZnO nanoparticles as nanofertilizers for sustainable growth of amaranthus hybridus. *Plants* 2022, 11(20), 2776.

Gautam, R.K.; Mudhoo, A.; Lofrano, G.; Chattopadhyaya, M.C. Biomass-derived biosorbents for metal ions sequestration: Adsorbent modification and activation methods and adsorbent regeneration. *J. Environ. Chem. Eng.* 2014, 2(1), 239–259.

Ghormade, V.; Deshpande, M.V.; Paknikar, K.M. Perspectives for nano-biotechnology enabled protection and nutrition of plants. *Biotechnol. Adv.* 2011, 29, 792–803.

Giraldo, J.P.; Landry, M.P.; Kwak, S.-Y.; Jain, R.M.; Wong, M.H.; Iverson, N.M.; Ben-Naim, M.; Strano, M.S. A ratiometric sensor using single chirality near-infrared fluorescent carbon nanotubes: Application to in vivo monitoring. *Small* 2015, 11, 3973–3984

Gottschalk, F.; Sonderer, T.; Scholz, R.W.; Nowack, B. Modelled environmental concentrations of engineered nanomaterials (TiO_2, ZnO, ag, CNT, fullerenes) for different regions. *Environ. Sci. Technol.* 2009, 43, 9216–9222.

Grillo, R.; Oliveira, H.C.; Lima, R.; Fraceto, L.F. Polymeric nanoparticles as carrier systems for herbicides. *J. Nanomed. Nanotechnol.* 2015, 6(4). https://doi.org/10.4172/2157-7439.S1.022.

Gross, M.S.; Bean, T.G.; Hladik, M.L.; Rattner, B.A.; Kuivila, K.M. Uptake, metabolism, and elimination of fungicides from coated wheat seeds in Japanese quail (*Coturnix japonica*). *J. Agric. Food Chem.* 2020, 68, 1514–1524.

Guang, X.Y.; Wang, J.J.; He, Z.G.; Chen, G.X.; Ding, L.; Dai, J.J.; Yang, X.H. Molluscicidal effects of nano-silver biological molluscicide and niclosamide. *Zhongguo Xue Xi Chong Bing Fang Zhi Za Zhi* 2013, 25(5), 503–505.

Islam, M.S.; Akter, N.; Rahman, M.M.; Shi, C.; Islam, M.T.; Zeng, H.; Azam, M.S. Mussel-inspired immobilization of silver nanoparticles toward antimicrobial cellulose paper. *ACS Sustain. Chem. Eng.* 2018, 6(7), 9178–9188.

Jhanzab, H.M.; Razzaq, A.; Jilani, G.; Rehman, A.; Hafeez, A.; Yasmeen, F. Silver nano-particles enhance the growth, yield and nutrient use efficiency of wheat. *Int. J. Agron. Agri. Res.* 2015, 7(1), 15–22

Jiang, M.; Song, Y.; Kanwar, M.K.; Ahammed, G.J.; Shao, S.; Zhou, J. Phytonanotechnology applications in modern agriculture. *J. Nanobiotechnol.* 2021, 19, 430.

Kale, A.; Gawade, S. Studies on nanoparticle induced nutrient use efficiency of fertilizer and crop productivity. *Green Chem.* 2016, 2, 88–92.

Kaziem, A.E.; Gao, Y.; Zhang, Y.; Qin, X.; Xiao, Y.; Zhang, Y.; You, H.; Li, J.; He, S. α-amylase triggered carriers based on cyclodextrin anchored hollow mesoporous silica for enhancing insecticidal activity of avermectin against Plutella xylostella. *J. Hazard. Mater.* 2018, 359, 213–21.

Kelly, K.L. Nanotechnology grows up. *Science* 2004, 304, 1732–17345.

Khan, F.; Pandey, P.; and Upadhyay, T.K. Applications of nanotechnology-based agrochemicals in food security and sustainable agriculture: An overview. *Agriculture* 2022, 12, 1672.

Khan, J.H.; Lin, J.; Young, C.; Matsagar, B.M.; Wu, K.C.; Dhepe, P.L.; Islam, M.T.; Rahman, M.; Shrestha, L.K.; Alshehri, S.M.; Ahamad, T. High surface area nanoporous carbon derived from Bangladeshi jute. *Mater Chem. Phys.* 2018, 216(1), 491–495.

Khan, M.; Khan, M.S.A.; Borah, K.K.; Goswami, Y.; Hakeem, K.R.; Chakrabartty, I. The potential exposure and hazards of metal-based nanoparticles on plants and environment, with special emphasis on ZnO NPs, TiO_2 NPs, and AgNPs: A review. *Environ. Adv.* 2021, 6, 100128.

Khiyami, M.A.; Almoammar, H.; Awad, Y.M.; Alghuthaymi, M.A.; Abd-Elsalam, K.A. Plant pathogen nanodiagnostic techniques: Forthcoming changes? *Biotechnol. Biotechnol. Equip.* 2014, 28(5), 775–785

Kong, W.J.; Li, J.Y.; Qiu, F.; Wei, J.H.; Xiao, X.H.; Zheng, Y.; Yang, M.H. Development of a sensitive and reliable high performance liquid chromatography method with fluorescence detection for high-throughput analysis of multi-class mycotoxins in Coix seed. *Analytica Chimica Acta*, 2013, 799, 68–76.

Kottegoda, N.; Sandaruwan, C.; Priyadarshana, G.; Siriwardhana, A.; Rathnayake, U.A.; Arachchige, D.M.B.; Kumarasinghe, A.R.; Dahanayake, D.; Karunaratne, V.; Amaratunga, G.A. Ureahydroxyapatite nanohybrids for slow release of nitrogen. *ACS Nano.* 2017, 11(2), 1214–1221.

Kumar, G.D.; Natarajan, N.; Nakkeeran, S. Antifungal activity of nanofungicide Trifloxystrobin 25% + Tebuconazole 50% against Macrophomina phaseolina. *Afr. J. Microbiol. Res.* 2016, 10(4), 100–105.

Kumar, M.; Shamsi, T.N.; Parveen, R.; Fatima, S. Application of nanotechnology in enhancement of crop productivity and integrated pest management. In: Prasad, R.; Kumar, M.; Kumar, V.; (Eds.), *Nanotechnology*. Springer: Singapore, 2017, pp. 361–371.

Lahiani, M.H.; Dervishi, E.; Chen, J.; Nima, Z.; Gaume, A.; Biris, A.S.; Khodakovskaya, M.V. Impact of carbon nanotube exposure to seeds of valuable crops. *ACS Appl. Mater. Interf.* 2013, 5(16), 7965–7973.

Lin, B.-S.; Diao, S.-Q.; Li, C.-H.; Qiao, S.-C.; Yu, M. Effect of tms (nanostructured silicon dioxide) on growth of changbai larch seedlings. *J. For. Res.* 2004, 15, 138–140.

Liu, R.; Lal, R. Potentials of engineered nanoparticles as fertilizers for increasing agronomic productions. *Sci. Total Environ.* 2015, 514, 131–139.

Liu, C.; Zhou, H.; Zhou, J. The Applications of Nanotechnology in Crop Production. *Molecules* 2021, 26(23), 7070.

Liu, R.; Lal, R. Synthetic apatite nanoparticles as a phosphorus fertilizer for soybean (Glycine max). *Sci. Rep.* 2014, 4, 5686

Lowry, G.V.; Avellan, A.; Gilbertson, L.M. Opportunities and challenges for nanotechnology in the agri-tech revolution. *Nat. Nanotechnol.* 2019, 14, 517–522.

Lü, S.; Feng, C.; Gao, C.; Wang, X.; Xu, X.; Bai, X.; Gao, N.; Liu, M. Multifunctional environmental smart fertilizer based on l-aspartic acid for sustained nutrient release. *J. Agric. Food Chem.* 2016, 64, 4965–4974.

Lv, Z.; Jiang, R.; Chen, J.; and Chen, W. Nanoparticle-mediated gene transformation strategies for plant genetic engineering. *Plant J.* 2020, 104, 880–891.

Ma, C.; White, J.C.; Zhao, J.; Zhao, Q.; Xing, B. Uptake of engineered nanoparticles by food crops: Characterization, mechanisms, and implications. *Annu. Rev. Food Sci. Technol.* 2018, 9, 129–153.

Mahakham, W.; Sarmah, A.K.; Maensiri, S.; Theerakulpisut, P. Nanopriming technology for enhancing germination and starch metabolism of aged rice seeds using phytosynthesized silver nanoparticles. *Sci. Rep.* 2017, 7, 8263.

Mahapatra, D.M.; Satapathy, K.C.; Panda, B. Biofertilizers and nanofertilizers for sustainable agriculture: Phycoprospects and challenges. *Sci. Total Environ.* 2022, 803, 149990.

Mahmoodi, P.; Yarnia, M.; Amirnia, R.; Tarinejad, A.; Mahmoodi, H. Comparison of the effect of nano urea and nono iron fertilizers with common chemical fertilizers on some growth traits and essential oil production of Borago officinalis L. *Dairy Vet. Sci. J.* 2017, 2(2), 555585.

Mahmoud, M.E.; Abdou, A.E.H.; Mohamed, S.M.S.; Osman, M.M. Engineered staphylococcus aureus via immobilization on magnetic Fe_3O_4 -phthalate nanoparticles for biosorption of divalent ions from aqueous solutions. *J. Environ. Chem. Eng.* 2016, 4, 3810–3824.

Malik, A.; Mor, V.S.; Tokas, J.; Punia, H.; Malik, S.; Malik, K.; Sangwan, S.; Tomar, S.; Singh, P.; Singh, N.; et al. Biostimulant-treated seedlings under sustainable agriculture: A global perspective facing climate change. *Agronomy* 2021, 11, 14.

Marchiol, L.; Filippi, A.; Adamiano, A.; Esposti, L.D.; Iafisco, M.; Mattiello, A.; Petrussa, E.; Braidot, E. Influence of hydroxyapatite nanoparticles on germination and plant metabolism of tomato (Solanum lycopersicum L.): Preliminary evidence. *Agronomy* 2019, 9, 161.

Mattiello, E.M.; Ruiz, H.A.; Neves, J.C.L.; Ventrella, M.C.; Araújo, W.L. Zinc deficiency affects physiological and anatomical characteristics in maize leaves. *J. Plant Physiol.* 2015, 183, 138–143.

McGehee, D.L.; Lahiani, M.H.; Irin, F.; Green, M.J.; Khodakovskaya, M.V. Multiwalled carbon nanotubes dramatically affect the fruit metabolome of exposed tomato plants. *ACS Appl. Mater. Interfaces* 2017, 9(38), 32430–32435.

Moghaddasi, S.; Fotovat, A.; Khoshgoftarmanesh, A.H.; Karimzadeh, F.; Khazaei, H.R.; Khorassani, R. Bioavailability of coated and uncoated ZnO nanoparticles to cucumber in soil with or without organic matter. *Ecotoxicol. Environ. Saf.* 2017, 144, 543–551.

Monreal, C.M.; DeRosa, M.; Mallubhotla, S.C.; Bindraban, P.S.; Dimkpa, C. Nanotechnologies for increasing the crop use efficiency of fertilizer-micronutrients. *Biol. Fertil. Soils* 2016, 52, 423–437.

Morales-Díaz, A.B.; Ortega-Ortíz, H.; Juárez-Maldonado, A.; Cadenas-Pliego, G.; GonzálezMorales, S.; Benavides-Mendoza, A. Application of nanoelements in plant nutrition and its impact in ecosystems. *Adv. Nat. Sci. Nanosci. Nanotechnol.* 2017, 8, 13001.

Murgueitio-Herrera, E.; Falconí, C.E.; Cumbal, L.; Gómez, J.; Yanchatipán, K.; Tapia, A.; Martínez, K.; Sinde-Gonzalez, I.; Toulkeridis, T. Synthesis of iron, zinc, and manganese nanofertilizers, using andean blueberry extract, and their effect in the growth of cabbage and lupin plants. *Nanomaterials* 2022, 12(11), 1921.

Naik, B.J.; Shimoga, G.; Kim, S.C.; Manjulatha, M.; Subramanyam, C.; Palem, R.R.; Kumar, M.; Kim, S.Y.; Lee, S.H. CRISPR/Cas9 and nanotechnology pertinence in agricultural crop refinement. *Front. Plant Sci.*, 2022, 13, 461.

Naseri, N.; Valizadeh, H.; Zakeri-Milani, P. Solid lipid nanoparticles and nanostructured lipid carriers: Structure, preparation and application. *Adv. Pharm. Bull.* 2015, 5(3), 305–313.

Neethirajan, S.; Gordon, R.; Wang, L. Potential of silica bodies (phytoliths) for nanotechnology. *Trends Biotechnol.* 2009, 27, 461–467.

Neme, K.; Nafady, A.; Uddin, S.; Tola, Y.B. Application of nanotechnology in agriculture, postharvest loss reduction and food processing: Food security implication and challenges. *Heliyon* 2021, 7(12), e08539.

Nongbet, A.; Mishra, A.K.; Mohanta, Y.K.; Mahanta, S.; Ray, M.K.; Khan, M.; Baek, K.H.; Chakrabartty, I. Nanofertilizers: A smart and sustainable attribute to modern agriculture. *Plants* 2022, 11(19), 2587.

Nuruzzaman, M.; Rahman, M.M.; Liu, Y.; Naidu, R. Nanoencapsulation, nano-guard for pesticides: A new window for safe application. *J. Agric. Food Chem.* 2016, 64(7), 1447–1483.

Oliveira, H.C.; Stolf-Moreira, R.; Martinez, C.B.R.; Grillo, R.; de Jesus, M.B.; Fraceto, L.F. Nanoencapsulation enhances the post-emergence herbicidal activity of atrazine against mustard plants. *PLoS One* 2015, 10(7), e0132971.

Omobhude, M.E.; Morenikeji, O.A.; Oyeyemi, O.T. Molluscicidal activities of curcumin-nisin polylactic acid nanoparticle on *Biomphalaria pfeifferi*. *PLoS Negl. Trop. Dis.* 2017, 11(8), e0005855.

Ouda, S.M. Antifungal activity of silver and copper nanoparticles on two plant pathogens, Alternaria alternata and Botrytis cinerea. *Res. J. Microbiol.* 2014, 9(1), 34–42.

Pandey, S.; Giri, K.; Kumar, R.; Mishra, G.; Rishi, R.R. Nanopesticides: Opportunities in crop protection and associated environmental risks. *Proceed. Nat. Acad. Sci. India. Sec. B Biol. Sci.* 2016, 2016, 1–22.

Pang, Y.; Qin, Z.; Wang, S.; Yi, C.; Zhou, M.; Lou, H.; Qiu, X. Preparation and application performance of lignin-polyurea composite micro controlled release of avermectin. *Colloid. Polym. Sci.* 2020, 298(8), 1001–1012.

Panichikkal, J.; Thomas, R.; John, J.C.; Radhakrishnan, E.K. Biogenic gold nanoparticle supplementation to plant beneficial pseudomonas monteilii was found to enhance its plant probiotic effect. *Curr. Microbiol.* 2019, 76, 503–509.

Pereira, A.E.S.; Sandoval-Herrera, I.E.; Zavala-Betancourt, S.A.; Oliveira, H.C.; Ledezma-Pérez, A.S.; Romero, J.; Fraceto, L.F. γ-Polyglutamic acid/chitosan nanoparticles for the plant growth regulator gibberellic acid: Characterization and evaluation of biological activity. *Carbohydr. Polym.* 2017, 157, 1862–1873.

Pourkhaloee, A.; Haghighi, M.; Saharkhiz, M.J.; Jouzi, H.; Doroodmand, M.M. Carbon nanotubes can promote seed germination via seed coat penetration. *Seed Technol.* 2011, 33, 155–169.

Prasad, R.; Bhattacharyya, A.; Nguyen, Q.D. Nanotechnology in sustainable agriculture: Recent developments, challenges, and perspectives. *Front. Microbiol.* 2017, 8, 1014.

Prasad, R.; Kumar, V.; Kumar, M.; Choudhary, D. *Nanobiotechnology in Bioformulations*. Springer International Publishing (ISBN 978-3-030-17061-5), 2019.

Rajput, V.D.; Minkina, T.; Upadhyay, S.K.; Kumari, A.; Ranjan, A.; Mandzhieva, S.; Sushkova, S.; Singh, R.K.; Verma, K.K. Nanotechnology in the restoration of polluted soil. *Nanomaterials* 2022, 12(5), 769.

Raliya, R.; Tarafdar, J.C.; Gulecha, K.; Choudhary, K.; Ram, R.; Mal, P.; Saran, R.P. Scope of nano science and nanotechnology in agriculture. *J. Appl. Biol. Biotechnol.* 2013a, 1, 41–44.

Raliya, R.; Saharan, V.; Dimkpa, C.; Biswas, P. Nanofertilizer for precision and sustainable agriculture: Current state and future perspectives. *J. Agric. Food Chem.* 2018, 66, 6487–6503.

Raliya, R.; Tarafdar, J.C. ZnO nanoparticle biosynthesis and its effect on phosphorous-mobilizing enzyme secretion and gum contents in clusterbean (*Cyamopsis tetragonoloba* L.). *Agric. Res*. 2013b, 2, 48–57

Rautela, I.; Dheer, P.; Thapliyal, P.; Shah, D.; Joshi, M.; Upadhyay, S.; Gururani, P.; Sinha, V.B.; Gaurav, N.; Sharma, M.D. Current scenario and future perspectives of nanotechnology in sustainable agriculture and food production. *Plant Cell Biotechnol. Mol. Biol*. 2021, 22, 99–121.

Rizwan, M.; Ali, S.; Ibrahim, M.; Farid, M.; Adrees, M.; Bharwana, S.A.; Zia-Ur-Rehman, M.; Qayyum, M.F.; Abbas, F. Mechanisms of silicon-mediated alleviation of drought and salt stress in plants: A review. *Environ. Sci. Pollut. Res. Int*. 2015, 22, 15416–15431.

Salinas, S.; Mosquera, N.; Yate, L.; Coy, E.; Yamhure, G.; González, E. Surface plasmon resonance nanosensor for the detection of arsenic in water. *Sens. Transducers* 2014, 183(12), 97–102.

Sathiyabama, M.; Manikandan, A. Foliar application of chitosan nanoparticle improves yield, mineral content and boost innate immunity in finger millet plants. *Carbohydrate Polymers* 2021, 258, 117691.

Schwartz, S.H.; Hendrix, B.; Hoffer, P.; Sanders, R.A.; Zheng, W. Carbon dots for efficient small interfering RNA delivery and gene silencing in plants. *Plant Physiol*. 2020, 184, 647–657.

Scott, N.R.; Chen, H.; Cui, H. Nanotechnology applications and implications of agrochemicals toward sustainable agriculture and food systems. *J. Agric. Food Chem*. 2018, 66, 6451–6456.

Sekhon, B.S. Nanotechnology in Agri-food production: An overview. *Nanotechnol. Sci. Appl*. 2014, 7, 31–53.

Shao, C.; Zhao, H.; Wang, P. Recent development in functional nanomaterials for sustainable and smart agricultural chemical technologies. *Nano. Convergence* 2022, 9(1), 1–17.

Sharonova, N.L.; Yapparov, A.K.; Khisamutdinov, N.S.; Ezhkova, A.M.; Yapparov, I.A.; Ezhkov, V.O.; Degtyareva, I.A.; Babynin, E.V. Nanostructured water-phosphorite suspension is a new promising fertilizer. *Nanotechnol. Russ*. 2015, 10, 651–661.

Si, Y.; Zhang, Z.; Wu, W.; Fu, Q.; Huang, K.; Nitin, N.; Ding, B.; Sun, G. Daylight-driven rechargeable antibacterial and antiviral nanofibrous membranes for bioprotective applications. *Sci. Adv.* 2022, 4(3), eaar5931.

Siddiqui, M.H.; Al-Whaibi, M.H.; Firoz, M.; Al-Khaishany, M.Y. Role of nanoparticles in plants. In: Siddiqui, M.H.; Al-Whaibi, M.H.; Firoz, M.; (Eds.), *Nanotechnology and Plant Sciences*. Springer: Switzerland, 2015, p. 303.

Singh, N.B.; Amist, N.; Yadav, K.; Singh, D.; Pandey, J.K.; Singh, S.C.; Zinc oxide nanoparticles as fertilizer for the germination, growth and metabolism of vegetable crops. *J. Nanoeng. Nanomanuf.* 2013, 3, 353–364.

Singh, H.; Sharma, A.; Bhardwaj, S.K.; Arya, S.K.; Bhardwaj, N.; Khatri, M. Recent advances in the applications of nanoagrochemicals for sustainable agricultural development. *Environ. Sci. Process. Impacts* 2021, 23, 213–239.

Suman, P.R.; Jain, V.K.; Varma, A. Role of nanomaterials in symbiotic fungus growth enhancement. *Curr. Sci.* 2010, 99, 1189–1191.

Tan, W.; Du, W.; Barrios, A.C.; Armendariz, R.J.; Zuverza-Mena, N.; Ji, Z.; Chang, C.H.; Zink, J.I.; Hernandez-Viezcas, J.A.; Peralta-Videa, J.R.; et al. Surface coating changes the physiological and biochemical impacts of nano-TiO(2) in basil (*Ocimum basilicum*) plants. *Environ. Pollut.* 2017, 222, 64–72.

Tarafdar, J.; Raliya, R.; Rathore, I. Microbial synthesis of phosphorous nanoparticle from tri-calcium phosphate using aspergillus tubingensis tfr-5. *J. Bionanosci.* 2012, 6, 1–6.

Thirugnanasambandan, T. Advances and trends in nano-biofertilizers. 2018.

Torney, F.; Trewyn, B.G.; Lin, V.S.Y.; Wang, K. Mesoporous silica nanoparticles deliver DNA and chemicals into plants. *Nat. Nanotechnol.* 2007, 2(5), 295.

Tungittiplakorn, W.; Cohen, C.; Kim, J.-Y. Engineered polymeric nanoparticles for soil remediation. *Environ. Sci. Technol.* 2004, 38, 1605–1610.

Venkatachalam, P.; Priyanka, N.; Manikandan, K.; Ganeshbabu, I.; Indiraarulselvi, P.; Geetha, N.; Muralikrishna, K.; Bhattacharya, R.C.; Tiwari, M.; Sharma, N.; et al. Enhanced plant growth promoting role of phycomolecules coated zinc oxide nanoparticles with P supplementation in cotton (Gossypium hirsutum L.). *Plant Physiol. Biochem. PPB* 2017, 110, 118–127.

Wu, S.G.; Huang, L.; Head, J.; Ball, M.; Tang, Y.J.; Chen, D.R. Electrospray facilitates the germination of plant seeds. *Aerosol. Air Qual. Res.* 2014, 14, 632–641

Xiang, Y.; Zhang, G.; Chi, Y.; Cai, D.; Wu, Z. Fabrication of a controllable nanopesticide system with magnetic collectability. *Chem. Eng. J.* 2017, 328, 320–30.

Xin, X.; Zhao, F.; Zhao, H.; Goodrich, S.L.; Hill, M.R.; Sumerlin, B.S.; Stoffella, P.J.; Wright, A.L.; He, Z. Comparative assessment of polymeric and other nanoparticles impacts on soil microbial and biochemical properties. *Geoderma* 2020, 367, 114278.

Yata, V.K.; Tiwari, B.C.; Ahmad, I. Research trends and patents in nano-food and agriculture. In: *Nanoscience in Food and Agriculture 5*. Ranjan, S.; Dasgupta, N.; Lichtfouse, E.; (Eds.), Springer International Publishing: Cham, Switzerland, 2017, pp. 1–20.

Ye, Y.; Cota-Ruiz, K.; Hernández-Viezcas, J.A.; Valdés, C.; Medina-Velo, I.A.; Turley, R.S.; Peralta-Videa, J.R.; Gardea-Torresdey, J.L. Manganese nanoparticles control salinity-modulated molecular responses in capsicum annuum L. through priming: A sustainable approach for agriculture. *ACS Sustain. Chem. Eng*. 2020, 8, 1427–1436.

Zambrano-Zaragoza, M.L.; González-Reza, R.; Mendoza-Muñoz, N.; Miranda-Linares, V.; Bernal-Couoh, T.F.; Mendoza-Elvira, S.; Quintanar-Guerrero, D. Nanosystems in edible coatings: A novel strategy for food preservation. *Int. J. Mol. Sci.* 2018, 19(3), 705.

Zaytseva, O.; Neumann, G. Carbon nanomaterials: Production, impact on plant development, agricultural and environmental applications. *Chem. Biol. Technol. Agric.* 2016, 3, 17.

Zhang, W. Environmental technologies at the nanoscale. *Environ. Sci. Technol.* 2003, 7, 103–108.

Zhao, X.; Cui, H.; Wang, Y.; Sun, C.; Cui, B.; Zeng, Z. Development strategies and prospects of nano-based smart pesticide formulation. *J. Agric. Food Chem.* 2017, 66(26), 6504–6512.

Zhao, F.; Xin, X.; Cao, Y.; Su, D.; Ji, P.; Zhu, Z.; He, Z. Use of Carbon Nanoparticles to Improve Soil Fertility, Crop Growth and Nutrient Uptake by Corn (*Zea mays* L.). *Nanomaterials* 2021, 11, 2717.

Zhu, H.; Shen, Y.; Cui, J.; Wang, A.; Li, N.; Wang, C.; Cui, B.; Sun, C.; Zhao, X.; Wang, C.; Gao, F.; Zhan, S.; Guo, L.; Zhang, L.; Zeng, Z.; Wang, Y.; Cui, H. Avermectin loaded carboxymethyl cellulose nanoparticles with stimuli-responsive and controlled release properties. *Ind. Crop Prod.* 2020, 152, 112497.

CHAPTER 7

Recent Advances in Postharvest Technology for Fresh Produce

AJIT KUMAR SINGH,[1,*] SINDHU P.M.,[2] RAGHAVENDRA H.R.,[2] and KRIPA SHANKAR[3]

[1]*College of Horticulture and Forestry, at Acharya Narendra Deva University of Agriculture and Technology, Kumarganj, Ayodhya, Uttar Pradesh, India*

[2]*Division of Food Science and Postharvest Technology, ICAR-Indian Agricultural Research Institute, New Delhi, India*

[3]*Division of Fruit and Horticultural Technology, ICAR-Indian Agricultural Research Institute, New Delhi, India*

*Corresponding author

ABSTRACT

Using postharvest technologies, horticulturists have been able to produce and distribute fresh produce that is high in nutritional value and sensory quality on a local and international scale. For horticultural products to retain their quality following harvest, the ripening and senescence processes must be controlled. A lack of effective management for these products can lead to significant nutritional and quality losses, outbreaks of foodborne pathogens, and financial losses for all stakeholders. It is imperative to slow down physiologic activity such as senescence and maturation in fresh produce. The occurrence and spread of physiological disorders can be minimized by using varied postharvest techniques, which will reduce contamination and inhibit microbial growth. In this chapter, we examine the current status of postharvest treatments of fresh produce as well as novel technologies such as plasma and ozone, nonthermal treatments that can help to maintain

Technological Advancements in Sustainable Crop Improvement and Production.
Devendra Kumar, Alok Kumar Singh, and Nisha Malik (Eds.)

DOI: 10.1201/9781779643155-7

quality and reduce postharvest losses of fresh produce. Furthermore, the authors attempt to assemble and discuss the postharvest biology, edible quality, nutritional profiles, and storage properties of genetically modified fruit and vegetables, along with the challenges associated with their use and consumer acceptance. In addition to that, the study highlights the need for further research with respect to fresh produce, which responds dynamically to a wider range of postharvest treatments.

7.1 INTRODUCTION

Mechanization and technological advancements in agriculture and horticulture have heralded a significant boost in productivity and overall crop yields. However, despite these strides, a substantial amount of food is lost annually throughout the supply chain, from field to consumer. The challenge of preserving food has persisted for centuries, with pest infestations and spoilage microbes being the primary culprits behind losses during storage, transportation, and marketing. Fruits and vegetables are among the most perishable food crops, with postharvest losses ranging from 15% to 20% depending on the crop. Due to numerous uncertainties across the supply chain, microbial contamination can lead to spoilage and postharvest losses at any stage, from farm to consumer. Hence, postharvest treatments play a crucial role in minimizing microbial spoilage and mitigating the risk of pathogen contamination for fruits and vegetables (Mostafavi et al., 2012). As a result, the country must prioritize preserving food (Fresh produce #15–20%) that would normally go to waste rather than attempting to turn garbage into valuable commodities after the food has been dumped. Hence, postharvest management holds a significant role in this context. The term "postharvest technology" refers to an interdisciplinary approach employed to preserve, refine, package, distribute, market, and use horticultural and agricultural commodities after harvest for community food and nutritional needs. For fresh produce to adhere to safety standards and maintain fresh-like qualities with high nutritional value, therefore, physical, chemical, and gaseous treatments may be applied after harvest. Implementation of these postharvest treatments often involves integrating them with careful management of storage temperatures to control the upsurge of their catabolic activity. This chapter provides an overview of the latest advancements in postharvest treatments and emerging technologies aimed at preserving the quality of fresh produce and minimizing postharvest waste.

This chapter provides a detailed overview of the different technologies that have been invented (Table 7.1).

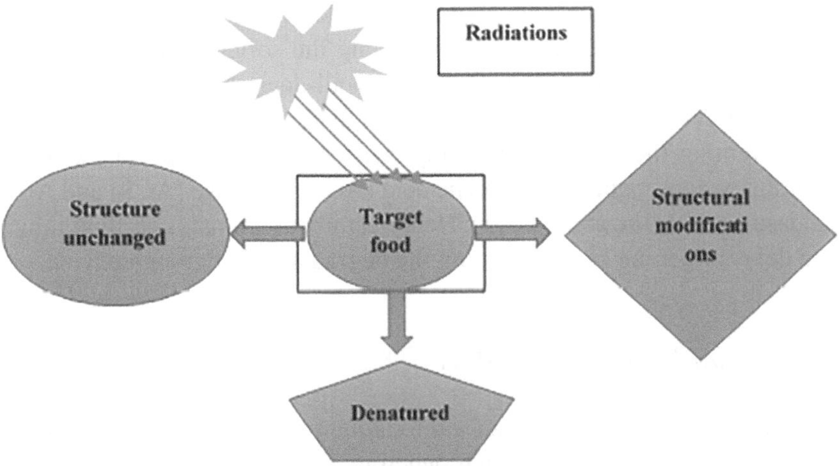

FIGURE 7.1 Diagrammatic representation of irradiation.

Source: Adapted from Khan et al., 2017

7.2 POSTHARVEST TREATMENT FOR THE SHELF-LIFE EXTENSION OF FRESH PRODUCE

7.2.1 PHYSICAL TREATMENT

7.2.1.1 HOT WATER TREATMENT (HWT)

The study has shown that heat treatment can be used as a substitute for chemical treatments for harvested fresh fruit and vegetables (FFV). The treatments encompass hot water dips (HWD), saturated water vapor, hot dry air, and hot water rinses combined with brushing. HWT acts as both a protective and curative agent and activates some phytoalexins that trigger the self-defense mechanisms of the treated fruits. Fawcett (1922) was the pioneer in reporting the effect of HWT on the prevention of decay of citrus fruits by thermal inhibition of pathogens by accumulation of phytoalexins (scoparone and scopoletin). Water is the best medium for eliminating microorganisms, and it has better heat transfer in comparison to air (Reddy et al., 2021). Thermal treatments have been applied, such as HWD, vapor heat, hot dry air, and hot water rinsing with brushing. The mode of action of heat treatment encompasses, for example, that (1) thermal treatment inhibits pathogen growth by activating phytoalexins like scoparone and scopoletin. (2) At elevated temperatures, sealing of cracks or natural openings due to meltdown of the stomata or lenticel leads to a significant reduction in weight loss, thus maintaining the firmness of the fruit during the prolonged storage period. (3) Delaying the softening of heat-treated fruit may be due to inhibition of the cell wall-degrading enzyme, such as pectin methyl esterase. IV: Heat treatment delays the emission of ethylene by preventing the action of the ethylene-forming enzyme, for example, 1-aminocyclopropane-1-carboxylic acid (ACC) synthase (ACS) and ACC oxidase (ACO) (Ilic et al., 2001). Heat treatment eliminates chilling injury and delays ripening by inactivating the degradative softening enzyme.

7.2.1.2 RADIATION TREATMENT

Irradiation involves subjecting food to radiant energy emitted by rays and electron beams (e-beams), which penetrate deeply and disrupt molecular bonds, including the DNA of living organisms. The shelf life of fresh produce can be extended by irradiating it with ionizing radiation from cobalt-60, caesium-137, or machine-generated electron beams. Irradiation's ability to inhibit cellular reproduction allows it to eliminate pests and address food safety issues. The effectiveness relies on the dosage, quantified in kilograys. Lower doses of irradiation (below 1 kGy) merely hinder cellular activity sufficiently to impede the sprouting of tubers, bulbs, and roots, as well as to postpone senescence. Moderate doses (1–10 kGy) diminish microbial populations, whereas higher doses (exceeding 10 kGy) eradicate a wide array of fungi, bacterial strains, and pests. Nevertheless, the majority of medium- and high-level doses are unsuitable for fresh produce as they can cause sensory defects (visual, textural, and flavor) and/or hasten senescence by causing irreversible damage to DNA and protein molecules. Irradiation emerges as a potent postharvest treatment method, effectively targeting bacteria, molds, and yeasts that are responsible for food spoilage. Additionally, it serves as a formidable tool against insect and parasite infestation, thereby preventing storage losses and extending the shelf life of fresh produce. By eliminating both parasitological and microbiological activity to ensure safe and healthy food In addition to that, irradiation plays a pivotal role in ensuring the quality and longevity of food products, as pointed out by Wang et al. (2016).

The commercialization of irradiation has led to the control of potato sprouting and strawberry decay. Low doses of γ-irradiation on mango (0.3–0.7 kGy) resulted in a delay in ripening and an extension of their shelf life by 3–4 days (Mahto and Das, 2013). The majority of irradiation use on FFV has been concentrated on extending shelf life and decreasing decay. The effectiveness of irradiation in killing and sterilizing has been known for many decades, as has preventing the development of a wide range of insect pests that are also of quarantine importance for perishable fruits and vegetables. Although there are some misconceptions, food exposed to irradiation does not become radioactive. The irradiation process produces very little chemical change in food and does not change the nutritional value of food. Irradiation has been shown to be safe and wholesome through extensive research and testing, even though there is always a fear among consumers when consuming it. Moreover, it is asserted that radiation processing is a residue-free technique that leaves no toxic chemical residue in food after processing (Figure 7.1).

TABLE 7.1 Food Items Approved for Radiation Preservation Under PFA Rules, 1955

Name of Produce	Purpose	Minimum Dose (kGy)	Maximum Dose (kGy)
Onion	Sprout inhibition	0.03	0.09
Potato		0.06	0.15
Ginger		0.03	0.15
Garlic		0.03	0.15
Mango	Disinfection (quarantine)	0.25	0.75
Spices	Microbial decontamination	6	14
Raisin, figs, dried dates	Shelf life extension and pathogen control	2.5	4

7.2.1.3 COLD PLASMA TECHNOLOGY

The major food processing technique used in the food industry is thermal processing, which has been in use for over two centuries. The nutritional and sensory quality of processed food can be negatively impacted by the use of harsh heat treatments. Cold plasma is an innovative, nonthermal technology that can effectively decontaminate food and increase its shelf life (Figure 7.2). Plasma is composed of ionized gas molecules that have been dissociated via an energy input. High or low temperatures can be generated, respectively, depending on the mode of particle activation and the excitation energy (Surowsky et al., 2014). At a radio frequency of 27 MHz, argon gas can be transformed into plasma to generate cold plasma at atmospheric pressure or by creating an electric current between two electrodes that are separated by dielectric barriers. Three basic mechanisms have been suggested for the inactivation of microbial spores in plasma environments: etching is achieved by reactive free radicals absorbing and eroding the surface atoms of microbial spores; direct destruction of DNA genetic molecules by UV irradiation; and volatilization of compounds from the spore surface by UV photons through intrinsic photo-desorption (Dey et al., 2016). The effectiveness of cold plasma against major food-contaminating pathogens has been reported, like *Escherichia coli*, *Salmonella typhimurium*, *Staphylococcus aureus*, and *Listeria monocytogenes* (Mollakhalili-Meybodi et al., 2021).

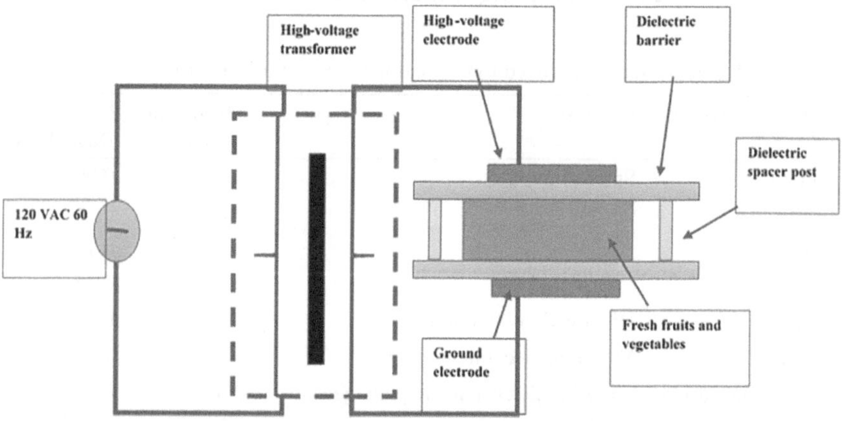

FIGURE 7.2 Diagrammatic representations of cold plasma.

7.2.1.4 EDIBLE COATING

The term "edible coating" refers to the application of thin layers onto the exterior of fresh produce, aiming to prolong shelf life. This can be accomplished by creating a waxy cuticle or substituting natural barriers, especially when the original cuticle is removed during the handling of fresh produce. Hence, fresh produce is protected from moisture migration from its surface by the application of edible coatings to reduce moisture loss during postharvest storage. As a result, the product is protected from respiration, senescence, and enzymatic oxidation, and its color and texture are preserved via a gas barrier that creates a modified atmosphere around the product. There are three main types of edible coatings: hydrophobic coatings, formed by lipid-based waxes; hydrocolloid/hydrophilic coatings, formed by polysaccharides or proteins; or a combination of these two types of coatings (Ghasemnezhad et al., 2013).

7.2.2 CHEMICAL TREATMENT

7.2.2.1 ETHYLENE

Aside from its beneficial effects, indigenous ethylene production and exogenous application have adverse effects on fresh produce. In some crops, especially bananas, avocados, kiwifruit, persimmon, tomato, mango, and citrus fruits, exogenously applied ethylene triggers the ripening, improves color and quality, and de-greens them. Ethylene has been demonstrated to have detrimental effects on postharvest horticultural commodities. These effects include reducing storage life, accelerating senescence, softening fruit, inducing discoloration such as browning and russet spots in lettuce, causing yellowing in leafy vegetables and cucumbers, and increasing susceptibility to decay. It is critical to manage ethylene in order to preserve postharvest life and maintain the quality of climacteric and nonclimacteric horticultural produce. Keeping horticultural commodities in good condition typically involves storing them at low temperatures, preventing ethylene biosynthesis and action, and controlling the temperature and composition of atmospheric gases during ripening, harvesting, storage, and transportation. The freshness of fruits and vegetables can be maintained through treatments like modified atmosphere packaging (MAP) or controlled atmosphere (CA) storage. Additionally, inhibitors of ethylene action, such as 1-methyl cyclopropane, can also be employed to downregulate their functional activity (Ziv and Fallik, 2021).

7.2.2.2 METHYLCYCLOPROPENE (1-MCP)

By inhibiting ethylene perception, cyclopropenes have made a significant contribution to controlling ethylene responses in horticulture. This advancement marked a significant milestone in the commercialization of 1-MCP, allowing for the release of the complex and its application to horticultural products. Regulatory approval for the use of 1-MCP has been secured in over 50 countries, with ongoing approval processes taking place globally. Physiological disorders are affected by 1-MCP in accordance with the role of ethylene. Treatment with 1-MCP leads to a reduction in senescence-related disorders, such as apple senescence breakdown and water soaking in watermelon induced by ethylene (Nock et al., 2010). A recent 1-MCP formulation approved for preharvest applications by the Environmental Protection Agency and other regulators is Harvista™ for fruit and vegetables. Semicommercial trials have been conducted in the United States, Argentina, Brazil, Canada, Chile, New Zealand, and South Africa. According to studies, Harvista ™ or Smart Fresh ™ retards fruit drop, slows fruit maturation and ripening, and improves postharvest quality (Li et al., 2022).

7.2.2.3 OZONE (O_3)

O_3 can replace traditional sanitizing agents, according to recent research and commercial applications. O_3 gas is a naturally occurring, highly pungent gas known for its potent and reactive oxidizing properties. Studies indicate that O_3 possesses an oxidizing potential 1.5 times higher than chlorine and 3000 times higher than hypochlorous acid (Horvitz and Cantalejo, 2014). Antimicrobial action contact times are typically four to five times shorter than those required for chlorine. O_3 swiftly targets bacterial cell walls and exhibits greater effectiveness than chlorine against the resilient spores of plant pathogens and animal parasites, at all the concentrations that are practical and safe. Its efficacy has been documented in various produce, including asparagus spears, apples, cherries, carrots, garlic, kiwi, onions, peaches, plums, potatoes, and table grapes. (Salvador et al., 2006). Nonetheless, O_3 exhibits limited penetration through natural openings or wounds. Further research is necessary to delineate the potential and constraints of O_3 effective application in postharvest treatments for ensuring the quality and safety of FFV. Furthermore, according to Huyskens-Keil et al. (2012), fruits exposed to 2.5 ppm O_3 exhibited elevated levels of total soluble solids, ascorbic acid, beta-carotene, and carotenoid content. Additionally, O_3 treatment aided in retaining higher levels of antioxidants, and treated fruits demonstrated reduced weight loss after 10 days compared to their untreated counterparts. The sensory qualities of O_3-treated fruit were notably enhanced in terms of sweetness and overall acceptability, suggesting that O_3 presents a viable, nonthermal, and safe method for preserving FFV.

7.2.2.4 CHEMICAL TREATMENT USED AS AN ANTIMICROBIAL AND ANTIBROWNING AGENT FOR FRUIT AND VEGETABLES

Sulfur dioxide (SO_2) is commonly employed in the preservation of table grapes to mitigate decay during storage. This is typically achieved through the initial fumigation of the fruit directly from the field, followed by the weekly fumigation of storage rooms. Alternatively, SO_2 may be gradually released from in-package pads containing sodium metabisulfite. A promising postharvest approach for fresh blueberries through SO_2 fumigation followed by CA storage (3% O_2 + 6 or 12% CO_2) to decrease decay, eventually expand market life, and maintain high nutritional value Yuan et al. (2022). Nitric oxide (NO) is an extremely reactive free radical gas that plays the role of a multifunctional signaling molecule in a wide range of plant physiological processes, such as the senescence and ripening of fruits and vegetables. The ideal levels of NO can delay the climacteric phase of numerous tropical fruits, prolonging postharvest storage life by impeding ripening and senescence and inhibiting ethylene biosynthesis, thereby slowing down fruit ripening. NO treatments decrease the rate of cellular metabolism, water loss, discoloration, and the occurrence of postharvest diseases. When NO fumigation is paired with cold storage, it synergistically influences and enhances the storage life of fruits such as plums and mangos (Corpas et al., 2022). Further, it was suggested that NO gas is utilized either directly as a fumigant or is released by compounds like sodium nitroprusside, s-nitrosothiols, and diazeniumdiolates employed in dipping treatments. Zaharah et al. (2011) pointed out that the diminished ethylene production observed during ripening in NO-fumigated fruit is attributed to NO's binding with ACC and ACO, forming a stable ternary complex, which consequently restrains ethylene production. Moreover, chlorination, peroxyacetic acid (PAA), organic acids, hydrogen peroxide (H_2O_2), and electrolyzed water are among the chemical-based agents utilized for the disinfectant of fresh produce. Due to their potent oxidizing properties, chlorine-based solutions like NaClO (100–150 ppm) have been extensively employed as disinfectants for fresh produce. PAA has demonstrated efficacy in controlling *E. coli* O157:H7 and *L. monocytogenes* on apples, strawberries, lettuce, and cantaloupe. Due to its oxidizing properties and capacity to generate additional cytotoxic oxidizing species like hydroxyl radicals, H_2O_2 is utilized for its bactericidal, sporicidal, and inhibitory properties. The implementation of H_2O_2 treatment on melons, oranges, apples, prunes, tomatoes, grapes, and fresh-cut produce effectively prolongs the shelf life of these items by inhibiting pathogenic microbial populations. Organic acids like ascorbic acid, acetic acid, malic acid, and calcium-based solutions are frequently employed to mitigate enzymatic and nonenzymatic browning, preserve texture, and inhibit microbial proliferation on fresh produce (Rico et al., 2007).

7.3 CONTROL ATMOSPHERE (CA) STORAGE

CA storage is a technology used in the postharvest management of fruits, vegetables, and other perishable commodities. It involves modifying the atmosphere surrounding the produce stored in a sealed environment to slow down physiological processes such as respiration and ripening, thereby extending shelf life and maintaining quality. In CA storage, the levels of oxygen (O_2), carbon dioxide (CO_2), and sometimes nitrogen are regulated to create optimal conditions for the specific commodity being stored. Typically, the O_2 levels are reduced to slow down respiration rate, while CO_2 levels are increased to inhibit ethylene production and microbial growth. The specific CA conditions required vary depending on the type of produce, its maturity stage, and other factors. CA storage can be customized to suit different commodities and storage durations. It is commonly used for fruits such as apples, pears, kiwifruit, and citrus, as well as certain vegetables. Overall, CA storage is an important tool in postharvest management, allowing for the preservation of quality, reduction of losses, and extension of marketable shelf life for a wide range of perishable products.

7.3.1 THE BENEFICIAL EFFECT OF CA STORAGE

The beneficial effect of a controlled atmosphere has been mentioned below:

1. Fresh produce will take longer to soften due to slow respiration, ethylene emissions, and biochemical changes associated with senescence.
2. Elimination sensitivity effect of ethylene against fruit and vegetables at <8% of O_2 and/or CO_2 levels > 1%.
3. Certain physiological disorders, such as avocado chilling injury and apple and pear superficial scald, can be alleviated.

4. Botrytis rot development on strawberries and cherries is suppressed when exposed to CO_2 levels of 10–15%, which may occur due to the direct or indirect influence of the pathogen.

7.4 PLANT EXTRACTS

Plant extracts of natural plant origin were proved to possess a substantial scope for use in a range of food applications and a majority of them enjoy GRAS status. The antimicrobial activities of the plant extracts might be due to the presence of a variety of phytochemical metabolites such as polyphenols. Polyphenols are known to function by adsorbing to the cell membranes of bacteria resulting in its disruption and subsequent leakage of cellular components and also through the generation of hydroperoxides from polyphenols. The plant extracts also exhibited antifungal activity against a wide range of fungi, antioxidant and antimutagenic activities (Boubaker et al., 2011; Cherdshewasart et al., 2009; Horn and Vargas, 2003), and inhibited lipid oxidation in foods (Table 7.2).

TABLE 7.2 Plant Extracts as a Natural Source of Antimicrobials

Plant	Phytochemical	Parts Used	Reference
Elephant foot yam (*Amorphophallus campanulatus*)	Amblyone	Root	Khan et al. (2008)
Aristolochia (*Aristolochia bracteata*)	Aristolochic acid	Root	Negi et al. (2003)
Jakaranda (*Jacaranda mimosifolia*)	Flavones, flavonoids, iridoids, triterpenes	Flowers/ leaves	Rojas et al. (2006)
Seabuckthorn (*Hippophae rhamnoides* L.)	Flavonoids, tannins, triterpenes	Berries	Gupta et al. (2011)
Kokum (*Garcinia* spp.)	Garcinol, α-mangostin, xanthones and xanthone derivatives, xantho-chymol rubraxanthone	Fruit	Mackeen et al. (2000)
Cinnamon (*Cinnamomum* spp.)	Cinnamaldehyde flavan-3-ols, *p*-coumaric acid, eugenol, camphor, α-pinene, β-pinene, β-caryophyllene, γ-cadinene	Bark/ leaves/root bark/fruit	Jayaprakasha et al. (2007)

7.5 ESSENTIAL OILS

Some essential oils have an antimicrobial property that enhances the storage life of perishable produce by controlling postharvest spoilage. Several reports have stated that effective usage of essential oils either by spraying or dipping can control the decay of several fruits and vegetables. For instance, an essential oil extracted from thyme (*Thymus capitatus L.*), that is, thymol has the ability to reduce the disease incidence of apricot gray mold and brown mold without phytotoxicity fumigation method at a lower concentration (2–4 mg/L) (Liu, 1999). Tripathi and Dubey (2004) reported that essential oils of *Zingiber officinale, Ocimum canum*, and *Mentha arvensis* have been found to enhance the storage life of citrus by preventing the blue mold incidence (Table 7.3).

TABLE 7.3 Efficacy of Different Essential Oils on The Microbial Load Reduction of Different Perishable Commodities

Essential Oils	Microorganism	Log10 Reduction	Produce	Reference
Thyme oil (5 ppm)	*Escherichia coli O157:H7*	1.0	Carrot	Tornuk et al. (2011)
Thyme oil (5 ppm)	*Salmonella typhimurium*	1.0	Apple	Tornuk et al. (2011)
Carvacrol (0.6 ppm)	*Aeromonas hydrophila*	3.0	Mixed vegetables	de Sousa et al. (2012)
Carvacrol (0.6 ppm)	*Listeria monocytogenes*	2.5	Mixed vegetables	de Sousa et al. (2012)
Oregano oil (100 ppm)	*S. typhimurium*	3.7	Tomato	Gunduz et al. (2010)
Oregano oil (250 ppm)	*Lactic acid bacteria*	1.1	Carrot	Berumen et al. (2015)

7.6 BIO AGENTS

Nowadays, a paradigm shifts from the application of synthetic fungicides to a green and eco-friendly approach to postharvest issues of horticultural produce. Out of these approaches, the use of biological agents for the management of postharvest diseases of fruits and vegetables is an outstanding emerging field. Direct parasitism, induction of resistance, antibiosis, and competition for nutrients are the main modes of action of the microbial antagonists which suppress the activity of postharvest pathogens during handling. Using several microbial antagonists, postharvest spoilage of horticultural produces has been successfully controlled. There were two approaches for the application of bioagents. The first approach is to promote and management of antagonistic microbes that are already hosted on the produce itself and the second approach is the artificial introduction of desired microbial against postharvest pathogens (Tables 7.4 and 7.5).

TABLE 7.4 Microbial Antagonists Used for the Successful Control of Postharvest Diseases of Fruits and Vegetables

Microbial Antagonist	Disease and Its Causal Agent	Fruit/ Vegetable	Reference
Aureobasidium Pullulans	Penicilliumrots (*Penicillium* spp.)	Citrus	Wilson and Chalutz (1989)
	Botrytis rot (*Botrytis cinerea*)	Grape	Schena et al. (2003)
	Soft rot (*Monilinia laxa*)	Grape	Barkai-Golan (2001)
Bacillus subtilis	Brown rot (*Lasiodiplodia theobromae*)	Peach, plum, and nectarine	Pusey and Wilson (1984)
	Green mold (*Penicillium digitatum*)	Citrus	Singh and Deverall (1984)
	Anthracnose (*Colletotrichum gloeosporioides*)	Mango	Govender et al. (2005)
Bacillus amyloliquefaciens	*Bacillus cinerea*	Pear	Qu et al. (2016)
Trichoderma atroviride	*Phomopsis vexans*	Brinjal	Javeria et al. (2014)

7.7 GENETIC MODIFICATION FOR SHELF-LIFE EXTENSION OF FRESH PRODUCE

FFVs are mostly composed of water, which makes up between 85% and 95% of their total weight. Even after harvest, they continue to respire, carry out metabolic activities, and undergo the senescence process. Their perishable nature and rapid deterioration cause the short postharvest storage lives of fruit and vegetables. During the handling, retailing, and storage of fruit and vegetables, there is a significant loss in physical appearance, including yellowing, wilting, shriveling, browning, overripening, and sprouting.

TABLE 7.5 Approaches for Maintaining the Postharvest Quality of Fresh Produce

Treatment	Produce	Effect	Reference
Hot water Treatment @ 52°C/5min	"Carabao" Mango	Killing the insect-pest, reduce the chilling injury, delay the ripening	Urquiola et al. (2022)
HWT @ 48°C/5 and 50°C/2	Apple Cv. Royal Delicious	Maintain fruit firmness and mitigate fungicide residues (90–98%)	Singh et al. (2024)
HWT @ 48°C and 50°C for 5 and 2 min, respectively	Royal Delicious Apple	Optimum retention of functional quality attributes.	Singh et al. (2023)
Irradiation 0.5–1.0, 1.5	Onion, potato, and apple	Inhibit sprouting of bulbs, tubers, and roots crops, delay ripening	Reddy et al. (2018)
Edible coating	Apple, pears, carrots, and strawberry	Act a partial barrier, reduce moisture loss, create modified atmosphere; preserve color and texture	Dhall (2013)
Ethylene @100ppm	Banana, mango, citrus fruit, apple, and tomato	Triggers ripening process, enhance fruit color, and quality attributes	Maduwanthi and Marapana (2019)
1-MCP	Apple, avocado, banana, broccoli, kiwifruit, mango, pear, and papaya	Act ethylene action inhibitors, delay fruit softening, development of aroma and flavor	Huber (2008)
Ozonation	Fruits and vegetables, strawberry @ 1.5 µL/L	Act as sanitizer, reduce decay caused by *B. cinerea*	Aslam et al. (2020)
NO	Application of SNP @ 0.5 mm, Nectarine, apple peach pear etc.	Reduce the respiration rate, water loss, inhibits ethylene biosynthesis	Jayarajan and Sharma (2018)
Sulfur dioxide @ 0.20%	Grape, litchi, fig	Prevent postharvest decay, color development	Sortino et al. (2017)
CA storage	Apple, pear peach kiwifruit, etc.	Maintain freshness by retard the senescence process, reduce the decay severity.	Prange et al. (2013)

The freshness of products affects not only visible quality traits but also hidden quality traits like vitamins, pigments, bioactive compounds, antioxidants, etc. There are various methods used by stakeholders (physical, chemical, and biological) to maintain the postharvest quality of produce. Among these approaches, genetic engineering is emerging as one of the potential tools for postharvest management and preserving produce quality for a longer duration. Friedman (2019) recently carried out a study to review candidate genes for extending the shelf life of tomatoes, melons, apples, peaches, bananas, and strawberries. A series of biochemical events delay the onset of ripening and senescence in GM crops, making them less perishable than traditional crops. Additionally, genetic modification targeting the enrichment of certain nutrients in the edible part can enhance nutritional value, which can help developing countries meet their nutrition needs (Singh et al., 2017). Additionally, developing countries suffer from very high postharvest losses of fruits and vegetables due to inadequate infrastructure, including cold chain logistics. Transgenic horticultural crops with delayed senescence may be very helpful to farmers and consumers in rural areas who lack a cold chain. Several authors have suggested that genetic modification may be an effective way to increase quality, yield, and shelf life beyond conventional breeding. In order to develop new varieties with enhanced phytochemical content, genetic modification may be useful, such as genome editing using CRISPR/Cas9 (clustered regularly interspaced palindromic repeats associated with 9) (Zhou et al., 2020).

7.7.1 FRUITS AND VEGETABLES THAT HAVE BEEN GENETICALLY MODIFIED

Genetically modified foods are those whose DNA has been modified by introducing recombinant DNA technology into one or more well-characterized genes of crops. Agrobacterium tumefaciens-based gene transfer technology was developed in the 1980s by Monsanto Corporation and the Max Planck Institute for Plant Breeding, Germany. As a result, China became the first country to commercialize transgenic tobacco and tomatoes in the early 1990s. The higher levels of bioactive compounds and nutritional content in genetically modified fruit and vegetables may lead to better functional quality as well as higher yields, better edible quality, and a longer shelf life after harvest. Using the transgenic technique is advantageous due to the abundance of unrestricted genetic resources and its quick, direct application to commercial cultivars. The biosynthesis of ethylene begins with L-methionine via S-adenosyl methionine (SAM) and ACC. An enzyme named ACS converts SAM to ACC, while an enzyme called ACO converts ACC to ethylene. A method for delaying ripening involves silencing the genes for ACS and ACO or expressing genes that encode enzymes involved in ethylene degradation (Guis et al., 1997). It has also been found that RNA antisense technology is very promising for regulating the expression of genes involved in fruit ripening. By inserting the entire gene under the cauliflower mosaic virus 35S promoter, they inhibited ACC synthase mRNA expression. A second strategy was used in tomatoes to suppress ethylene production by inserting antisense oxidized ACC (Hamilton et al., 1990). Another transgenic broccoli with enhanced shelf life was developed by *Agrobacterium rhizogenes* co-transformation by modifying ACC synthase 1 and ACO 1 and 2 cDNAs in sense and anti-sense orientations. In the transformed broccoli, ethylene production was lower and chlorophyll loss was delayed (Higgins et al., 2006). Similarly, transgenic melons (Malus domestica L.) expressing antisense ACO showed a 39-fold increase in ethylene inhibition at harvest compared with wild-type melons (Silva et al., 2004). Delaying fruit ripening (by suppressing polygalacturonase (PG) and ethylene control technology) improves the flavor, color, texture, and shelf-life of fruits, as well as their transportation and handling characteristics (Thayer, 1994). It was possible to create a transgenic apple by silencing either the ethylene-producing enzymes ACS or ACO in plants. This plant produced fruit that had reduced autocatalytic ethylene production, was firmer, and had a longer shelf life than the control. It was also observed that ethylene regulates fruit quality attributes, and these transgenic apple trees allow researchers to investigate the role of ethylene and other factors that regulate fruit development. Another company, DNA Plant Technology Corporation, created event 1345-4 (trade name: Endless Summer) through inserting a construct containing the ACS gene, thereby suppressing the conversion of SAM to ACC.

7.7.2 POSTHARVEST QUALITY MODULATION

The quality of fruit and vegetables after harvest involves both visible and hidden traits that determine consumer preference. It is important to understand that visual characteristics include size, shape, color, and appearance; however, hidden characteristics include sugar, acids, vitamins, minerals, bioactive compounds, and flavors. While most consumers purchase fruit and vegetables based on their visual appearance, reliable traders consider both external and internal characteristics. Following are details about the postharvest quality attributes of fruit and vegetables modulated by genetic engineering.

7.7.3 THE MODULATION OF FIRMNESS

Genetic engineering has also been used to modulate the expression of cell wall-degrading enzymes such as PG, pectin methylesterase (PME), glucanase, and galactosidase. Each of these factors plays a crucial role in diminishing firmness. PG primarily depolymerizes pectin, subsequently rendering it soluble, while PME serves as a mediator in its depolymerization process. An effort in strawberry (*Fragaria ananassa* L.) has been made to investigate the effects of endo-1,4-glucanase (EGase) studied by Lee and Kim (2011). It was found that lowering the expression of the EGase gene on strawberries through down-regulation mediated by the APX promoter was an effective technique to increase fruit firmness. By suppressing PL gene expression, tomato firmness was also enhanced, thereby increasing shelf life (Uluisik et al., 2016).

7.7.4 COLOR

In the plant kingdom, carotenoids and anthocyanins are widely distributed pigments. The antioxidant activity of these compounds is particularly valuable, as they protect against several age-related degenerative diseases (Fraser and Bramley, 2004). In fruit and vegetables, carotenoids are synthesized in the plastids and accumulated in the chromoplasts. It is produced as a result of the condensation of geranylgeranyl diphosphate molecules into colorless phytoene by the enzyme phytoene synthase (PSY). In the carotenoid biosynthesis pathway, PSY is the rate-limiting enzyme and ORANGE (OR) is the main regulator gene involved, as reported by Tzuri et al. (2015). A CRISPR/Cas9 site-specific genome editing technique was used to enhance lycopene content in tomatoes by targeting genes for SGR1 (Stay-Green 1), LCY-E (lycopene β-cyclase 1), Blc (beta lycopene cyclase 1), and LCY-B2 (lycopene β-cyclase 2). With the help of metabolic engineering, they were developed a transgenic sweet orange (*Citrus sinensis* L.) that is endowed with enhanced levels of beta-carotene (Pons et al., 2014).

7.7.5 CONTENT OF VITAMINS

Vegetables can also be genetically engineered to increase their vitamin levels (vitamins A, C, and E). Due to their antioxidant properties, these vitamins can slow or prevent biological oxidation, which is a harmful chain reaction that contributes to cancer, heart disease, and blindness (Cho et al., 2005). The lettuce plant *(Lactuca sativa* L.) cv. In Chungchima, tocopherol accumulation is enhanced by over-expressing *Arabidopsis thaliana* cDNA, which encodes the tocopherol methyl transferase. Meanwhile increasing ascorbate levels by twofold when the same gene is overexpressed in potatoes was noticed by Hemavathi et al. (2010).

7.7.6 TOTAL SOLUBLE SOLID CONTENT

Genetic engineering has also increased the soluble solids content in some fruits and vegetables. Increasing the soluble solids content of the produce, especially carbohydrates, improves its processing characteristics. In potato, recombinant DNA technology increased the soluble solids content, which makes it suitable for French fries. Monsanto Company created a potato with a high solids content by introducing a carbohydrate producing gene into the plant. These potatoes absorb less oil when made into French fries.

7.7.7 CONTENT OF MINERALS

Genetic engineering has made it possible to develop varieties with high mineral nutrient content, which was crucial in meeting the nutritional needs of large populations around the world. An edible part of the genetically modified carrot has a calcium content 1.5-fold higher than that of the unmodified carrot due to overexpression of the Cax1 enzyme from the Arabidopsis plant (Cation Exchanger 1) was studied by Park et al. (2004). It was also found that expressing the iron-binding gene ferritin from soybeans increased lettuce's iron content by about 1.2–1.7-fold. Transgenic lettuce was also reported to grow faster than the control because of its high photosynthesis rate as pointed out by Macknight et al. (2017).

7.7.8 FLAVOR CONTENT

A characteristic flavor comes from a distinctive blend of sugar, acids, and volatile compounds. It has been observed that transgenes can change the flavoring compounds of tomatoes in novel ways. To increase the volatile terpenoid compounds in tomatoes, the basil (*Ocimum basilicum* L.) *geraniol synthase* gene was expressed under the control of tomato ripening-specific PG promoter. The metabolic engineering of the terpenoid pathway led to an increased amount of the flavor compound S-Linalool in tomatoes reported by Davidovich-Rikanati et al. (2007).

7.8 CONCLUSION AND PROSPECTS FOR THE FUTURE

Fresh and processed foods have been preserved using different traditional methods since ancient times. As one of the basic methods of food preservation, canning emerged as the most widely used until Louis Pasteur's discovery of pasteurization in 1864, which revolutionized the food industry and remains today for perishable foods like fruit juices and milk. These preservation methods effectively controlled spoilage and pathogenic microbes and enzymes, but they adversely affected the sensory and nutritional quality parameters of the fresh produce. As a consequence, the perishable food industry needed to develop alternative thermal and nonthermal technologies that would satisfy the food safety requirements without compromising the quality of their nutrition or sensory characteristics. The shelf life of fresh produce can be extended and maintained through a variety of physical and chemical treatments. It is critical to maintain the physical, nutritional, and sensory attributes of the fresh produce after harvest and reduce decay incidence by combining postharvest treatments such as CA and MAP. Thus, the perishable food industry greatly needs to develop alternative, thermal, and nonthermal technologies that could be used to meet food safety measures without compromising their nutritional and sensory quality attributes. Aside from improving edible quality and postharvest shelf life, GM fruits and vegetables have the potential to offer higher retention of nutritional and functional quality attributes during the storage period. In order to overcome fears and reduce risks with genetically modified produce, systematic research is needed with a precautionary and rational approach. Among the major horticultural crops, banana, papaya, guava, and others, which are mainstays for large populations worldwide, urgent attention is needed. Human health and the alleviation of hunger and malnutrition should be promoted by the benefits of GM technologies while addressing concerns about possible adverse effects (Tables 7.6 and 7.7).

TABLE 7.6 Regulation of Ethylene Biosynthesis in Fresh Produce

Regulation of Ethylene Biosynthesis in Fresh Produce Through Genetic Engineering			
Name of the Crop	**Outcome**	**Targeted Gene/Inserted Gene**	**References**
Tomato	Decline of 90% C_2H_4 production during the peak of ripening	ACC deaminase from pseudomonas Chloraphis 6G5	Klee et al. (1991)
	Decline of 44% in C_2H_4 production	ADC gene from oat	Gupta et al. (2019)
Musk melon	C_2H_4 production reduce	By RNA antisense technology, targeted ACO (CMAO-1)	Nunez-Palenius et al. (2007)
Melon	Reduction of C_2H_4 production by 39-fold at harvesting	Insertion of ACC oxidase (ACO) gene from apple by antisense orientation	Silva et al. (2004)
Mango	6-benzylaminopurine (BA) delay the ripening and senescence by downregulating the C_2H_4 biosynthesis.	ACO was suppressed in mango fruit after 6-BA application, which resulted in decreased ethylene production.	Zhang et al. (2022)
Improvement of Pigment Content in Fruits and Vegetables			
Sweet potato	Carotenoids as well as anthocyanin levels have both increased	The *IbMYB1* gene has been upregulated.	Park et al. (2015)
Orange	Up to a 36-fold increase in β-carotene levels	Endogenous beta-carotene hydroxylase (Cs-CHX) gene expression	Pons et al. (2014)
Tomato	Anthocyanin content has increased	Gene S1MYB75 overexpression	Jian et al. (2019)
Pepper	Carotene levels go up 7–10-fold	Upregulation of the Arabidopsis lycopene beta-cyclase (-Lcy) gene	El Nagar (2018)
Tomato	Lycopene content might have increased by 5.1.	Editing of the *SGR1, LCY-E, Blc, LCY-B1*, and *LCY-B2* genes using CRISPR/Cas9.	Li et al. (2018)

TABLE 7.6 *(Continued)*

Name of the Crop	Outcome	Targeted Gene/Inserted Gene	References
	Enrichment of Nutritional Level in Fruit and Vegetable		
Lettuce	Vitamin C, phenolic content, and antioxidant properties all have increased	The *AtGLDH* gene has been upregulated.	Guo et al. (2013)
	Calcium content was increased by 25–32% times	Arabidopsis H+/Ca^{2+} transporter sCAX1 is expressed in a deregulated state.	Park et al. (2009)
	Iron content was increased by 1.2–1.7-fold.	Soyabean iron-binding gene ferritin expression.	Goto et al. (2000)
Strawberry	Elevated synthesis of folic acid	*ADCS* gene is overexpressed.	Diaz de la Garza et al. (2004)
Potato	Ascorbate has increased.	The *GLOase* gene is overexpressed.	Hemavathi et al. (2010)

TABLE 7.7 Fruits and Vegetables with Genetically Modified Crop are Commercially Available

Name of the Crop	Targeted Characteristic	Parental Source	Gene of Interest	Innovator	Commercial Name
Apple	Polyphenoloxidase activity is suppression in ordertoprevent browning	Golden Delicious	This is a sequence of four PPO fragments in sense orientation (pineapple).	Okanagan Specialty Fruit Inc, Canada	Artic* Golden
Pineapple	Fruit color modification due to high lycopene accumulation (Pink flesh), as well as delay ripening might be reducing the ethylene production	Golden extract sweet	PSY gene (tangerine), lcyb and lcye gene fragment for antisense whereas ACS gene fragment in sense and antisense (Pineapple)	DelMonte Fresh Produce Company.	Rose/Extrac sweet pink flesh
Melon	Ripening delayed	–	SAM-K (E. c bacteriophage T3	Agritope Inc.	–
Tomato	Loss of firmness delayed	T-7	Partial PG gene (tomato)	Calgene Inc.	MacGregor's
	Ripening delayed	91103-114	ACS (Tomato)	DNA Plant Technology	Endless Summer

KEYWORDS

- **fresh produce**
- **heat treatment**
- **irradiation**
- **cold plasma**
- **plant extracts**
- **bio-agent**
- **genetic modification**

REFERENCES

Aslam, R.; Alam, M.S.; Saeed, P.A. Sanitization potential of ozone and its role in postharvest quality management of fruits and vegetables. *Food Eng. Rev.* 2020, 12(1), 48–67.

Barkai-Golan, R. *Postharvest Diseases of Fruits and Vegetables: Development and Control*. Elsevier, 2001.

Berumen, G.; Partida, L.D.C.; Jimenez, V.A.O.; Lopez, M.A.C.; Martinez, P.G. Efecto del quitosanoen la induccion de resistencia contra *Colletotrichum* sp. mango (*Mangifera indica* L.) cv. Tommy Atkins. *Investigacióny Ciencia* 2015, 23(66), 16–21.

Boubaker, J.; Mansour, H.B.; Ghedira, K.; Chekir-Ghedira, L. Antimutagenic and free radical scavenger effects of leaf extracts from *Accacia salicina*. *Ann. Clin. Microbiol. Antimicrob.* 2011, 10(1), 1–10.

Cherdshewasart, W.; Sutjit, W.; Pulcharoen, K.; Chulasiri, M. The mutagenic and antimutagenic effects of the traditional phytoestrogen-rich herbs, *Pueraria mirifica* and *Pueraria lobata*. *Braz. J. Med. Biol.* 2009, 42, 816–823.

Cho, E.A.; Lee, C.A.; Kim, Y.S.; Baek, S.H.; Reyes, B.G.; Yun, S.J. Expression of γ-tocopherol methyltransferase transgene improves tocopherol composition in lettuce (*Latuca sativa* L.). *Mol. Cells* 2005, 19(1), 16–22.

Corpas, F.J.; Rodriguez-Ruiz, M.; Muñoz-Vargas, M.A.; Gonzalez-Gordo, S.; Reiter, R.J.; Palma, J.M. Interactions of melatonin, reactive oxygen species, and nitric oxide during fruit ripening: An update and prospective view. *J. Exp. Bot.* 2022, 73(17), 5947–5960.

Davidovich-Rikanati, R.; Sitrit, Y.; Tadmor, Y.; Iijima, Y.; Bilenko, N.; Bar, E.; Lewinsohn, E. Enrichment of tomato flavor by diversion of the early plastidial terpenoid pathway. *Nat. Biotechnol.* 2007, 25(8), 899–901.

de Sousa, J.P.; de Azeredo, G.A.; de Araújo Torres, R.; da Silva Vasconcelos, M.A.; da Conceição, M.L.; de Souza, E.L. Synergies of carvacrol and 1,8-cineole to inhibit bacteria associated with minimally processed vegetables. *Int. J. Food Microbiol.* 2012, 154(3), 145–151.

Dey, A.; Rasane, P.; Choudhury, A.; Singh, J.; Maisnam, D.; Rasane, P. Cold plasma processing: A review. *J. Chem. Pharm. Sci.* 2016, 9, 2980–2984.

Dhall, R.K. Advances in edible coatings for fresh fruits and vegetables: A review. *Crit. Rev. Food. Sci. Nutr.* 2013, 53(5), 435–450.

Diaz de la Garza, R.; Quinlivan, E.P.; Klaus, S.M.; Basset, G.J.; Gregory III, J.F.; Hanson, A.D. Folate biofortification in tomatoes by engineering the pteridine branch of folate synthesis. *Proc. Natl., Acad. Sci. USA*. 2004, 35, 345–354

El Nagar, M.M. Genetic engineering to improve β-carotene content in pepper. *Ann. Agric. Sci. Moshtohor* 2018, 56(4th ICBAA), 119–126.

Fawcett, H.S. Packing house control of brown rot. *Citrograph* 1922, 7, 232–234.

Fraser, P.D.; Bramley, P.M. The biosynthesis and nutritional uses of carotenoids. *Prog. Lipid Res.* 2004, 43(3), 228–265.

Friedman, H. Candidate genes to extend fleshy fruit shelf life. *Plant Breed. Rev.* 2019, 43, 61–94.

Ghasemnezhad, M.; Zareh, S.; Rassa, M.; Sajedi, R.H. Effect of chitosan coating on maintenance of aril quality, microbial population and PPO activity of pomegranate *(Punica granatum L. cv. Tarom)* at cold storage temperature. *J. Sci. Food Agric.* 2013, 93(2), 368–374.

Goto, F.; Yoshihara, T.; Saiki, H. Iron accumulation and enhanced growth in transgenic lettuce plant expressing the iron-binding protein ferritin. *Theor. Appl. Genet. THEOR.* 2000, 100(5), 658–664.

Govender, V.; Korsten, L.; Sivakumar, D. Semi-commercial evaluation of *Bacillus licheniformis* to control mango postharvest diseases in South Africa. *Postharvest Biol. Technol.* 2005, 38(1), 57–65.

Guis, M.; Botondi, R.; Ben-Amor, M.; Ayub, R.; Bouzayen, M.; Pech, J.C.; Latche, A. Ripening-associated biochemical traits of Cantaloupe Charentais melons expressing an antisense ACC oxidase transgene. *J. Am. Soc. Hortic. Sci.* 1997, 122(6), 748–751.

Gunduz, G.T.; Gonul, S.A.; Karapinar, M. Efficacy of sumac and oregano in the inactivation of *Salmonella Typhimurium* on tomatoes. *Int. J. Food Microbiol.* 2010, 141(1–2), 39–44.

Guo, X.; Liu, R.H.; Fu, X.; Sun, X.; Tang, K. Over-expression of l-galactono-γ-lactone dehydrogenase increases vitamin C, total phenolics and antioxidant activity in lettuce through bio-fortification. *Plant Cell, Tissue Organ Cult. (PCTOC)*, 2013, 114(2), 225–236.

Gupta, S.M.; Gupta, A.K.; Ahmed, Z.; Kumar, A. Antibacterial and antifungal activity in leaf, seed extract and seed oil of seabuckthorn (*Hippophae salicifolia* D. Don) plant. *J. Plant Pathol. Microbiol.* 2011, 2(2), 1–4.

Gupta, P., et al. Biotechnological interventions for modulating ethylene production: The role of ADC gene expression in reducing ethylene biosynthesis. In Asrey, R. (Ed.), *Biological and Postharvest Interventions to Manage the Ethylene in Fruit: A Review*, Royal Society of Chemistry, 2019, pp. 1–54. https://doi.org/10.1039/d3fb00037k

Hamilton, A.J.; Lycett, G.W.; Grierson, D. Antisense gene that inhibits synthesis of the hormone ethylene in transgenic plants. *Nature* 1990, 346(6281), 284–287.

Hemavathi; Upadhyaya, C.P.; Akula, N.; Young, K.E.; Chun, S.C.; Kim, D.H.; Park, S.W. Enhanced ascorbic acid accumulation in transgenic potato confers tolerance to various abiotic stresses. *Biotechnol. Lett.* 2010, 32(2), 321–330.

Higgins, J.D.; Newbury, H.J.; Barbara, D.J.; Muthumeenakshi, S.; Puddephat, I.J. The production of marker-free genetically engineered broccoli with sense and antisense ACC synthase 1 and ACC oxidases 1 and 2 to extend shelf-life. *Mol. Plant Breed.* 2006, 17, 7–20.

Horn, R.C.; Vargas, V.M.F. Antimutagenic activity of extracts of natural substances in the Salmonella/microsome assay. *Mutagenesis* 2003, 18(2), 113–118.

Horvitz, S.; Cantalejo, M.J. Application of ozone for the postharvest treatment of fruits and vegetables. *Criti. Rev. Food Sci. Nutrit.* 2014, 54(3), 312–339.

Huber, D.J. Suppression of ethylene responses through the application of 1-methylcyclopropene: A power tool for elucidating ripening and senescence mechanisms in climacteric and non-climacteric fruits and vegetables. *Hort. Sci.* 2008, 43(1), 106–111.

Huyskens-Keil, S.; Hassenberg, K.; Herppich, W.B. Impact of postharvest UV-C and ozone treatment on textural properties of white asparagus (*Asparagus officinalis* L.). *J. Appl. Bot. Food Qual.* 2012, 84(2), 229.

Ilic, Z.; Polevaya, Y.; Fallik, E. The effectiveness of a short prestorage hot water rinse and hot water dips on tomato ripening. *Contemporary Agriculture* 2001.

Javeria, S.; Kumar, H.; Gangwar, R.K.; Tyagi, S.; Yadav, R.S. Isolation of stem rot disease causing organism of Brinjal and their in-vitro inhibition with fungicides and bio-control agents. *Eur. Res.* 2014, 9(2), 1662–1670.

Jayaprakasha, G.K.; Negi, P.S.; Jena, B.S.; Rao, L.J.M. Antioxidant and antimutagenic activities of Cinnamomum zeylanicum fruit extracts. *J. Food Compos. Anal.* 2007, 20(3), 330–336.

Jayarajan, S.; Sharma, R.R. Impact of nitric oxide on shelf life and quality of nectarine (*Prunus persica var. nucipersica*). *Acta Physiol. Plant.* 2018, 40(12), 1–9.

Jian, W.; Cao, H.; Yuan, S.; Liu, Y.; Lu, J.; Lu, W.; Li, Z. SlMYB75, an MYB-type transcription factor, promotes anthocyanin accumulation and enhances volatile aroma production in tomato fruits. *Hort. Res.* 2019, 6(23), 2–8.

Khan, A.; Rahman, M.; Islam, M.S. Antibacterial, antifungal and cytotoxic activities of amblyone isolated from *Amorphophallus campanulatus*. *Indian J. Pharmacol.* 2008, 40(1), 41.

Khan, Muhammad & Ahmad, Khalil & Hassan, Sadia & Imran, Muhammad & Ahmad, Nazir & Xu, Changmou. (2017). Effect of novel technologies on polyphenols during food processing.

Klee, H.J.; Van Der Knaap, E.; Giovannoni, J.J. The tomato fruit-ripening mutant ripening-inhibitor (RIN) demonstrates roles for ethylene and gene expression in fruit development. *Proc. Natl. Acad. Sci. USA*, 1991, 88(8), 2342–2346. https://doi.org/10.1073/pnas.88.8.2342

Lee, Y.K.; Kim, I.J. Modulation of fruit softening by antisense suppression of endo-β-1, 4-glucanase in strawberry. *Mol. Plant Breed.* 2011, 27, 375–383.

Li, F.; Zhang, X.; Wang, J.; Jiang, Y.; Zhang, X.; Li, X. Preharvest application of 1-methylcyclopropene and Ethephon altered cuticular wax biosynthesis and fruit quality of apples at harvest and during cold storage *Hortic. Plant J.* 2022, 8(2), 143–152.

Li, X.; Wang, Y.; Chen, S.; Tian, H.; Fu, D.; Zhu, B.; Zhu, H. Lycopene is enriched in tomato fruit by CRISPR/Cas9-mediated multiplex genome editing. *Front. Plant Sci.* 2018, 9, 559.

Liu, K. Biotech crops: Products, properties, and prospects. *Food Tech., (Chicago)*, 1999, 53(5), 42–49.

Mackeen, M.M.; Ali, A.M.; Lajis, N.H.; Kawazu, K.; Hassan, Z.; Amran, M.; Mohamed, S.M. Antimicrobial, antioxidant, antitumour-promoting and cytotoxic activities of different plant part extracts of *Garcinia atroviridis*. *J. Ethnopharmacol.* 2000, 72(3), 395–402.

Macknight, R.C.; Laing, W.A.; Bulley, S.M.; Broad, R.C.; Johnson, A.A.; Hellens, R.P. Increasing ascorbate levels in crops to enhance human nutrition and plant abiotic stress tolerance *Curr. Opin. Biotechnol.* 2017, 44, 153–160.

Maduwanthi, S.D.T.; Marapana, R.A.U.J. Induced ripening agents and their effect on fruit quality of banana. *Int. J. Food Sci.* 2019, 2019, 2520179.

Mahto, R.; Das, M. Effect of gamma irradiation on the physico-chemical and visual properties of mango (*Mangifera indica L.*), cv. 'Dushehri' and 'Fazli'stored at 20°C. *Postharvest Biol. Technol.* 2013, 86, 447–455.

Mollakhalili-Meybodi, N.; Yousefi, M.; Nematollahi, A.; Khorshidian, N. Effect of atmospheric cold plasma treatment on technological and nutrition functionality of protein in foods. *Eur. Food Res. Technol.* 2021, 247(7), 1579–1594.

Mostafavi, H.A.; Mirmajlessi, S.M.; Fathollahi, H. The potential of food irradiation: Benefits and limitations. *Trends Food Sci. Technol.* 2012, 5, 43–68.

Negi, P.S.; Anandharamakrishnan, C.; Jayaprakasha, G.K. Antibacterial activity of *Aristolochia bracteata* root extracts. *J. Med. Food* 2003, 6(4), 401–403.

Nock, J.F.; Watkins, C.B.; James, H.; Reed, N.; Oakes, R.L. Preharvest application of 1-methylcyclopropene (1-MCP) to control fruit drop of apples, and its effects on postharvest quality. In: *VI International Postharvest Symposium 877,* 2010, 1, 365–374.

Núñez-Palenius, H.G.; Huber, D.J.; Gasser, C.S. Characterization of ripening behavior in transgenic melons expressing an antisense 1-aminocyclopropane-1-carboxylate (ACC) oxidase gene from apple. *Plant Sci.* 2007, 173(2), 161–171. https://doi.org/10.1016/j.plantsci.2007.03.012

Park, S.C.; Kim, Y.H.; Kim, S.H.; Jeong, Y.J.; Kim, C.Y.; Lee, J.S.; Kwak, S.S. Overexpression of the IbMYB1 gene in an orange-fleshed sweet potato cultivar produces a dual-pigmented transgenic sweet potato with improved antioxidant activity. *Physiol. Plant* 2015, 153(4), 525–537.

Park, S.; Elless, M.P.; Park, J.; Jenkins, A.; Lim, W.; Chambers IV, E.; Hirschi, K.D. Sensory analysis of calcium-biofortified lettuce. *Plant Biotechnol. J.* 2009, 7(1), 106–117.

Park, S.; Kim, C.K.; Pike, L.M.; Smith, R.H.; Hirschi, K.D. Increased calcium in carrots by expression of an Arabidopsis H+/Ca^{2+} transporter. *Mol. Plant Breed.* 2004, 14(3), 275–282.

Pons, E.; Alquezar, B.; Rodriguez, A.; Martorell, P.; Genoves, S.; Ramon, D.; Pena, L. Metabolic engineering of β-carotene in orange fruit increase its in vivo antioxidant properties. *Plant Biotechnol. J.* 2014, 12(1), 17–27.

Prange, R.K.; Wright, A.H.; DeLong, J.M.; Zanella, A. History, current situation and future prospects for dynamic controlled atmosphere (DCA) storage of fruits and vegetables, using chlorophyll fluorescence. *VII International Postharvest Symposium* 2013, 1012, 905–915.

Pusey, P.L.; Wilson, C. Postharvest biological control of stone fruit brown rot by *Bacillus subtilis*. *Plant Dis.* 1984, 68(9), 753–756.

Qu, H.; Wei, Y.; Xue, J.; Wang, Y.; Li, C.; Zhang, J. Biocontrol of gray mold decay in pear by Bacillus amyloliquefaciens strain BA3 and its effect on postharvest quality parameters. *Polish J. Microbiol.* 2016, 65(2), 171–177. https://doi.org/10.33073/pjm-2016-029

Reddy, S.V.R.; Sharma, R.R.; Raghavendra, H.R.; Prajapati, U. Microbiological efficacy of decontamination methodologies for fresh produce. *Food Security Plant Dis. Manag.* 2021, 16, 323–355.

Reddy, S.; Rakesh, V.; Sharma, R.R.; Gundewadi, G. Use of irradiation for postharvest disinfection of fruits and vegetables. In: *Postharvest Disinfection of Fruits and Vegetables*. Academic Press, 2018, pp. 121–136.

Rico, D.; Martin-Diana, A.B.; Barat, J.M.; Barry-Ryan, C. Extending and measuring the quality of fresh-cut fruit and vegetables: A review. *Trends Food Sci. Technol.* 2007, 18(7), 373–386.

Rojas, J.J.; Ochoa, V.J.; Ocampo, S. M.; Muñoz, J.F. Screening for antimicrobial activity of ten medicinal plants used in Colombian folkloric medicine: Part I. *Memór. Inst. Oswaldo Cruz* 2006, 101(7), 739–742. https://doi.org/10.1590/s0074-02762006000700006

Salvador, A.; Abad, I.; Arnal, L.; Martinez-Javega, J.M. Effect of ozone on postharvest quality persimmon. *J. Food Sci.* 2006, 71(6), S443-S446.

Schena, L.; Nigro, F.; Pentimone, I.; Ligorio, A.; Ippolito, A. Control of postharvest rots of sweet cherries and table grapes with endophytic isolates of *Aureobasidium pullulans*. *Postharvest Biol. Technol.* 2003, 30(3), 209–220.

Silva, J.A.; da Costa, T.S.; Lucchetta, L.; Marini, L.J.; Zanuzo, M.R.; Nora, L.; Rombaldi, C.V. Characterization of ripening behavior in transgenic melons expressing an antisense 1-aminocyclopropane-1 carboxylate (ACC) oxidase gene from apple. *PostharvestBiol. Technol.* 2004, 32(3), 263–268.

Singh, A.K.; Banerjee, T.; Sethi, S.; Tippannanavar, M.; Joshi, A.; Kumar, R.; Pandey, R. Fungicide residue degradation in hot water treated apple. *Appl. Fruit Sci.* 2024, 1–13.

Singh, A.K.; Sethi, S.; Asrey, R.; Kumar, R. Influence of hot water treatment on nutritional quality attributes of cold stored apple (*Malus* × *domestica* Borkh). *Indian J. Agric. Sci.* 2023, 93(9), 1025–1031.

Singh, A.; Rai, A.; Rai, A.; Singh, M. Biotechnological applications in post-harvest management of vegetable crops. In: Singh, B.; Singh, S.; Koley, T.K.; (Eds.), *Advances in Postharvest Technologies in Vegetable Crops*. Apple Academic Press and CRC Press: USA, 2017, pp. 171–202.

Singh, V.; Deverall, B.J. Bacillus subtilis as a control agent against fungal pathogens of citrus fruit. *Trans. Br. Mycol. Soc.* 1984, 83(3), 487–490.

Sortino, G.; Allegra, A.; Passafiume, R.; Gianguzzi, G.; Gullo, G.; Gallotta, A. Postharvest application of Sulphur dioxide fumigation to improve quality and storage ability of "Red Globe" grape cultivar during long cold storage. *Chem. Eng. Trans.* 2017, 58, 403–408.

Surowsky, B.; Froehling, A.; Gottschalk, N.; Schlüter, O.; Knorr, D. Impact of cold plasma on *Citrobacter freundii* in apple juice: Inactivation kinetics and mechanisms. *Int. J. Food Microbiol.* 2014, 174, 63–71.

Thayer, A. FDA gives go-ahead to bioengineered tomato. *Chem. Eng. News* 1994, 72(21), 7–8.

Tornuk, F.; Cankurt, H.; Ozturk, I.; Sagdic, O.; Bayram, O.; Yetim, H. Efficacy of various plant hydrosols as natural food sanitizers in reducing *Escherichia coli* O157: H7 and

Salmonella Typhimurium on fresh cut carrots and apples. *Int. J. Food Microbiol.* 2011, 148(1), 30–35.

Tripathi, P.; Dubey, N. K. Exploitation of natural products as an alternative strategy to control postharvest fungal rotting of citrus fruits. *Postharvest Biol. Technol.* 2004, 32(3), 235–245. https://doi.org/10.1016/j.postharvbio.2003.11.001

Tzuri, G.; Zhou, X.; Chayut, N.; Yuan, H.; Portnoy, V.; Meir, A.; Tadmor, Y.A. 'golden' SNP in CmOr governs the fruit flesh color of melon (*Cucumis melo*). *Plant J.* 2015, 82(2), 267–279.

Uluisik, S.; Chapman, N.H.; Smith, R.; Poole, M.; Adams, G.; Gillis, R.B.; Seymour, G.B. Genetic improvement of tomato by targeted control of fruit softening. *Nat. Biotechnol.* 2016, 34(9), 950–952.

Urquiola, M.A.J.; Bayogan, E.R.V.; De Faveri, S.G.; Johnson, P. Effect of hot water treatment and prochloraz on the postharvest quality of 'Carabao mango'. *Acta Hortic.* 2022, 1336, 363–370.

Wang, C.; Zhang, Y.; Zhan, Y.; Luo, C.; Li, Y.; Qiu, D.; Hua, Y. The effect of 17 α-ethynylestradiol induced intrahepatic cholestasis of pregnancy on placental P-glycoprotein in mice: Implications in the individualized transplacental digoxin treatment for fetal heart failure. *Placenta* 2016, 44, 91–97.

Wilson, C.L.; Chalutz, E. Postharvest biological control of Penicillium rots of citrus with antagonistic yeasts and bacteria. *Sci. Hortic.* 1989, 40(2), 105–112.

Yuan, Y.; Wei, J.; Xing, S.; Zhang, Z.; Wu, B.; Guan, J. Sulfur dioxide (SO2) accumulation in postharvest grape: The role of pedicels of four different varieties. *Postharvest Biol. Technol.* 2022, 190, 111–953.

Zaharah, S.S.; Singh, Z. Mode of action of nitric oxide in inhibiting ethylene biosynthesis and fruit softening during ripening and cool storage of 'Kensington Pride' mango. *Postharvest Biol. Technol.* 2011, 62(3), 258–266.

Zhang, Y.; Gao, Z.; Hu, M.; Pan, Y.; Xu, X.; Zhang, Z. Delay of ripening and senescence in mango fruit by 6-benzylaminopurine is associated with inhibition of ethylene biosynthesis and membrane lipid catabolism. *Postharvest Biol. Technol.* 2022, 185, 111797.

Zhou, J.; Li, D.; Wang, G.; Wang, F.; Kunjal, M.; Joldersma, D.; Liu, Z. Application and future perspective of CRISPR/Cas9 genome editing in fruit crops. *J. Integr. Plant Biol.* 2020, 62(3), 269–286.

Ziv, C.; Fallik, E. Postharvest storage techniques and quality evaluation of fruits and vegetables for reducing food loss. *Agronomy* 2021, 11(6), 1133.

CHAPTER 8

Molecular Markers and Their Effectiveness in Crop Improvement to Select Superior Genotypes

PRASANTA KUMAR MAJHI,[1,*] TANMAYA KUMAR BHOI,[2] IPSITA SAMAL,[3] PARTHA PRATIM BEHERA,[4] PRACHI PATTNAIK,[5] SUBRAT KUMAR SENAPATI,[6] SUDIPTA PADHAN,[7] and SARITA PRADHAN[8]

[1]*Department of Plant Breeding and Genetics, Regional Research and Technology Transfer Station (RRTTS), Odisha University of Agriculture and Technology (OUAT), Keonjhar, Odisha, India*

[2]*Forest Protection Division, ICFRE-Arid Forest Research Institute (AFRI), Jodhpur, Rajasthan, India*

[3]*Department of Entomology, Faculty of Agriculture, Sri Sri University, Cuttack, Odisha, India*

[4]*Department of Plant Breeding and Genetics, Assam Agricultural University, Jorhat, Assam, India*

[5]*Department of Horticulture, Institute of Agricultural Sciences, Banaras Hindu University (BHU), Varanasi, Uttar Pradesh, India*

[6]*Department of Floriculture and Landscaping, Odisha University of Agriculture and Technology (OUAT), Bhubaneswar, Odisha, India*

[7]*Department of Entomology, Lovely Professional University, Jalandhar, Punjab, India*

[8]*Department of Seed Science and Technology, Odisha University of Agriculture and Technology, Bhubaneswar, Odisha, India*

*Corresponding author

Technological Advancements in Sustainable Crop Improvement and Production.
Devendra Kumar, Alok Kumar Singh, and Nisha Malik (Eds.)

DOI: 10.1201/9781779643155-8

ABSTRACT

Improved food grain production is mostly attributable to plant breeding activities. Given the current yield patterns, projected population growth, and environmental effects, it is crucial for plant breeding efforts to prioritize traits that ensure the stability of yield and sustainability. While the impact of marker-assisted selection (MAS) on variety development has been limited so far, it has the potential to assist plant breeders greatly in the near future. Greater integration with breeding programs, as well as a thorough understanding of existing hurdles and the development of relevant solutions, are necessary if MAS is to fulfil its potential. Benefiting from the potential of MAS over traditional breeding methods could have a major effect on crop development. Incorporating quantitative resistance controlled by quantitative trait locus (QTL) expression is an additional viable technique for creating long-lasting disease resistance. Utilizing DNA markers during the backcrossing process significantly improves the quality of the selection process. Marker-assisted backcrossing can be broken down into three distinct stages. The process of merging several genes to form a single phenotype is referred to as "pyramiding." Therefore, marker-assisted gene pyramiding can be employed to simultaneously select many genes, leading to improved resistance to environmental stresses.

8.1 INTRODUCTION

To continually satisfy the requirements of changing markets for crop varieties, plant breeders must devise methods to accelerate the breeding process and enhance the quality of their choices. The progress of agriculture and the enhancement of human life can be directly attributed to the efforts of plant breeders (Cobb et al., 2019). Thousands of years ago, humans used artificial selection to domesticate wild plant species and turn them into farmable crops. The domestication and continued selection of crop species have played a crucial role in advancing human civilization. In modern plant breeding programs supported by established genetic principles and cutting-edge techniques, the breeders' experience and individual understanding remain crucial to the achievement of desirable results in cultivar development. As a result, plant breeding is often viewed as a fusion of scientific method, creative expression, and modern technologies (Acquaah, 2012). The following are some of the difficulties encountered in plant breeding: how do we choose which individuals will become parents, and how do we organize populations for breeding? How can I uncover the variations I want and pick the ones I need? What methods can predict breeding material expression and performance across generations and contexts? What strategies may be implemented to increase the efficiency and speed of selection trials? How can we speed up crop variety development to meet changing market demands? A breeder's job is to find answers to these kinds of issues, yet doing so with only traditional breeding techniques is sometimes unrealistic.

In practical breeding, selection for desired features can be made directly or indirectly. Direct selection evaluates the fitness of candidates for further breeding depending on how well they perform or how phenotypically they value the features of interest. The use of direct selection for a complex trait with quantitative inheritance has limited implications in many scenarios. If an alternative characteristic is closely connected with the target trait and is inherited in a simpler way, indirect selection may be preferable (Richards et al., 2010). However, there is a severe lack of features that can be employed for yield-focused indirect selection. Phenotypic value is determined by both genetic control and environmental factors. Opting for the desired gene(s) would be the most straightforward kind of selection, and so might offer some guarantee of effectiveness. While many breeding programs might benefit from selecting for target genes, doing so is often impractical due to a lack of suitable infrastructure and funding (Ober et al., 2021). Conventional plant breeding relies on phenotypic evaluation of the desired qualities to make decisions (Figure 8.1). The heritability of a quantitative trait determines the phenotypic value of the trait for selection purposes. As a result, quantitative qualities require extensive, preferably independent, testing. This raises the expense of evaluation and restricts trials to times and places where the relevant features can be well expressed. As a result, qualities like yield cannot be selected for in off-season nursery and greenhouse operations. Furthermore, it is only after a plant reaches maturity that characteristics such as the quality of its fruit or seeds and its yield can be evaluated. Due to the inability to employ the selected plants for in-generation hybridization, they will be lost. Marker-assisted selection (MAS) is an effective and novel method currently utilized in practical breeding and selection for crop enhancement. Molecular markers, such as DNA markers, may not directly select genes, but they are the most effective means of indirectly selecting target genes at the DNA level. As a result, they are a powerful and very practical tool in the field of plant breeding (Lema, 2018). Molecular marker-assisted breeding

(MAB) is the process of using DNA markers, linkage maps, and genomics to selectively breed plants and animals for specific desirable traits (Jiang, 2013). Many contemporary breeding methods can be categorized under this overarching concept. These methods consist of MAS, marker-assisted backcrossing (MABC), marker-assisted recurrent selection (MARS), and genome-wide selection or genomic selection (GS) (Ribaut et al., 2010). MAB has been extensively utilized in several crop species to enhance their genetic composition since its establishment (Jiang, 2013; Xu, 2010). This chapter specifically addressed the different categories of markers employed in crop enhancement, the predominant molecular marker systems, the prerequisites and procedure of MAB, as well as the utilization of MAS in crop improvement. Additionally, it discussed MABC, gene pyramiding, and recurrent selection as means to augment both qualitative and quantitative traits.

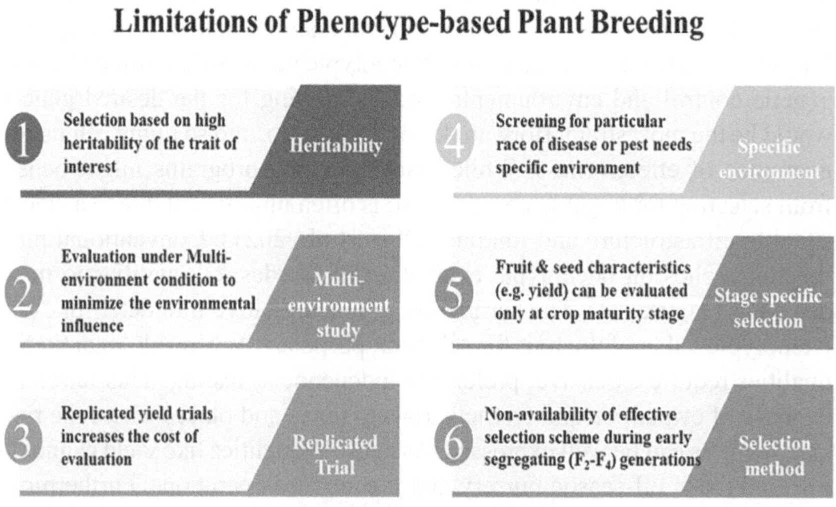

FIGURE 8.1 Limitations of phenotype-based plant breeding.

8.2 VARIOUS CATEGORIES OF MARKERS EMPLOYED IN THE FIELD OF PLANT BREEDING

There are primarily four categories of markers utilized in the process of plant breeding selection (Figure 8.2). The following are the four types of markers used in biological research: (1) morphological markers, which involve the study of morphological characteristics; (2) biochemical markers, which involve the analysis of chemical substances; (3) cytological markers, which involve the examination of cellular structures; and (4) DNA markers, which involve the study of genetic material.

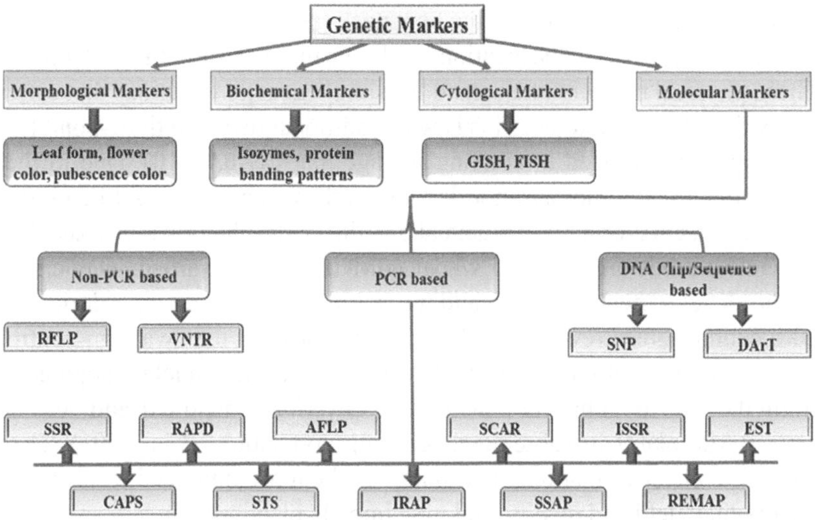

FIGURE 8.2 Classification of genetic markers.

8.2.1 MORPHOLOGICAL MARKERS

The early genetic markers utilized in scientific investigations were morphological qualities, hence the alternative name "visible markers" for these characteristics. Form of the leaf, color of the blossom, color of the pubescence, color of the pod, color of the seed, shape of the seed, color of the hilum, type and length of the awn, shape of the fruit, color and stripes on the exocarp (rind), flesh color, stem length, etc. are all observable morphological traits that breeders use as markers. "Naked eye polymorphisms" are another name for these characteristics because they can be observed and scored visually (Singh and Singh, 2015). These characteristics are what plant breeders care about because they reflect the true phenotypes of plants. Protein and DNA markers, however, generally represent unspecified genomic locations that may or may not have a direct connection to observable characteristics. Morphological marker assays often do not require complex apparatus or preliminary procedures. As a result, these markers are easy to score, quick to analyze, and cheap, and often the scores can even be transferred to preserved specimens (Stussey, 1990).

The leaf-tip-burning signal in wheat is associated with the leaf rust resistance gene "*Lr34*," while the pigmented seedling/black chaff marker is associated with the stem rust resistance gene "*Sr2*." Certain signs can be useful for distinguishing different crop kinds, such as the black glumes of the wheat variety HD-2329 or the bent neck (peduncle) of the wheat variety Kalyan Sona. All available locations will keep making use of these identifiers. These physical markers often stand for simple genetic variations that can be easily detected and altered. This is why traditional two- and three-point tests are commonly utilized in creating linkage maps (Xu, 2010). Due to their connections with other agronomic factors, these markers might be used as indirect selection criteria in practical breeding. Morphological markers help generate semidwarf rice and wheat cultivars. Semidwarfism was crucial to the development of high-yielding rice and wheat cultivars during the 20th-century green revolution. Backcrossing brought the dwarfism-controlling *Rht10* gene into male-sterile wheat from the Taigu nuclear population, strengthening its connection with *Ta1*. Dwarfism identified and selected male-sterile plants in subsequent breeding populations (Liu, 1991). There are a few physical markers, and many of them are unrelated to economic factors like yield and quality, and may impede plant growth.

8.2.1.1 LIMITATIONS OF MORPHOLOGICAL MARKERS

1. The number of highly reliable morphological or visual indications for a given species is small.
2. Because it is difficult to identify the phenotypes of distinct features in a single plant, only a small number of these markers may typically be investigated in a single cross or mapping population.
3. In most cases, they can only be scored on fully grown plants at specific stages of their development.
4. It is possible that there is a minimum expression requirement for several traits, including disease resistance.
5. The characteristic of interest, or the trait with which the marker connection is to be studied, may be impacted by specific genes that influence the marker traits in a pleiotropic way. Because of this, mistakes in gene mapping would result from an inflated segregation ratio.
6. Lastly, it would be necessary to keep genetic groups that exhibit all of the marker traits.

8.2.2 BIOCHEMICAL OR PROTEIN MARKERS

Electrophoretic protein variations are used to identify protein-based markers, such as enzymes. These markers are made when there are small changes in the relevant genes' coding sequences, which change the corresponding proteins' amino acid sequences. Consequently, differential electrophoretic mobility reveals that mutant protein molecules have a different electrical charge than wild type. Individuals have many isozymes of an enzyme with the same catalytic function. They move at various rates in electrophoresis according to net charge or conformation. Gene duplication or polyploidization can create isozymes, which are enzyme variations encoded by different genes. However, allozymes are enzymes encoded by several copies of the same gene. This indicates that only allozymes can be marker locus alleles in linkage analysis. Strict isozymes are inherited separately at various loci. Molecular weight or isoelectric point variations of nonenzymatic proteins are also identified. Markers like seed storage proteins and isozymes are common (Tanksley and Orton, 1983). Plant breeding employs biochemical indicators, such as the high molecular weight glutenin subunit (HMW-GS) in wheat. Payne et al. (1987) discovered that the presence of specific HMW-GS is positively associated with gluten strength, as measured by the SDS-sedimentation volume test. A numerical scale called the Glu-1 quality score was created to rank breads based on the listed components (Payne et al., 1987; Rogers et al., 1989). The predictive ability of bread-making quality can be determined by summing the scores of the alleles found in the given line, under the assumption that the alleles have an additive effect. Evidence indicates that variations in the alleles at the Glu-D1 locus have a more significant impact on the bread's quality compared to variations at the other Glu-1 loci. Different types' bread manufacturing quality cannot be explained solely by HMW-GS composition. The low molecular weight glutenin subunit, gliadins in lower amounts, and their interactions with the HMW-GS affect gluten strength and bread quality (Jiang, 2013).

8.2.2.1 THE BENEFITS THAT PROTEIN-BASED MARKERS HAVE OVER MORPHOLOGICAL INDICATORS

1. Only need a tiny bit of tissue to locate them.
2. They can often be found when the plants are young or even when they are just seeds.

3. Most of the time, looking at one marker does not change the way you look at other protein-based markers.
4. Isozymes usually work together.
5. Analysis is not too hard.
6. Many reference data make it easier to figure out what the data mean.

8.2.2.2 LIMITATIONS OF PROTEIN MARKERS

1. Only a few protein-based markers can distinguish parents.
2. Isozymes are just a small, nonrandom sample of an organism's structural genes.
3. They only look for mutations that change the way an enzyme moves in an electrophoretic gel. Also, a single band can stand for two different isozymes that move the same way.
4. They vary based on tissue type, development stage, and environment. Researchers have uncovered isozyme markers connected to genomic areas or genes that create several fascinating features.
5. Indirect selection of traits using markers is known as "MAS." MAS has been applied with the tomato acid phosphatase locus asp1, which is connected to nematode resistance.

8.2.3 CYTOLOGICAL OR CHROMOSOMAL MARKERS

Cytogenetic traits such as aneuploidy, chromosomal pairings, and defective chromosomes are discovered by comparing the quantity, shape, and structure of chromosomes across species. These can be employed for genetic mapping by replacing chromosomes or as markers to find other genes and determine their relative placements (Nadeem et al., 2018). Using a karyotype and bands, we can see the intricate structure of a chromosome. The bands' hues, sizes, orders, and placements reveal the dispersed natures of euchromatin and heterochromatin. The Giemsa stain creates the Q bands, whereas the reversed Giemsa stain creates the G bands and the reversed Giemsa stain creates the R bands (Jiang, 2013). These chromosome markers help with normal chromosome characterization and also with detecting chromosome mutations. Linkage groups within specific chromosomes can be identified using cytological markers, which have found widespread usage in physical mapping. Low number and resolution limit their use in genetic diversity studies, genetic mapping, and MAS.

8.2.4 DNA OR MOLECULAR MARKERS

DNA markers, segments of DNA with mutations/variations, are needed to identify gene genotypes or alleles in a population or gene pool. Since such fragments are related to certain genome sites, molecular technology can detect them. Base deletions, insertions, and substitutions cause DNA markers to vary between individuals (Jiang, 2013). Southern blotting, a nucleic acid hybridization method (Southern, 1975), and polymerase chain reaction (PCR) are frequently used to determine variation (Mullis, 1990). PCR and molecular hybridization followed by electrophoresis can identify DNA sample variation or polymorphisms in a given DNA sequence. There have been advances in various detection methods. DNA hybridization with tagged nucleotides in novel array chip technologies and novel sequencing methods for DNA polymorphism identification are examples. Marker techniques enable simultaneous selection of desirable features in F2 and backcross populations, near isogenic lines, doubling haploids, and recombinant inbred lines. DNA markers, or molecular markers, are crucial to molecular breeding.

8.2.4.1 FEATURES OF AN IDEAL MOLECULAR (DNA) MARKER

The optimal DNA markers for effective utilization in MAB should include the following attributes (Jiang, 2013). However, these properties may differ depending on the specific application and species involved.

1. A significant degree of polymorphism was observed.
2. The distribution is uniform throughout the entire genome, without any clustering in specific areas.
3. Codominance refers to the simultaneous expression of both alleles in heterozygotes, allowing them to be easily recognized from homozygotes.
4. The allelic traits should be distinguishable to facilitate easy identification of different alleles.
5. This refers to a situation when there is only one copy of a gene present, and it does not have any additional effects on other traits or characteristics.
6. Cost-effective marker creation and genotyping with low expenses.
7. The assay/detection process is simple and may be easily automated.
8. The system should have high availability, meaning it may be used without restrictions, and it should be suitable for duplication or

multiplexing, allowing data to be accumulated and transferred between laboratories.

9. Specifically tailored to a particular genome, particularly when dealing with polyploids.
10. There is no negative impact on the observable characteristics of an organism.

8.3 TYPES OF DNA MARKER SYSTEMS

DNA-based markers represent variations in individuals' genomic DNA sequences. Fragment mobility disparities in a gel, hybridization to an array, PCR, and direct sequencing of DNA are all techniques utilized for identifying genetic variations. The process of creating these markers began in 1974 when restriction enzyme digestion of DNA was used to physically map an adenovirus gene. Human geneticists discovered in 1980 that restriction enzyme digestion of genomic DNA produced DNA fragments of different lengths from the same genomic regions. This fragment length variation can be seen by gel electrophoresis of DNA digests and hybridization with a probe that matches the genomic region. This process created unique band patterns for everyone. The first DNA-based marker, restriction fragment length polymorphism (RFLP), measures small differences in DNA fragment length. Over time, many DNA-based marker systems have been developed to address challenges such as high-throughput, low cost, reproducibility, availability, and ease of use. These marker systems can detect single-nucleotide variants, insertion/deletion (InDel) mutations ranging from one to several bases, and tandem repeat variations spanning a few to several nucleotides. The molecular markers are categorized according to many criteria, including the sequence of their creation, the use of PCR reaction, the dominant effect, and their high-throughput efficiency (Table 8.1).

There are multiple marker systems, and their performance in molecular plant breeding has been compared in Table 8.2.

1. Restriction fragment length polymorphism
2. Random amplified polymorphic DNA (RAPD)
3. Amplified fragment length polymorphism (AFLP)
4. Sequence-characterized amplified region (SCAR)
5. Sequence-tagged sites (STS)
6. Simple sequence repeat (SSR) polymorphisms

TABLE 8.1 Classification of DNA Markers Based on Different Criteria

Sl. No.	Basis of Classification	Categories	Example
1.	Based on the chronology of development	1st generation	RFLP, RAPD, and their modifications
		2nd generation	SSRS, AFLPs, and their modifications
		3rd generation	ESTS and SNPS markers
2.	Based on the use of the PCR reaction	PCR-based markers	RAPD, AFLP, and SSR
		Non-PCR-based markers depending	RFLP
3.	Based on the dominance effect	Dominant marker	RAPD, AFLP, and SCAR
		Codominant marker	RFLP, SSR, SNP, and CAPS
4.	Based on location and functional significance of markers	Random markers	Obtained from polymorphic sequences found at arbitrary locations in the genome. These markers may or may not be situated within genes, and their role in the formation of a phenotype is uncertain
		Gene-based markers	The gene-targeted markers show different parts of genes, but we do not know how they connect to the important trait phenotypes
		Functional markers	The functional markers are obtained from polymorphism locations within genes that are causally linked to certain phenotypes of the traits under question
5.	Based on throughput criteria	Low-throughput hybridization-based markers	RFLP
		Medium-throughput PCR-based markers	RAPD, AFLP, and SSR
		High-throughput sequence-based markers	SNPs

7. Arbitrary primed PCR
8. DNA amplification fingerprinting
9. Allele-specific associated primers
10. Single primer amplification reactions
11. SSR-anchored PCR
12. Cleaved amplified polymorphic sequences (CAPS)
13. Allele-specific PCR

14. Allele-specific ligation
15. Single-strand conformation
1. Single-nucleotide polymorphism (SNP)
17. Single-strand conformation polymorphism
18. Diversity array technology
19. Inter-SSR (ISSR) markers
20. Amplicon length polymorphism
21. Target region amplification polymorphism
22. Sequence related amplified polymorphism (SRAP)
23. Transposable element-based marker

8.3.1 RESTRICTION FRAGMENT LENGTH POLYMORPHISM MARKERS

The utilization of DNA polymorphisms as genetic markers was initially implemented by Botstein et al. (1980) in the process of human linkage mapping. It is widely recognized that all genomes contain many areas of neutral variation at the DNA level. These variation sites have no discernible effect on phenotype. A neutral site can refer to either a single nucleotide alteration within or between genes, or it can denote the position of a variable number of repeated sections of nonfunctional DNA located between genes. RFLP markers have sped up molecular linkage map production, improved gene pinpointing, and shortened the time needed to create a comprehensive map for many organisms. Restriction enzymes break pure DNA at recognition site sequences (usually four to eight base pairs [bps]) to create RFLPs. A personalized molecular fingerprint may result from this technique. Cleavage of DNA by an enzyme with a six-base recognition site occurs on average once per 4096 bases if the bases are randomly distributed throughout the genome. About 250,000 restriction fragments of varying lengths might be generated from a genome with 109 bases. When a substantial amount of genomic DNA digestion products is run *via* gel electrophoresis, the resulting image is a continuous smear. Only molecular probes employing the Southern method can disentangle specific pieces that are homologous between many people and perhaps allelic (Southern, 1975).

TABLE 8.2 Comparison of Most Widely Used DNA Marker Systems in Crop Improvement

Sl. No.	Features and Description	RFLP	RAPD	AFLP	SSR	SNP
1.	Genomic abundance	High	High	High	Moderate to high	Very high
2.	Genomic coverage	Low copy coding region	Whole genome	Whole genome	Whole genome	Whole genome
3.	Expression/inheritance	Codominant	Dominant	Dominant/ codominant	Codominant	Codominant
4.	Number of loci	Small (<1000)	Small (<1000)	Moderate (1000s)	High (1000s–10000s)	Very high (>100,000)
5.	Level of polymorphism	Moderate	High	High	High	High
6.	Type of polymorphism	Single-base changes, InDels	Single-base changes, InDels	Single-base changes, InDels	Changes in the length of repeats	Single-base changes, InDel s
7.	Type of probes/primers	Low-copy DNA or cDNA clones	10 bp random nucleotides	Specific sequence	Specific sequence	Allele-specific PCR primers
8.	Cloning and/or sequencing	Yes	No	No	Yes	Yes
9.	PCR-based	Usually no	Yes	Yes	Yes	Yes
10.	Radioactive detection	Usually yes	No	Yes or no	Usually no	No
11.	Reproducibility/ reliability	High	Low	High	High	High
12.	Effective multiplex ratio	Low	Moderate	High	High	Moderate to high
13.	Marker index	Low	Moderate	Moderate to high	High	Moderate
14.	Genotyping throughput	Low	Low	High	High	High
15.	Amount of DNA required	Large (5 –50 μg)	Small (0.01–0.1 μg)	Moderate (0.5–1.0 μg)	Small (0.05–0.12 μg)	Small (≥ 0.05 μg)
16.	Quality of DNA required	High	Moderate	High	Moderate to high	High
17.	Technically demanding	Moderate	Low	Moderate	Low	High

TABLE 8.2 *(Continued)*

Sl. No.	**Features and Description**	**RFLP**	**RAPD**	**AFLP**	**SSR**	**SNP**
18.	Time demanding	High	Low	Moderate	Low	Low
19.	Ease of use	Not easy	Easy	Moderate	Easy	Easy
20.	Ease of automation	Low	Moderate	Moderate to high	High	High
21.	Development/start-up cost	Moderate to high	Low	Moderate	Moderate to high	High
22.	Cost per analysis	High	Low	Moderate	Low	Low
23.	Number of polymorphic loci per analysis	1.0–3.0	1.5–5.0	20–100	1.0–3.0	1.0
24.	Primary application	Genetics	Diversity	Diversity and genetics	All purposes	All purposes

Source: Adapted from Collard et al., 2005; Jiang, 2013; Xu, 2010

8.3.1.1 STEPS OF RESTRICTION FRAGMENT LENGTH POLYMORPHISM ANALYSIS

1. To use a restriction enzyme, a significant amount of DNA must be extracted from multiple individuals of the target genotype (e.g., parents, segregating populations, germplasm survey, and garden blot) and purified to a sufficient degree.
2. Restriction digestion: For the second step, called restriction digestion, isolated genomic DNA is mixed with a restriction enzyme in a buffered environment. The enzyme makes hundreds of thousands of cuts in the DNA at recognition sites.
3. Gel electrophoresis allows DNA restriction fragments to be visualized as smears on an agarose gel.
4. NaOH solution is used to denature the agarose gel, and subsequently, the gel is neutralized.
5. After that, Southern blotting is used to transfer the DNA fragments to a nitrocellulose membrane.
6. The membrane-bound genomic DNA is hybridized with a cloned segment of the target genome or a closely similar species for probe visualization.
7. Then wash the membrane to eliminate any DNA that has hybridized to it by accident. Radiometric techniques are typically used to determine the sizes of the particles.
8. One or more pieces of varying sizes may be identified using the probe-restriction enzyme combinations. When comparing pieces have different lengths, polymorphism becomes apparent.

8.3.1.2 ADVANTAGES OF RESTRICTION FRAGMENT LENGTH POLYMORPHISM

1. The high density of RFLP loci in a mapping population allows for scoring and mapping even small regions of chromosomes.
2. The expression of the gene linked to an RFLP marker is not necessary for its mapping.
3. The third is that they have a high rate of reproduction and are equally dominant in the wild.
4. Support mapping even QTLs.
5. RFLP maps are constructed faster than conventional linkage maps.

8.3.1.3 LIMITATIONS OF RESTRICTION FRAGMENT LENGTH POLYMORPHISM

1. The RFLP method is quite time-consuming and costly.
2. The technology used radioactive probes, requiring specialized disposal arrangements and handling risks. This difficulty can be solved with nonradioactive labelling.
3. Building genomic/cDNA libraries to identify acceptable probes is a time-consuming and complex step in the RFLP development process that requires a high level of expertise.
4. Because RFLP analysis relies on restriction digestion, the DNA utilized must be of very high quality.
5. RFLP scoring takes longer and requires more effort than other molecular markers like SSRs.
6. This marker system cannot automate or do high-throughput analysis. Therefore, RFLPs are no longer commonly used.

8.3.2 RANDOM AMPLIFIED POLYMORPHIC DNA MARKERS

Welsh and McClelland (1990) and Williams et al. (1990) independently described RAPD and AP-PCR as low-stringency PCR (35–45°C) methods for amplification of multiple unique DNA fragments using a single, random-sequence oligonucleotide primer. A PCR on the subject's DNA with a short, unknown-sequence primer is the core of RAPD. Amplifying random sequences from a complex DNA template with a primer that binds to multiple loci or has few mismatches. The length and size of the primer and target genome affect the PCR-generated fragments (Datta et al., 2011). Ten-base oligomers with 40–100% GC. PCR requires homologous (at least 3000 bp) and opposing hybridization sites for amplification. Ethidium bromide staining shows amplified products (up to 3.0 kb) on agarose gels. Target RAPD products can also be cloned and sequenced to yield STS, SCAR, and other PCR-based markers. RAPD marker information must be reproducible and standardized for comparison across laboratories, samples, and trials to be collected and distributed.

8.3.2.1 ADVANTAGES OF RANDOM AMPLIFIED POLYMORPHIC DNA MARKERS

1. Specific primers can be designed without either DNA probes or prior knowledge of the target sequence.
2. Since neither blotting nor hybridization is required, the process is rapid, easy, and effective.
3. RAPDs operate automatically, use only 10 ng of DNA per reaction, and detect higher levels of polymorphism than RFLPs.
4. No special markers need to be developed, and the technology can be used on almost any organism with only a little bit of setup.
5. Primers need not be species-specific; rather, a single set of primers can be utilized across all species.

8.3.3 AMPLIFIED FRAGMENT LENGTH POLYMORPHISM MARKERS

Selective PCR amplification turns DNA RFLPs into observable patterns. AFLP uses selective PCR amplification of restriction fragments from a genomic DNA double-digest by coupling polymorphism at restriction sites with arbitrary primer hybridization (Xu, 2010). Some call AFLP "selective restriction fragment amplification." AFLP primers are 17–21 nucleotides long and consist of a synthetic adaptor sequence, restriction endonuclease recognition sequence, and nondegenerate selection sequence. The primers must anneal appropriately to their target sequences, which include the adapter, restriction sites, and a few nucleotides near the restriction sites, for this approach to operate. AFLP begins with two restriction enzymes digesting 500 ng of genomic DNA, one of which has a 6-bp recognition site (Eco RI, Pts I, or Hind III). For PCR amplification using known sequences, adaptors are ligated to both fragment ends. The double-stranded oligonucleotide adaptors are designed to prevent ligation from reestablishing the restriction site. Thus, the magnified pieces will only be those cut by the frequent or unusual cutter. AFLP methods generate dominant rather than codominant markers, but their time and cost efficiency, replicability, and resolution can match or exceed those of allozymes, RAPD, RFLP, and microsatellites. Replicability and ease of application make AFLP markers popular in systematics, pathotyping, population genetics, DNA fingerprinting, and QTL mapping (Datta et al., 2011).

8.3.4 SIMPLE SEQUENCE REPEAT MARKERS

SSRs, also known as sequence-tagged microsatellite sites or short tandem repeats, are multiallelic PCR-based markers that cover the genome effectively. They are multiallelic, have locus identity, and are hence automatable (Datta et al., 2011). Typically, between 2 and 6 bps in length, these motifs occur in a random tandem fashion. Plant and animal genomes contain a large number of di-, tri-, and tetranucleotide repeats, such as (GT)n, (AAT)n, and (GATA)n. Plants display polymorphism as a result of variations in the number of copies of these repetitions. Primers designed for microsatellite regions can be generated for use in the PCR due to the frequent conservation of DNA sequences surrounding these sites. Microsatellite loci are useful genetic markers because of the high level of allelic variation they often exhibit. Primers that amplify SSR alleles use the distinct sequences flanking SSR motifs as templates. PCR. Each SSR locus is amplified independently by PCR with a set of oligonucleotide primers that recognize a different DNA sequence on either side of the SSR. High-resolution electrophoresis methods like agarose gel electrophoresis (AGE) and polyacrylamide gel electrophoresis (PAGE) can separate PCR-amplified products, and fluorescent labelling or silver-staining can record the bands (Xu, 2010). SSR markers have hypervariability, repeatability, codominance, locus specificity, and an apparently random genome distribution. SSR markers are easy to test by PCR and detect by PAGE or AGE. Automation, high-throughput genotyping, and multiplexing are possible using SSR markers due to their simplicity. Manual SSR testing requires only 100 ng of DNA per person and a small initial investment. SSR manufacturers are the most breeder-friendly marker system, despite the need for nucleotide information for primer production, a lengthy marker generating procedure, and a large initial cost for automated detections. Since the 1990s, SSR markers have been used in plant genetic linkage mapping, QTL mapping, MAS, and germplasm analysis (Jiang, 2013).

8.3.5 SINGLE-NUCLEOTIDE POLYMORPHISM MARKERS

The SNP in genetics is a bp variability. Nucleotide substitutions (SNPs) can be categorized as straight swaps (C/T or G/A) or reverse swaps. Xu (2010) defines SNPs as insertions and deletions of a single base in the genome and differences in the number of bases in a molecule of RNA (cDNA). A single nucleotide base is the smallest unit of inheritance; hence, SNPs provide the most markers most simply and effectively. SNPs are everywhere in plants and animals. SNPs occur 1–3 per 100–300 bps in plants (Edwards et al., 2007; Xu, 2010). SNPs are found in gene coding sequences, noncoding sections, and intergenic regions (Sobrino et al., 2005). Each form of allelic discrimination reaction is detected by gel-electrophoresis, mass spectrophotometry, chromatography, fluorescence polarization, arrays or chips, etc. Sequencing now finds SNPs (Jiang, 2013). SNPs are codominant, connected with genes, and the simplest/ultimate form for polymorphism, making them the best genetic markers. SNPs are increasingly used in plant genome-wide association studies. Commercial SNP assays are available for many species; however, the high costs for start-up or marker generation, high-quality DNA, and high technical/equipment needs limit their use in some laboratories and breeding efforts. SNP chips, a high-density genotyping array with thousands, tens or hundreds of thousands of SNP markers, allow scientists to characterize genetic diversity and linkage disequilibrium and create high-resolution linkage maps to improve whole-genome sequence assembly.

8.3.6 SIGNIFICANCE OF GENETIC MARKERS

According to Helentjaris (1992), DNA markers possess the following advantageous characteristics:

1. The number of distinct marker loci is extremely great, which enables all of the genomic areas to be mapped at extremely high marker densities.
2. The sequence variation being discovered by these markers is, for the most part, innocuous, except in instances in which it is situated in coding regions and has an impact on the functioning of the genes in question.
3. The assessment of a single DNA marker typically does not have any impact on the scores of the other markers, which allows for the evaluation of many markers at the same time.
4. Molecular markers demonstrate a straightforward Mendelian inheritance pattern.
5. The genotyping of markers is not dependent on the environment that is now present.
6. Neither the developmental stage of the plant nor the marker assays is damaging to the plant. Due to the fact that the characteristic phenotypes are not examined, MAS can be utilized successfully in every setting and at any stage of genetic development.

7. MAS can also be utilized for an allele that is not expressed in the available genotypes, such as a recessive allele in heterozygotes.
8. It is possible to preserve the DNA samples for use at a later time.
9. It is not necessary to have predetermined marker stocks.

8.3.7 UTILIZATIONS OF GENETIC MARKERS

Molecular markers possess a diverse range of applications, encompassing the following:

1. Identification of strains and varieties through the use of fingerprints (unambiguous).
2. The mapping of genes and QTLs is the second step.
3. Cost-effective methods for identifying tightly related QTLs and other oligogenes, for which direct selection may be prohibitively expensive or troublesome.
4. Positional cloning of genes and QTLs.
5. The identification of chromosomal segments that would participate in the enhancement of the features that are being investigated.
6. Establishing phylogenetic links between the many strains and species of organisms under study.
7. Determine which parents will be used for hybridization.
8. Assessing the basis of somaclonal variance.
9. The determination of the races and biotypes of the parasites.
10. Make predictions on heterotic cross pairings.
11. Identifying wide hybrids.
12. Gene pyramiding.
13. Management and utilization of genetic resources.
14. The use of off-season nursery and greenhouse facilities is permitted by MAS, which facilitates a reduction in the amount of time required for the development of varieties

8.4 NEEDS OF MARKER-ASSISTED BREEDING PROGRAM

A more sophisticated infrastructure is required for DNA MAB than for conventional breeding approaches. According to Jiang (2013), MAB in plants to be successful, all the required ingredients are listed below.

8.4.1 EFFECTIVE MARKER SYSTEM AND RELIABILITY MARKERS

To initiate a MAB program for a plant species or crop, it is essential to have access to a suitable marker system and dependable markers. Based on the information we have discussed so far, effective markers should possess the following characteristics: (1) affordability and simplicity in analysis; (2) requiring a small amount of DNA; (3) exhibiting codominance; (4) being able to produce consistent results; (5) having a high level of genetic variation; and (6) being widely present and distributed (Jiang, 2013). Additionally, it would be highly advantageous if the markers were closely associated with the genes of interest. The selection of markers can ensure the successful identification of the gene, provided that they are located in close proximity to the target gene or are integral components of the gene itself.

8.4.2 RAPID EXTRACTION OF DNA AND EFFICIENT DETECTION OF GENETIC MARKERS AT A LARGE SCALE

A quick DNA extraction method and high-throughput marker detection apparatus are needed to screen multiple markers in many tissue samples (Jiang, 2013). Many labs and programs utilize 96- or 384-well plates for DNA extraction, which is more efficient and processes smaller tissue samples. High-throughput PAGE and AGE equipment are standard for marker detection. Several labs offer marker detection services using automated detection methods like SNP chips with thousands of markers.

8.4.3 UNDERSTANDING MARKER-TRAIT ASSOCIATION

Markers that strongly correspond to attributes that are more effective and efficient. Gene mapping, QTL analysis, association mapping, genome-wide association research, classical mutant analysis, linkage or recombination analysis, bulked segregants analysis, etc., can provide this data. Additionally, it is crucial to grasp the linkage situation, or the markers that are coupled or repelled (cis/trans) to the trait of interest allele. Even if numerous markers are strongly linked to a QTL, a plant breeder must determine allele linkage in his breeding material. QTLs' effects vary by context, unlike genes. This makes QTL information transmission between materials difficult.

8.4.4 GENETIC MAP CONSTRUCTION

Linkage maps facilitate the identification of marker-trait correlations and aid in the selection of markers for MAB. Thus, MAB must possess a genetic linkage map, with a specific emphasis on a high-density linkage map (Jiang, 2013). If a trait-related region is discovered, precise mapping using additional markers can pinpoint the marker(s) most closely connected with the gene(s) responsible for that trait. An ideal genetic map would include enough uniformly spaced polymorphic markers to pinpoint the locations of target QTLs and genes (Babu et al., 2004).

8.4.5 RAPID AND EFFECTIVE PROCESSING OF DATA AND MONITORING

Breeders can derive advantages through assessments that are prompt and valuable, provided that data is handled and maintained with speed and efficiency. Having a fast and efficient system for classifying, storing, retrieving, handling, and examining large data sets, as well as merging sets of data from other systems, is of utmost importance. Recent advancements in bioinformatics and statistical tools provide valuable assistance in this regard (Jiang, 2013).

8.4.6 ECONOMIC COMPONENTS

Setting up an MAB laboratory with essential equipment such as a DNA extraction instrument, PCR machine, electrophoresis system, gel imaging system, etc., can be expensive in comparison to conventional breeding operations. If there is no MAB laboratory or equipment available within the breeding program, an alternative option is to send DNA samples or plant tissue samples to a specialized laboratory that provides MAB services, such as marker screening or genotyping.

8.5 THE PROCESS OF MARKER-ASSISTED BREEDING

MAB encompasses the following actions, assuming that the necessary requirements are adequately provided:

1. The initial stage involves planting the breeding populations that likely exhibit segregation for desired features or polymorphism for the markers used.
2. Extracting a tissue specimen from a plant, usually during its early stages of growth, ranging from emergence to the seedling phase.
3. The process involves extracting DNA from tissue samples obtained from each individual or family in the communities to prepare DNA samples for PCR and marker screening.
4. The molecular markers associated with the trait of interest are amplified using PCR or other methods.
5. Utilizing suitable separation and detection methods, such as PAGE, AGE, etc., to isolate and quantify PCR/amplified products.
6. Identifying families and individuals who possess the optimal marker alleles.
7. Employing marker data and additional selection criteria to identify the most suitable individuals and families that possess the marker alleles associated with the intended target qualities, while also displaying the preferred performance and phenotypes for other aspects of interest.
8. The aforementioned steps are iterated until stable superior or elite lines with improved traits are established. The duration of this process is contingent upon the correlation between the markers and the attributes, as well as the condition of the marker alleles (homozygous or heterozygous).

8.6 THE UTILIZATION OF MARKER-ASSISTED SELECTION IN CROP BREEDING

Molecular markers have a wide range of applications in improving field crops, fruit, and vegetable crops through innovative breeding schemes such as MAS. These applications include studying genetic diversity and identifying varieties, tagging genes, identifying disease resistance, analyzing pedigrees, detecting hybrids, differentiating sexes, and assisting in selection processes (Figure 8.3). With the use of MAS, the breeder can make earlier selections by examining a smaller number of plants for selection. Furthermore, once marker information becomes accessible, it would be possible to selectively breed for disease resistance even in the absence of the pathogen. QTLs have been provisionally identified and located for the majority of significant characteristics in various fruits. The polygenic character and extended juvenile stage of fruit trees provide significant challenges for traditional breeding programs. The utilization of molecular markers and genetic maps in conjunction with traditional approaches offers an alternate strategy to expedite the breeding process. The extensive utilization of molecular markers in crop enhancement has been examined in Table 8.3.

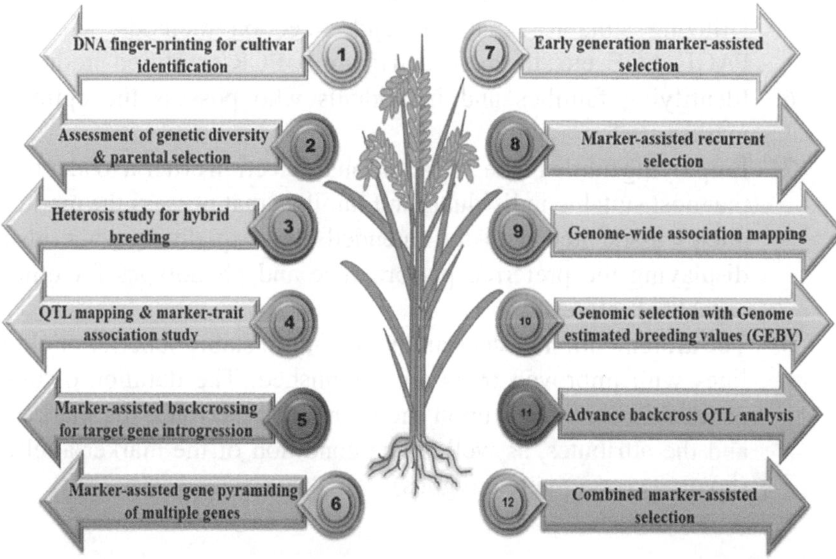

FIGURE 8.3 The novel breeding strategies of marker-assisted selection (MAS) for enhancing crop quality and productivity.

8.6.1 MARKER-ASSISTED SELECTION FOR THE ENHANCEMENT OF QUALITATIVE TRAITS

Key genes/QTLs in agricultural plants control a wide range of economically important characteristics, including disease and insect resistance,

TABLE 8.3 Different Molecular Markers and Their Application in Vegetable and Fruit Crops Improvement

Sl. No.	Crop	Trait Improved	Gene/QTLs	Types of Markers Used	References
			Field Crops		
1.	Rice	Black-streaked dwarf virus disease and small brown planthopper resistance	Black-streaked dwarf virus: *qRBSDV3, qRBSDV6, qRBSDV9, and qRBSDV11* BPH: *qSBPH1, qSBPH5, qSBPH8,* and *qSBPH9*	SSR	Sun et al. (2017)
		Blast resistance	*Pi68*	SNP and SSR	Devi et al. (2020)
		Drought tolerance	$DTY_{1.1}$, $DTY_{2.1}$, and $DTY_{3.1}$	SSR	Majhi et al. (2022b); Singh et al. (2022)
2.	Maize	Maize stem borers	The Embu site has a 3.37 LOD score, while Kakamega has 2.68. A suggestive QTL is also on chromosome 1 at the Kakamega location	SNPs	Munyiri and Mugo (2017)
3.	Sorghum	Sorghum shoot fly resistance	qDH9.1 (dead-heart) and qEC9.1 (oviposition) shared chromosome 9	SSR	Vikal et al. (2020)
4.	Cotton	Resistance to bollworm and spider mite in cotton	*GhPAP1D*	SSR and SNP	Li et al. (2019)
5.	Chickpea	Pod borer resistance	*qUlwv01* and *qLpp03*	SNP	Barmukh et al. (2021)
6.	Mungbean	Bruchid tolerance	*qBrc5.1* and *qBrm5.1*	SSR	Kaewwongwal et al. (2017); Majhi et al. (2022a)
7.	Wheat	Wheat Aphid Resistance	*Dn1, Dn2, Dn5, Dn6,* and *Dnx*	SSR	Tolmay et al. (2020)
		Stripe Rust Resistance	*Yr* genes	SSR	Liu et al. (2020)

TABLE 8.3 *(Continued)*

Sl. No.	Crop	Trait Improved	Gene/QTLs	Types of Markers Used	References
8.	Sugarcane	Sugarcane aphid resistance	*RMES1*	SNPs	Cuevas et al. (2022)
		red rot resistance	*Chitinase A gene*	SNP	O'Connell et al. (2022)
9.	Mustard	Alternaria blight resistance	*NBS-LRR gene*	SNP	Inturrisi et al. (2018)
10.	Lentil	Ascochyta Blight Resistance	*AbR1*	SNP	Sudheesh et al. (2016)
			Vegetable Crops		
11.	Tomato	Late blight	*Ph-3*	CAPS	Robbins et al. (2010);
		and yield		SNP	Brekke et al. (2019)
		Yellow leaf curl virus	*Ty-2*	SCAR	Garcia et al. (2007);
				ACY (InDel)	Sahu et al. (2019
		Bacterial wilt	*Bwr-6* and *Bwr-12*	CAPS SNP	Kim et al. (2018); Tiwari et al. (2022)
		Root knot Nematode resistance	*Mi-1*	SCAR	Arens et al. (2010)
		Leaf mold	*Cf-4/Cf-9*	SNP/InDel	Tiwari et al. (2022)
		Powdery Mildew	*ol-2*	RAPD	De Giovanni et al. (2004)
		ToMV	*Tm-1*	SCAR	Arens et al. (2010)
		Tomato spotted wilt virus	*Sw-5*	SCAR/CAPS/ RAPD	Zhang and Panthee (2021)
		Male sterility	*ms-15* (*TM6 gene*)	SNP	Tiwari et al. (2022)
		Early flowering	*QTL*	SNP	Ruangrak et al. (2018)
12.	Potato	Nematode resistance to *M. chitwood*	*QTL*	SNP	Bali et al. (2022)
		Late blight	*R11*	RAPD	Taoutaou et al. (2022)

TABLE 8.3 *(Continued)*

Sl. No.	Crop	Trait Improved	Gene/QTLs	Types of Markers Used	References
13.	Cucumber	Angular leaf spot	*psl*	SNP08	Wang et al. (2019)
		Downy mildew	*dm*	SNP08	
		Powdery mildew	*Pm-s*	pmsSSR27 pmSSR17	Liu et al. (2017)
14.	Brinjal	Resistance to Ralstonia solanacearum phylotype I	*ERs1* (*EBWR9*)	SNPs	Salgon et al. (2017)
		Fusarium wilt	*Fm1*	SSR	Miyatake et al. (2016)
		Parthenocarpy	*Cop8.1*	SSR	Miyatake et al. (2012)
15.	Onion	Purple blotch resistance	*ApR1*	SSRs and STS	Chand et al. (2018)
16.	Pepper	CMV resistance	*Cmr1*	CAPS and HRM	Kim et al. (2017b)
		Bacterial spot	*Bs2*	AFLP and SCAR	Kim et al. (2017b)
		Anthracnose	*AnRGO5*	SCAR	Zhao et al. (2020)
17.	*Brassica rapa* spp. pekinensis	Clubroot resistance	*CRb*	TCR74 and TCR79	Zheng et al. (2022)
18.	Lettuce	Leafminer resistance	*qSTP*	QTLs	Kandel et al. (2021)
19.	Pumpkin	Early flowering	*FLT* and *FLT2*	InDels and SSR (InDel2507 and InDel6146)	Abbas et al. (2022)
			Fruit Crops		
20.	Apple	Fire blight resistance	–	SCAR and SSR	Keller-Przybyłkowicz et al. (2009)

TABLE 8.3 *(Continued)*

Sl. No.	Crop	Trait Improved	Gene/QTLs	Types of Markers Used	References
21.		Scab	*Vb*	SSR	Erdin et al. (2006)
22.	Citrus	Citrus leprosis virus resistance	–	AFLP and RAPD	Bastianel et al. (2009)
		Cold tolerance	–	SSR and AFLP	Abouzari et al. (2020)
		Citrus nematode	*Tyr 1*	RAPD	Ling et al. (2000)
		Citrus tristeza virus	*Ctv-R*	RAPDs	–
23.	Pear	Fruit traits	–	AFLP and SSR	Zhang et al. (2013)
		Fruit skin colour	–	RAPD	Inoue et al. (2006)
24.	Peach	Gummosis resistance	–	SNPs	Hoyos-Villegas et al. (2022)
		Root-knot nematode resistance	*Mj*	STS-834b	Yamamoto et al. (2004)
		Resistance to green peach aphid	*Rm2*	SNP	Lambert et al. (2016)
25.	Prunus species	Sharka disease (plum pox virus [PPV])	*PPVres* locus	SSR	Soriano et al. (2012)
26.	Mango	Anthracnose	*β-1* and *3-GLU2*	SNP	Felipe et al. (2022)
27.	Guava	High vitamin C content	–	SRAP	Youssef and Ibrahim (2016)
28.	Grapes	Seed-lessness, berry size, and ripening date	–	AFLP, SSR, RAPDs, ISSRs, and SCARs	Mejía et al. (2007)
29.	Papaya	Papaya ring spot virus	*PRSV-P*	CAPS	Dillon et al. (2006)
30.	Coconut	Root wilt	*PR1* and *PR4*	SNP/InDel/ SSR	Verma et al. (2017)

male sterility, self-incompatibility, as well as qualities relating to the overall shape, color, and structure of plants or specific plant parts. These traits are frequently inherited from one generation to the next in a household with either one or two parents. However, many quality metrics exhibit significant phenotypic variation that can be attributed to a small number of critical QTLs or genes (Bilyeu et al., 2006; Pham et al., 2012). Therefore, the transfer of this gene to a specific lineage might greatly augment the trait in the growing crop. Indirect selection at marker loci that have functional connections to target genes may, in certain situations, be more efficient than direct selection at the target genes. A potential reason for the increase in efficiency when a marker is situated within a gene is that the marker mRNA is expressed at a greater magnitude. When there is difficulty in detecting a SNP between two alleles of a gene of interest, using an external marker with a more easily detectable polymorphism may be the most practical alternative. Consider the soybean cyst nematode (SCN; *Heterodera glycines*) as an illustration of a major-gene MAS technique. This fungal infection is responsible for the greatest amount of harm to soybeans in the commercial sector compared to any other recognized pathogen. Utilizing pest-resistant cultivars is the main approach for reducing or eradicating the harm caused by this bug (Cregan et al., 1999). Detecting resistant individuals in breeding populations is a difficult and expensive endeavor. Processing 100 plant samples using a regularly used phenotypic assay (Young, 1999) requires 5 weeks, a significant amount of greenhouse space, and approximately 5–10 hours of work. The identification of the SSR marker Satt309, located within a range of 1–2 cM from the *rhg1* resistance gene, has led to the development of numerous breeding programs in both public and commercial sectors (Cregan et al., 1999). According to Young (1999), the application of genotypic selection with Satt309 accurately predicted SCN-susceptible lines in later greenhouse testing for two test populations from the same source of SCN resistance. However, for a third population from a different source of SCN resistance, the accuracy was only 80%. Furthermore, Shi et al. (2009) employed molecular markers in a J05 V94-5152 cross to generate five F4:5 lines that exhibited homozygosity for all eight marker alleles linked to the genes/loci responsible for resistance to soybean mosaic virus (SMV). The lines exhibited resistance to SMV strains G1 and G7 and were likely to possess all three resistance genes (*Rsv1, Rsv3,* and *Rsv4*), which could confer broad and long-lasting resistance to SMV (Shi et al., 2009).

8.6.2 MARKER-ASSISTED SELECTION FOR ENHANCING QUANTITATIVE ATTRIBUTES

The majority of important agronomic variables are affected by a multitude of QTLs or are determined by many genes. The inclusion of a substantial number of genes or QTLs, along with the consideration of QTLE interaction and epistasis, renders the use of MAS for improving these qualities a difficult and intricate undertaking (Jiang, 2013). The manifestation of these features is often regulated by a mix of environmental variables and the minute influence that each of these genes exerts on the phenotypic expression of the characteristic. The effectiveness of MAS is reduced in the presence of QTLE interaction, as epistasis might lead to a distorted influence of QTL on the trait. Defining the relationship between marker phenotypic and quantitative factors is inherently problematic. Iterative field investigations are necessary to accurately define the effects of the QTLs and assess their consistency under different situations. Although there has been extensive experimentation in QTL mapping in the past ten years, the practical use of QTL mapping information in plant breeding has been restricted due to several issues. These challenges consist of a substantial interaction between QTLs and the environment, which complicates phenotyping due to variations in expression across different locations and years. Additionally, there is a dearth of generally applicable relationships between QTLs and markers that can be used across different populations. Some breeders lose interest in MAS because each time new germplasm is utilized, a new project of QTL mapping must be initiated to identify new QTL markers. Additionally, deficiencies in QTL statistical analysis can result in either overestimation or underestimation of the number of QTLs involved and their effect on the trait. Furthermore, it is common to find that there are no QTLs with significant effects on the trait, necessitating the identification of a large number of QTLs. Accomplishing this process can be difficult at times and increases the intricacy of determining the relationship between marker QTLs.

Practical breeding for enhancing a quantitative feature involves using more minor QTLs than are typically addressed in MAS. In certain cases, the utilization of the gene pyramiding strategy may be required to incorporate several QTLs, particularly those with minimal effects. This technique is favored by breeders in such circumstances. Fusarium head blight (FHB), also referred to as scab, is a very destructive disease that affects wheat and barley. Since the 1990s, scientists from various fields such as plant breeding, genetics, molecular genetics, genomics, plant pathology, and integrated management have been diligently collaborating to discover a resolution to this issue. QTLs for FHB resistance in wheat and barley have been identified in various germplasm collections (Buerstmayr et al., 2009; Jiang et al., 2007a, 2007b). For numerous breeding efforts, MAS has emerged as a feasible method for enhancing resistance. The wheat QTL, *Fhb1*, was consistently identified in several habitats and populations, and it was determined to explain 20–40% of the phenotypic variation in wheat (Buerstmayr et al., 2009; Jiang et al., 2007b). Thus, wheat breeders aim to utilize this prominent QTL in order to develop new cultivars that possess resistance to FHB. The mean decrease in disease severity among NIL pairs was 23% for disease severity and 27% for kernel infection. These values were obtained by comparing 19 NIL pairs for *Fhb1* from a continuous breeding project conducted by Pumphrey et al. (2007). Follow-up investigations conducted by the same group of researchers validated the effectiveness of MAS for this specific QTL (Anderson et al., 2007). In addition, MAS was employed to incorporate several QTLs in order to enhance resistance to FHB. Miedaner et al. (2006) demonstrated that targeting three FHB resistance QTLs simultaneously in MAS resulted in a considerable enhancement in FHB resistance in German spring wheat. The most FHB-resistant recombinant lines were created by combinations of different QTLs, particularly 3B and 5A.

Jiang et al. (2007a, 2007b) conducted a multiple-locus combination analysis on a recombinant inbred line (RIL) population obtained by crossing Veery with CJ 9306. Except for the nonreciprocal comparisons, the average degrees of resistance in genotypes for the three loci varied from the lowest to the greatest as follows: genotypes with no advantageous allele, genotypes with one favorable allele, genotypes with two favorable alleles, and genotypes with three favorable alleles. The coefficients of determination between the combined effects of alleles for various combinations and the average number or percentage of diseased spikelets for the corresponding RILs ranged from 0.33 to 0.41 ($P < 0.01$) when only four or five loci containing beneficial alleles from the resistant parent CJ 9306 were taken into account (Jiang et al., 2007a). The scientists hypothesized that by deliberately choosing beneficial genetic markers at many locations in breeding programs, the effects of QTLs resistant to FHB could be combined and the resistance may be significantly enhanced.

Asoro et al. (2013) recently conducted a comparison of two cycles of selection for glucan concentration in oat. They used marker-based selection methods and the best linear unbiased prediction (BLUP) based phenotypic selection (PS). In Cycle 2, marker-based selection procedures exhibited significantly higher averages compared to PS methods. Furthermore, the offspring resulting from marker-based selection approaches demonstrated the highest amounts of glucan. Furthermore, the utilization of marker-based selection strategies showed a higher retention of genetic variability compared to BLUP PS. This suggests that marker-based strategies have the potential to yield greater selection advantages in the future, as reported by Asoro et al. (2013).

8.6.3 MARKER-ASSISTED BACKCROSSING FOR GENE INTROGRESSION

MABC is the prevailing and efficient technique in contemporary molecular breeding. It is also the most straightforward type of MAS (Jiang, 2013). The objective of MAB for improvement is to augment a desirable trait by transferring one or more genes/QTLs from a genetic source (the donor parent, which may have inferior agronomic qualities or lack comprehensive performance) into a superior cultivar or elite breeding line (the recurrent parent). Unlike conventional backcrossing, MABC does not select markers based on their phenotypic performance. Instead, markers are chosen for their association with the specific gene(s) or QTL(s) of interest. In order to effectively execute MABC, breeders need to initially ascertain the extent to which markers offer substantial enhancements compared to conventional backcrossing or act as valuable complements to traditional breeding endeavors (Jiang, 2013). MABC is crucial and beneficial in situations where phenotyping is challenging, costly, or unfeasible; the heritability of the desired trait is minimal; the trait is manifested in the later stages of plant growth and development, such as flowers, fruits, and seeds; the traits are governed by genes that necessitate specific conditions for expression; the traits are controlled by recessive genes; and gene pyramiding is necessary for one or more traits.

So far, molecular analysis-based selection (MABC) has shown to be the most widely used and successful molecular breeding approach for plants. Several species, such as rice, wheat, maize, barley, pearl millet, soybeans, and tomatoes, have experienced positive effects from its application.

This has led to improvements in various characteristics, including disease and pest resistance, drought tolerance, and overall quality (Collard et al., 2005; Dwivedi et al., 2007; Xu, 2010). *Bacillus thuringiensis*, a bacterium present in maize, synthesizes insecticidal toxins that, upon ingestion by maize borer larvae, result in their mortality (Ragot et al., 1995). MABC has been employed to deliver the Bt transgene into various maize genetic backgrounds. The recessive genetics of rice aroma can be attributed to an eight-base-pair deletion and three SNP in the *betaine aldehyde dehydrogenase-2* gene (Bradbury et al., 2005a). This discovery differentiates between individuals with homozygous recessive and dominant traits, as well as those with heterozygous traits, in populations that are segregating for the specific characteristic. Additionally, it aids in the identification of rice varieties that possess aromatic and nonaromatic properties. The application of MABC for the purpose of aroma selection in rice was described by Bradbury et al. (2005b). The high *lysine opaque-2* gene was introduced into maize using the method of MABC as described by Babu et al. (2005). When endeavoring to incorporate a substantial number of QTLs, the likelihood of achieving success tends to diminish. Sebolt et al. (2000) found two QTLs for soybean seed protein content using MABC. However, only one QTL was validated in BC3 F4:5. Out of the three genetic backgrounds that were evaluated, only one exhibited a response to the insertion of the QTL. Advanced backcross-QTL (AB-QTL) is a method developed by Tanksley and Nelson (1996) to transfer resistance genes from wild or unadapted tomato genotypes into high-quality germplasm. This approach has had a positive impact on some critical characteristics of tomatoes, such as fruit quality and resistance to black mold (Fulton et al., 2002). The use of AB-QTL has shown the successful transfer of advantageous genes from wild/unadapted germplasm to elite germplasm in many crop species such as rice, barley, wheat, maize, cotton, and soybean (Wang and Chee, 2010). Resistance to barley yellow dwarf virus was achieved by employing a backcrossing approach, wherein a marker closely associated (0.7 cM) with the *Yd2* gene was utilized (Jefferies et al., 2003). Zhao et al. (2012) reported the successful integration of the key QTL (qHSR1) for resistance to head smut in maize into ten high-yielding inbred lines. (Prone to head smut infection). After crossing with the donor parent Ji 1037, which contains the *qHSR1* gene and is completely resistant to head smut, and performing five generations of backcrossing to the original parents, we successfully obtained ten lines with high yield.

Wan et al. (2016) employed DNA markers to develop a new strain of rice (KR022) that contains a combination of BPH-resistance genes *Bph14* and *Bph15*, *Bt* gene *cry1C*, and glufosinate-resistance gene bar.

8.6.4 MARKER-ASSISTED GENE PYRAMIDING

A key use of DNA markers in plant breeding is gene pyramiding. Marker-assisted gene pyramiding (MAGP) selects three or more genes simultaneously to improve disease and pest resistance (Jiang, 2013). Ice pyramids defend against bacterial blight and blast (Luo et al., 2012; Singh et al., 2001). Castro et al. (2003) pyramided qualitative genes and QTLs to boost barley's stripe rust resistance. Selecting QTL-allele-linked markers with the same phenotype is possible. Pyramiding multiple genes or QTLs may improve a quantitatively inherited characteristic in plant breeding (Richardson et al., 2006). Multiple-QTL pyramiding increases wheat, barley, and soybean yields (Jiang et al., 2007a; Li et al., 2010; Richardson et al., 2006; Wang et al., 2012). Pyramiding several genes/QTLs can be done using multiple-parent crossing, difficult crossing, backcrossing, and recurrent selection. The optimal MAGP breeding scheme depends on the heritability of traits of interest, the number of genes/QTLs needed for improvement, the number of parents carrying them, and other factors like marker-gene association, expected duration to complete the plan, and relative cost. If three or four desirable genes/QTLs exist independently in three or four lines, pyramiding can be performed through three-way, four-way, or double crossing. They can be combined using convergent or stepwise backcrossing. If pyramiding more than four genes/QTLs, a breeder may use sophisticated crossover or recurrent selection. The genes *xa13* and *Xa21* for BB resistance, *Pi54* for blast resistance, and *qSBR11-1* for a major QTL have been combined through marker-assisted backcross breeding in rice for the first time, resulting in the first transfer of genes against three diseases in a single crop (Singh et al., 2012). Pyramided lines will donate BB, blast, and ShB genes and perhaps become cultivars in Basmati rice breeding.

Three breeding systems may be used for MABC-based gene pyramiding:

1. Step-by-step transfer
2. Synchronized/simultaneous transfer
3. Convergent transfer or backcrossing

Assuming cultivar "W" is superior in overall performance but lacks a trait of interest, four genes/QTLs (desired gene/marker alleles designed as A, B, C, and D) contributing to the trait have been identified in four germplasm lines (P1, P2, P3, and P4).

8.6.4.1 STEPWISE GENE PYRAMIDING

The step-wise MABC method entails sequentially introducing four target genes/ QTLs into the repeating parent "W." To begin, parent W is mated with each of the four donor parents, resulting in the production of four single-cross F1 hybrids. These hybrids are then used in the backcrossing process for the first gene/QTL. When two out of the four single-cross F1s are crossed with each other, two double-cross F1s are produced. Crossing these two double-cross F1s again results in a hybrid that contains all four target genes/QTLs in a heterozygous manner. Once all four target genes/ QTLs have been effectively incorporated into the RP, the hybrid and/or offspring that possess heterozygous markers for all four are bred with the RP "W." Due to the fact that only one gene/QTL is being examined at a time, the necessary population size and the number of genotypes to be analyzed are reduced, thereby making gene pyramiding a more viable and precise technique. If two or three of the integrated genes/QTLs meet the criteria before the final stage, the improved recurrent parent can be approved for release (Jiang, 2013). Time is a concern as it prolongs the procedure.

8.6.4.2 SIMULTANEOUS OR SYNCHRONIZED MARKER-ASSISTED BACKCROSSING

The process of self-fertilization that occurs repeatedly and is completed by a single generation in the synchronized or simultaneous MABC method (Jiang, 2013). This approach has the benefit of requiring the least amount of time to finish. Backcrossing, on the other hand, involves all target genes/ QTLs simultaneously, necessitating a sizable population and extensive genotyping.

8.6.4.3 CONVERGENT MARKER-ASSISTED BACKCROSSING

The novel method, convergent MABC, combines the benefits of stepwise and synchronized backcrossing. After single-crossing donors bearing the four-target gene/QTLs into the recurrent parent W, markers related to the genes/QTLs are used to back-cross four better lines (WAA, WBB, WCC, and WDD). Pyramiding all four genes/QTLs, crossing two lines, then crossing the hybrids again creates a better line (e.g., WAABBCCDD). Convergent MABC is recommended because it saves time (compared to stepwise transfer) and ensures gene fixation and/or pyramiding (compared to simultaneous transfer) (Jiang, 2013). We investigated theoretical issues and MABC's gene pyramiding efficacy using computational simulations (Servin et al., 2004; Ye and Smith, 2008). For practical gene pyramiding, rice, wheat, barley, cotton, soybean, common bean, and pea have been used in MABC. Creating long-lasting crop stress tolerance is extremely important (Jiang, 2013). This strategy introduced commercial cultivars, although specifics are scarce. Joshi and Nayak (2010) and Xu (2010) studied marker-based gene pyramiding and provided examples. Somers et al. (2005) molecularly bred Canadian wheat with pest-resistant genes. They back-crossed and selected six FHB resistance QTLs, the orange blossom wheat midge resistance gene *Sm1*, and the leaf rust resistance gene *Lr21* using high-throughput SSR genotyping and half-seed analysis (Jiang, 2015).

8.6.5 MARKER-ASSISTED RECURRENT SELECTION

The process of selecting or pyramiding numerous genes/QTLs using markers is challenging and lacks substantial evidence to support its effectiveness. Recurrent selection is widely acknowledged as an effective strategy for enhancing polygenic traits. PS heavily depends on environmental circumstances, while genotypic selection takes longer (at least two to three crop seasons for one selection cycle). Consequently, the effectiveness and efficiency of selection may not always be sufficient (Jiang, 2013). The technique known as MARS was devised to simultaneously achieve genotypic selection and intercrossing during a single crop season. Consequently, MARS can enhance the effectiveness of recurrent selection and accelerate the process (Jiang et al., 2007a), particularly in the integration of several advantageous genes/QTLs from different sources through recurrent selection using a multiparent population. MAS is suggested for the advancement of native genes and the combination of many QTLs to enhance complex traits such as grain yield, tolerance to biotic and abiotic factors. Intrapopulation breeding involving selected parents and their children (Crosbie et al., 2006; Eathington, 2005; Ribaut et al., 2010). Molecular MAS utilizes molecular markers to identify and select specific genomic regions that contribute to the expression of complex traits, to construct the most high-performing genotype within a single population or across related populations (Ribaut et al., 2010). Johnson (2004) provided an example of the effectiveness of MARS in analyzing quantitative features. Implementing markers in biparental populations on a large scale can enhance the efficiency of long-term selection in maize MARS programs. This involves using markers for QTL detection and subsequently employing MARS on yield, which entails rapid cycles of recombination and selection based on associated markers for yield. The primary benefit of this approach is the increased frequency of favorable alleles. Studies conducted by Crosbie et al. (2006) and Eathington (2005) have shown that the genetic gain achieved through the use of MARS in maize is about double that of PS in certain reference populations. The efficacy of MARS in providing resistance to Helicoverpa armigera in upland cotton has been demonstrated by Yi et al. (2004). Based on their research, the average levels of resistance in improved populations after recurrent selection were significantly higher than those of previous groups.

8.7 ADVANTAGES OF MARKER-ASSISTED SELECTION

1. Selection for features whose phenotypic evaluation is difficult, time-consuming, damaging, and/or dependent on threshold circumstances is considerably aided by foreground selection.
2. The use of MAS enables the sequential backcrossing process to be carried out while incorporating recessive genes.
3. The use of MAS allows for the selection of desirable traits to be conducted in nurseries and greenhouses that operate outside of the regular growing season. This leads to the production of two to four generations of harvestable crops every year.
4. The practice of MAS can be implemented during the seedling stage of the crop. This facilitates the crossbreeding of chosen plants within the same generation, particularly when the desired traits are present in the fruits and seeds.
5. In the context of MABC, the use of MAS might expedite the reestablishment of the genotype of the parent that occurs most frequently (known as background selection).

6. Recombinant selection helps to decrease or eliminate the negative effects of gene transfer from nonadopted germplasm, known as linkage drag.
7. Horizontal resistance selection is facilitated by the use of MAS, which also allows for the combination of many genes and traits.
8. Combining oligogenic and polygenic resistance through MAS results in a more durable and enduring resistance to diseases. Without the implementation of multiagent systems, this objective will remain unattainable.
9. It is possible to successfully accumulate QTLs for the target traits by combining MAS with recurrent selections as MARS.
10. It is possible that the most far-reaching subfield of MAS is GS. The use of GS could prove to be a very efficient method for enhancing low-heritability characteristics.

8.8 LIMITATIONS OF MARKER-ASSISTED SELECTION

1. There are a few target features for which there are tightly related, broadly applicable, and consistently diagnostic indicators.
2. It is essential to independently confirm and validate the marker-trait associations discovered in one community by conducting the same analysis in unrelated populations. This is vital for the finding of QTLs.
3. Selecting for broad genomic areas can be difficult during QTL introgression. This is due to the fact that QTL confidence intervals span 10 cM.
4. The expression of QTLs introduced into different genetic backgrounds is frequently ambiguous. The incorporation of QTL for complex traits such as yield often results in unsatisfactory enhancements, and in certain cases, even negative effects are observed.
5. As a result of interactions between QTL and the surrounding environment, QTL transmission is more difficult.
6. A seven-fold increase in data generation from a breeding program due to the use of MAS means more work for breeders.
7. When using MAS, breeding program decisions are more involved and made more frequently (three to four times yearly vs once or twice annually).

8. To accomplish marker genotyping during a single growing season, it may be necessary to have high-throughput genotyping facilities.
9. MAS generally raises the cost of the breeding program as a whole. The high price of marker genotyping has been the primary barrier to implementing MAS in plant breeding efforts, especially in less economically developed regions of the world.
10. Unfortunately, many breeding operations may not have access to reliable off-season/greenhouse facilities necessary to reap the full benefits of MAS.

The utilization of molecular technology in plant breeding continues to encounter several limitations:

1. It is crucial to acknowledge that not all markers are conducive to breeding. The solution may lie in replacing the problematic markers with more beneficial ones for breeders, as transitioning from RFLP to STS and from RAPD to SCAR.
2. Furthermore, it should be noted that not all markers exhibit dependable marker-trait relationships or possess sufficient polymorphism to be applicable across diverse populations. Enhanced understanding of the variety of genetic markers and the impact of genetic background can be achieved by employing different mapping populations. In addition, breeders should reassess the placements and effects of QTLs in their specific germplasm of interest to ensure precision and correctness (Heffner et al., 2009).
3. Additionally, recombination between the markers and the genes/QTLs of interest can occur, resulting in erroneous selection. Flanking markers or additional markers can be used to provide a reliable selection for the target gene/QTL.
4. Furthermore, the development process is delayed due to the imprecise forecasts of QTL locations and their impacts. QTL identification efficiencies are influenced by various aspects such as algorithms, mapping procedures, polymorphism marker concentrations, and population characteristics (Wang et al., 2012). Improved accuracy in identifying QTL locations and their effects can be achieved by doing fine mapping using a high density of genetic markers in large populations, together with well-designed phenotyping across a range of settings.

5. Additionally, to be effective, MAB schemes and methods must be easily accessible to plant breeders. Computer-generated multi-armed bandit techniques that disregard real-world limitations are unlikely to be extensively adopted in practical breeding initiatives.
6. Plant breeders often work with numerous segregating populations, each consisting of hundreds or even thousands of individuals. Completing tissue sample collection, DNA extraction, marker screening, and data processing within a specific timeframe is highly difficult when resources are scarce. Limited finances and inadequate infrastructure hinder the extensive adoption of MAB in numerous breeding operations.
7. Furthermore, numerous organizations are apprehensive about the escalating expenses associated with staff and initial investments. Therefore, similar to other innovative methods of plant breeding, such as transgenic breeding or genetic manipulation, MAB is not capable of completely replacing conventional breeding. Instead, it functions as a supplementary tool to conventional breeding (Jiang, 2013).
8. The high cost of MAB and the requirement for specialized training and equipment will hinder its widespread usage shortly, particularly in impoverished countries (Collard and Mackill, 2008; Ribaut et al., 2010). Hence, the incorporation of MAB into conventional breeding programs holds great potential for augmenting agricultural improvement in the coming years (Jiang, 2013).

As MAB theory, technology, and application advance, its limitations are likely to be diminished, if not eliminated entirely. This bodes well for the widespread implementation of MAB in practical breeding projects for a wider variety of crop species and in more nations.

8.9 CONCLUSION AND FUTURE PROSPECTS

The process of domesticating agricultural species initiated a lengthy and productive era in the field of plant breeding. A fundamental component of contemporary agricultural technology is modern plant breeding, which utilizes the fundamental principles of heredity shared by all other breeding methods. Conventional methods of plant breeding have been highly effective in developing novel plant varieties and genetic resources. Instances comprise semidwarf, high-yielding grain varieties that were produced during the green revolution, as well as hybrid rice that was developed in the 1970s. In contrast, traditional breeding continues to heavily depend on subjective evaluation and empirical selection. Scientific breeding requires less subjectivity and more reliance on practical and precise evaluation, as well as effective and efficient selection methods. Molecular MAB presents both challenges and prospects for conventional breeding. Advancements in molecular biotechnology have led to the development of more sophisticated molecular markers that can be utilized in crop plants. Due to the swift progress of molecular markers, particularly DNA markers, and the continuous improvement of molecular assays, molecular MAB has emerged as a transformative innovation in plant breeding. The use of molecular technology has been demonstrated to be a reliable and efficient method for manipulating important traits in crop plants. It is widely used in various areas such as evaluating germplasm, mapping genes, discovering genes through mapping, characterizing traits, and introducing genes for both biotic and abiotic stresses. The effectiveness of commonly used molecular approaches in modern molecular breeding attempts is apparent. Some crop species and plant breeding programs in developing nations will continue to struggle due to the high cost of MAS. Tailoring MAS methods to individual crops, characteristics, and financial resources may be necessary. The introduction of cutting-edge marker technology has the potential to significantly lessen the expense of MAS. MAS has the potential to become more generally applied for crop breeding projects if the success of the new methods can be verified and the equipment is readily available.

KEYWORDS

- **crop improvement**
- **genomic selection**
- **marker-assisted selection**
- **marker-assisted backcrossing**
- **quantitative trait locus mapping**

REFERENCES

Abbas, H.M.K.; Zhou, Y.Y.; Huang, H.X.; Qamar, M.T.U.; Wang, M.M.; Lu, S.; Zhong, Y.J. QTL mapping, whole genome resequencing, and marker-assisted selection provide basics of early flowering in pumpkin. *Plant Breed.* 2022, 141(2), 266–276.

Abouzari, A.; Solouki, M.; Golein, B.; Fakheri, B.A.; Sabouri, A.; Dadras, A.R. Screening of molecular markers associated to cold tolerance-related traits in Citrus. *Sci. Hortic.* 2020, 263, 109145.

Acquaah, G. *Principles of Plant Genetics and Breeding, Chapter: Molecular Marker*. John Wiley and Sons Ltd., 2012, pp. 385–407.

Anderson, J.A.; Chao, S.; Liu, S. Molecular breeding using a major QTL for Fusarium head blight resistance in wheat. *Crop Sci.* 2007, 47, S-112–S-119.

Arens, P.; Mansilla, C.; Deinum, D.; Cavellini, L.; Moretti, A.; Rolland, S.; Vosman, B. Development and evaluation of robust molecular markers linked to disease resistance in tomato for distinctness, uniformity and stability testing. *Theor. Appl. Genet.* 2010, 120(3), 655–664.

Asoro, F.G.; Newell, M.A.; Beavis, W.D.; Scott, M.P.; Tinker, N.A.; Jannink, J.L. Genomic, marker-assisted, and pedigree-BLUP selection methods for β-glucan concentration in elite oat. *Crop Sci.* 2013, 53(5), 1894–1906.

Babu, R.; Nair, S.K.; Kumar, A.; Venkatesh, S.; Sekhar, J.C.; Singh, N.N.; Srinivasan, G.; Gupta, H.S. Two-generation marker-aided backcrossing for rapid conversion of normal maize lines to quality protein maize (QPM). *Theor. Appl. Genet.* 2005, 111, 888–897.

Babu, R.; Nair, S.K.; Prasanna, B.M.; Gupta, H.S. Integrating marker-assisted selection in crop breeding prospects and challenges. *Curr. Sci.* 2004, 87, 607–619.

Bali, S.; Brown, C.; Majtahedi, H.; Yilma, S.; Ingham, R.E.; Cimrhakl, L.; Sathuvalli, V. Genomic markers linked to *Meloidogyne chitwoodi* resistance introgressed from *Solanum bulbocastanum* to cultivated potato and their utility in marker-assisted selection. *Mol. Breed.* 2022, 42(3), 1–12.

Barmukh, R.; Roorkiwal, M.; Jaba, J.; Chitikineni, A.; Mishra, S.P.; Sagurthi, S.R.; Munghate, R.; Sharma, H.C.; Varshney, R.K. Development of a dense genetic map and QTL analysis for pod borer *Helicoverpa armigera* (Hübner) resistance component traits in chickpea (*Cicer arietinum* L.). *Plant Genome* 2021, 14(1), e20071.

Bastianel, M.; Cristofani-Yaly, M.; de Oliveira, A.C.; Freitas-Astúa, J.; Garcia, A.A.F.; de Resende, M.D.V.; Machado, M.A. Quantitative trait loci analysis of citrus leprosis resistance in an interspecific backcross family of (*Citrus reticulata* Blanco × C. *sinensis* L. Osbeck) × C. *sinensis* L. Osb. *Euphytica* 2009, 169(1), 101–111.

Bilyeu, K.L.; Palavalli, D.A.; Sleper; Beuselinck, P. Molecular genetic resources for development of 1% linolenic acid soybeans. *Crop Sci.* 2006, 46, 1913–1918.

Botstein, D.; White, R.L.; Skolnick, M. and Davis, R.W. Construction of a genetic linkage map in man using restriction fragment length polymorphisms. *Am. J. Hum. Genet.* 1980, 32, 314–331.

Bradbury, L.M.T.; Fitzgerald, T.L.; Henry, R.J.; Jin, Q.; Waters, D.L.E. The gene for fragrance in rice. *Plant Biotech. J.* 2005a. 3, 363–370.

Bradbury, L.M.T.; Henry, R.J.; Jin, Q.; Reinke, R.F.; Waters, D.L.E. A perfect marker for fragrance genotyping in rice. *Mol. Breed.* 2005b, 16, 279–283.

Brekke, T.D.; Stroud, J.A.; Shaw, D.S.; Crawford, S.; Steele, K.A. QTL mapping in salad tomatoes. *Euphytica* 2019, 215(7), 1–12.

Buerstmayr, H.; Ban, T.; Anderson, J.A. QTL mapping and marker-assisted selection for Fusarium head blight resistance in wheat: A review. *Plant Breed.* 2009, 128, 1–26.

Castro, A.J.; Capettini, F.; Corey, A.E.; Filichkina, T.; Hayes, P.M.; Kleinhofs, A.; Kudrna, D.; Richardson, K.; Sandoval-Islas, S.; Rossi, C.; Vivar, H. Mapping and pyramiding of qualitative and quantitative resistance to stripe rust in barley. *Theor. Appl. Genet.* 2003, 107, 922–930.

Chand, S.K.; Nanda, S.; Joshi, R.K. Genetics and molecular mapping of a novel purple blotch-resistant gene ApR1 in onion (*Allium cepa* L.) using STS and SSR markers. *Mole. Breed.* 2018, 38(9), 1–13.

Cobb, J.N.; Juma, R.U.; Biswas, P.S.; Arbelaez, J.D.; Rutkoski, J.; Atlin, G.; Hagen, T.; Quinn, M.; Ng, E.H. Enhancing the rate of genetic gain in public-sector plant breeding programs: Lessons from the breeder's equation. *Theor. Appl. Genet.* 2019, 132(3), 627–645.

Collard, B.C.Y.; Jahufer, M.Z.Z.; Brouwer, J.B.; Pang, E.C.K. An introduction to markers, quantitative trait loci (QTL) mapping and marker-assisted selection for crop improvement: The basic concepts. *Euphytica* 2005, 142, 169–196.

Collard, B.C.Y.; Mackill, D.J. Marker-assisted selection: An approach for precision plant breeding in the twenty-first century. *Phil. Trans. R. Soc.* 2008, 363, 557–572.

Cregan, P.B.; Mudge, J.; Fickus, E.W.; Danesh, D.; Denny, R.; Young, N.D. Two simple sequence repeat markers to select for soybean cyst nematode resistance conditioned by the rhg1 locus. *Theor. Appl. Genet.* 1999, 99, 811–818.

Crosbie, T.M.; Eathington, S.R.; Johnson, G.R.; Edwards, M.; Reiter, R.; Stark, S.; Mohanty, R.G.; Oyervides, M.; Buehler, R.E.; Walker, A.K.; Dobert, R.; Delannay, X.; Pershing, J.C.; Hall, M.A. and Lamkey, K.R. Plant breeding: Past, present and fu- ture. In: Lamkey, K.R.; Lee, M.; (Eds.), *Plant Breeding: The Arnel R. Hallauer International Symposium*. Blackwell Publishing: Oxford, UK, 2006, pp. 3–50.

Cuevas, H.E.; Knoll, J.E.; Harris-Shultz, K.R.; Punnuri, S.M. Genetic mapping of sugarcane aphid resistance in sorghum line SC112-14. *Crop Sci.* 2022, 62(6), 2267–75.

Datta, D.; Gupta, S.; Chaturvedi, S.K.; Nadarajan, N. *Molecular Markers in Crop Improvement*. Indian Institute of Pulses Research: Kanpur 208024, 2011, 201–260.

De Giovanni, C.; Dell'Orco, P.; Bruno, A.; Ciccarese, F.; Lotti, C.; Ricciardi, L. Identification of PCR-based markers (RAPD, AFLP) linked to a novel powdery mildew resistance gene (ol-2) in tomato. *Plant Sci.* 2004, 166(1), 41–48.

Devi, S.J.; Singh, K.; Umakanth, B.; Vishalakshi, B.; Rao, K.; Suneel, B.; Sharma, S.K.; Kadambari, G.K.; Prasad, M.S.; Senguttvel, P.; Syamaladevi, D.P. Identification and characterization of a large effect QTL from *Oryza glumaepatula* revealed Pi68 (t) as putative candidate gene for rice blast resistance. *Rice* 2020, 13(1), 1–3.

Dillon, S.; Ramage, C.; Ashmore, S.; Drew, R.A. Development of a codominant CAPS marker linked to PRSV-P resistance in highland papaya. *Theor. Appl. Genet.* 2006, 113(6), 1159–1169.

Dwivedi, S.L.; Crouch, J.H.; Mackill, D.J.; Xu, Y.; Blair, M.W.; Ragot, M.; Upad- hyaya, H.D.; Oritiz, R. The molecularization of public sector crop breeding: Progress, problems and prospects. *Adv. Agron.* 2007, 95, 163–318.

Eathington, S.R. Practical applications of molecular technology in the development of commercial maize hybrids. In: *Proceedings of the 60th Annual Corn and Sorghum Seed Research Conferences, American Seed Trade Association*, Washington, D.C., 2005. pp. 1–25.

Edwards, D.; Forster, J.W.; Chagne, D.; Batley, J. What is SNPs? In: Oraguzie, N.C.; Rikkerink, E.H.A.; Gardiner, S.E.; de Silva, H.N.; (Eds.), *Association Mapping in Plants*. Springer: Berlin, 2007, pp. 41–52.

Erdin, N.; Tartarini, S.; Broggini, G.A.; Gennari, F.; Sansavini, S.; Gessler, C.; Patocchi, A. Mapping of the apple scab-resistance gene Vb. *Genome* 2006, 49(10), 1238–1245.

Felipe, J.E.L.; Lachica, J.A.P.; Cueva, F.M.D.; Laurel, N.R.; Alcasid, C.E.; Sison, M.L.J.; Ocampo, E.T.M. Validation and molecular analysis of β-1, 3-GLU2 SNP marker associated with resistance to *Colletotrichum gloeosporioides* in mango (*Mangifera indica* L.). *Physiol. Mol. Plant Pathol.* 2022, 118, 101804.

Fulton, T.M.; Bucheli, P.; Voirol, E.; Lopez, J.; Peetiard, V.; Tanksley, S.D. Quantitative trait loci (QTL) affecting sugars, organic acids and other biochemical properties possibly contributing to flavor, identified in four advanced backcross populations of tomato. *Euphytica* 2002, 127, 163–177.

Garcia, B.E.; Graham, E.; Jensen, K.S.; Hanson, P.; Mejía, L.; Maxwell, D.P. Co-dominant SCAR marker for detection of the begomovirus-resistance Ty-2 locus derived from *Solanum habrochaites* in tomato germplasm. *Tomato Genetic Cooperative Report.* 2007, 57, 21–24.

Heffner, E.L.; Sorrells, M.E.; Jannink, J.L. Genomic selection for crop improvement. *Crop Sci.* 2009, 49, 1–12.

Helentjaris, T.G. RFLP analysis for manipulating agronomic traits in plants. In: Stalker, H.T.; Murphy, J.P.; (Eds.), *Plant Breeding in the 1990s*. CAB International: Wallingford, 1992, pp. 357–372

Hoyos-Villegas, V.; Chen, J.; Mastrangelo, A.M.; Raman, H. Advances in breeding for quantitative disease resistance. *Front. Plant Sci.* 2022, 13, 890002–890002.

Inoue, E.; Kasumi, M.; Sakuma, F.; Anzai, H.; Amano, K.; Hara, H. Identification of RAPD marker linked to fruit skin color in Japanese pear (*Pyrus pyrifolia* Nakai). *Sci. Hortic.* 2006, 107(3), 254–258.

Inturrisi, F.C.; Bayer, P.E.; Yang, H.; Chan, C.K.; Edwards, D.; Batley, J. Genome-wide analysis of NBS-LRR genes in Indian mustard (*Brassica juncea*) and prediction of candidate disease resistance genes. *Phytopathology* 2018, 108(10), 111.

Jefferies, S.P.; King, B.J.; Barr, A.R.; Warner, P.; Logue, S.J.; Langridge, P. Marker-assisted backcross introgression of the Yd2 gene conferring resistance to barley yellow dwarf virus in barley. *Plant Breed.* 2003, 122, 52–56.

Jiang, G.L. *Advances in Plant Breeding Strategy: Breeding Biotechnology and Molecular Tools, Chapter. Molecular Marker Assisted Breeding: A Plant Breeder's Review*. Springer International Publishing: Switzerland, 2015, pp. 431–471.

Jiang, G.L. Molecular markers and marker-assisted breeding in plants. In: *Plant Breeding from Laboratories to Fields*. Intech Open Science, 2013, pp. 1–40. http://dx.doi.org/10.5772/52583

Jiang, G.L.; Dong, Y.; Shi, J.; Ward, R.W. QTL analysis of resistance to Fusarium head blight in the novel wheat germplasm CJ 9306. II. Resistance to deoxy- nivalenol accumulation and grain yield loss. *Theor. Appl. Genet.* 2007b, 115, 1043–1052.

Jiang, G.L.; Shi, J.; Ward, R.W. QTL analysis of resistance to Fusarium head blight in the novel wheat germplasm CJ 9306. I. Resistance to fungal spread. *Theor. Appl. Genet.* 2007a, 116, 3–13.

Johnson, R. Marker assisted selection. In: Jannick, J.; (Ed.), *Plant Breed. Rev.* 2004, 24(1), 293–309.

Joshi, R.K.; Nayak, S. 2010. Gene pyramiding—A broad spectrum technique for developing durable stress resistance in crops. *Biotechol. Mol. Biol. Rev.* 2010, 5, 51–60.

Kaewwongwal, A.; Chen, J.; Somta, P.; Kongjaimun, A.; Yimram, T.; Chen, X.; Srinives, P. Novel alleles of two tightly linked genes encoding polygalacturonase-inhibiting proteins (*VrPGIP1* and *VrPGIP2*) associated with the Br locus that confer bruchid resistance to mungbean accession V2709. *Front. Plant Sci.* 2017, 8, 1692.

Kandel, R.; Lu, H.; Sandoya, G.V. Identification and mapping of quantitative trait loci for resistance to *Liriomyza trifolii* in romaine lettuce cultivar 'Valmaine'. *Sci. Rep.* 2021, 11(1), 1–9.

Keller-Przybyłkowicz, S.; Lewandowski, M.; Korbin, M. Molecular screening of apple (*Malus domestica*) cultivars and breeding clones for their resistance to fire blight. *J. Fruit Ornam. Plant Res.* 2009, 17(2), 31–43.

Kim, B.; Hwang, I.S.; Lee, H.J.; Lee, J.M.; Seo, E.; Choi, D.; Oh, C.S. Identification of a molecular marker tightly linked to bacterial wilt resistance in tomato by genome-wide SNP analysis. *Theor. Appl. Genet.* 2018, 131(5), 1017–1030.

Kim, H.; Yoon, J.B.; Lee, J. Development of fluidigm SNP type genotyping assays for marker-assisted breeding of chili pepper (*Capsicum annuum* L.). *Hortic. Sci. Technol.* 2017b, 35, 465–479

Lambert, P.; Campoy, J.A.; Pacheco, I.; Mauroux, J.B.; Da Silva Linge, C.; Micheletti, D.; Arús, P. Identifying SNP markers tightly associated with six major genes in peach [*Prunus persica* (L.) Batsch] using a high-density SNP array with an objective of marker-assisted selection (MAS). *Tree Genet. Genom.* 2016, 12(6), 1–21.

Lema, M. Marker assisted selection in comparison to conventional plant breeding. *Agric. Res. Technol.* 2018, 14, 555914.

Li, X.; Ouyang, X.; Zhang, Z.; He, L.; Wang, Y.; Li, Y.; Zhao, J.; Chen, Z.; Wang, C.; Ding, L.; Pei, Y. Over-expression of the red plant gene R1 enhances anthocyanin production and resistance to bollworm and spider mite in cotton. *Mole. Genet. Genom.* 2019, 294(2), 469–478.

Li, X.; Han, Y.; Teng, W.; Zhang, S.; Yu, K.; Poysa, V.; Anderson, T.; Ding, J.; Li, W. Pyramided QTL underlying tolerance to Phytophthora root rot in mega-environment from soybean cultivar 'Conrad' and 'Hefeng 25'. *Theor. Appl. Genet.* 2010, 121, 651–658.

Ling, P.; Duncan, L.W.; Deng, Z.; Dunn, D.; Hu, X.; Huang, S.; Gmitter Jr, F.G. Inheritance of citrus nematode resistance and its linkage with molecular markers. *Theor. Appl. Genet.* 2000, 100(7), 1010–1017.

Liu, L.; Wang, M.; Zhang, Z.; See, D.R.; Chen, X. Identification of stripe rust resistance loci in US spring wheat cultivars and breeding lines using genome-wide association mapping and Yr gene markers. *Plant Dis.* 2020, 104(8), 2181–2192.

Liu, B.H. Development and prospects of dwarf male-sterile wheat. *Chin. Sci. Bull.* 1991, 36(4), 306.

Liu, P.N.; Miao, H.; Lu, H.W.; Cui, J.Y.; Tian, G.L.; Wehner, T.C.; Zhang, S.P. Molecular mapping and candidate gene analysis for resistance to powdery mildew in Cucumis sativus stem. *Genet. Mol. Res.* 2017, 16(3).

Luo, Y.; Sangha, J.S.; Wang, S.; Li, Z.; Yang, J. Yin, Z. Marker-assisted breeding of Xa4, Xa21 and Xa27 in the restorer lines of hybrid rice for broad-spectrum and enhanced disease resistance to bacterial blight. *Mol. Breed.* 2012, 30, 1601–1610, DOI 10.1007/s11032-012-9742-7.

Majhi, P.K.; Bhoi, T.K.; Mogali, S.C.; Shiv, A.; Sahoo, K.C.; Saini, V. Advances in molecular breeding for bruchid (*Callosobruchus* spp.) resistance in mungbean [*Vigna radiata* (L.) Wilczek]: A review. *Legume Res.* 2022a, 45(8), 933–941. https://doi.org/10.18805/LR-4695.

Majhi, P.K.; Singh, S.K.; Anandan, A.; Khaire, A.R.; Korada, M.; Habde, S.V.; Singh, D.K.; Bhagvan, A.P.; Jayasudha, S.; Singh, A. Parental polymorphism survey for evaluation and selection of contrasting parents for drought tolerance in rice (*Oryza sativa* L.) by using SSR markers. *Int. J. Environ. Clim. Chang.* 2022b, 2(11), 2507–2519. https://doi.org/10.9734/IJECC/2022/v12i1131244

Mejía, N.; Gebauer, M.; Muñoz, L.; Hewstone, N.; Muñoz, C.; Hinrichsen, P. Identification of QTLs for seedlessness, berry size, and ripening date in a seedless x seedless table grape progeny. *Am. J. Enol. Vitic.* 2007, 58(4), 499–507.

Miedaner, T.; Wilde, F.; Steiner, B.; Buerstmayr, H.; Korzun, V.; Ebmeyer, E. Stacking quantitative trait loci (QTL) for Fusarium head blight resistance from nonadapted sources in an European elite spring wheat background and assessing their effects on deoxynivalenol (DON) content and disease severity. *Theor. Appl. Genet.* 2006, 112, 562–569.

Miyatake, K.; Saito, T.; Negoro, S.; Yamaguchi, H.; Nunome, T.; Ohyama, A.; Fukuoka, H. Development of selective markers linked to a major QTL for parthenocarpy in eggplant (*Solanum melongena* L.). *Theor. Appl. Genet.* 2012, 124(8), 1403–1413.

Miyatake, K.; Saito, T.; Negoro, S.; Yamaguchi, H.; Nunome, T.; Ohyama, A.; Fukuoka, H. Detailed mapping of a resistance locus against Fusarium wilt in cultivated eggplant (*Solanum melongena*). *Theor. Appl. Genet.* 2016, 129(2), 357–367.

Mullis, K. The unusual origin of the polymerase chain reaction. *Sci. Am.* 1990, 262 (4), 56–61.

Munyiri, S.W.; Mugo, S.N. Quantitative trait loci for resistance to spotted and African maize stem borers (*Chilo partellus* and *Busseola fusca*) in a tropical maize (Zea mays L.) population. *African J. Biotechnol.* 2017, 16(28), 1579–1589.

Nadeem, M.A.; Nawaz, M.A.; Shahid, M.Q.; Comertpay, Y.D.G.; Hatipoğlu, M.Y.R.; Fiaz A.; Alsaleh, A.; Labhane, N.; Özkan, H.; Chung, G.; Baloch, F.S. DNA molecular markers in plant breeding: Current status and recent advancements in genomic selection and genome editing. *Biotechnol. Biotechnol. Equip.* 2018, 32(2), 261–285, https://doi.org/10.1080/13102818.2017.1400401

O'Connell, A.; Deo, J.; Deomano, E.; Wei, X.; Jackson, P.; Aitken, K.S.; Manimekalai, R.; Mohanraj, K.; Hemaprabha, G.; Ram, B.; Viswanathan, R. Combining genomic selection with genome-wide association analysis identified a large-effect QTL and improved selection for red rot resistance in sugarcane. *Front. Plant Sci.* 2022, 13, 1021182.

Ober, E.S.; Alahmad, S.; Cockram, J.; Forestan, C.; Hickey, L.T.; Kant, J.; Maccaferri, M.; Marr, E.; Milner, M.; Pinto, F.; Rambla, C.; Reynolds, M.; Salvi, S.; Sciara, G.;

Snowdon, R.J.; Thomelin, P.; Tuberosa, R.; Uauy, C.; Voss-Fels, K.P.; Wallington, E.; and Watt, M. Wheat root systems as a breeding target for climate resilience. *Theor. Appl. Genet.* 2021, 134(6), 1645–1662.

Payne, P.I.; Nigtingale, M.A.; Krattiger, A.F.; Holt, L.M. The relationship between HMW glutenin subunit composition and the bread-making quality of British-grown wheat varieties. *J. Sci. Food Agric.* 1987, 40, 51–65.

Pham, A.T.; Shannon, J.G.; Bilyeu, K.D. Combinations of mutant FAD2 and FAD3 genes to produce high oleic acid and low linolenic acid soybean oil. *Theor. Appl. Genet.* 2012, 125, 503–515.

Pumphrey, M.O.; Bernardo, R.; Anderson, J.A. Validating the Fhb1 QTL for Fusarium head blight resistance in near-isogenic wheat lines developed from breeding populations. *Crop Sci.* 2007, 47, 200–206.

Ragot, M.; Biasiolli, M.; Dekbut, M.F.; Dell'orco, A.; Margarini, L.; Thevenin, P.; Ver noy, J.; Vivant, J.; Zimmermann, R.; Gay, G. Marker-assisted backcrossing: A practical example. In: Berville, A.; Tersac, M.; (Eds.), *Les Colloques, No. 72.* INRA: Paris, 1995, pp. 45–46.

Ribaut, J.M.; de Vicente, M.C.; Delannay, X. Molecular breeding in develop- ing countries: Challenges and perspectives. *Curr. Opin. Plant Biol.* 2010, 13, 1–6.

Richards, R.A.; Rebetzke, G.J.; Watt, M.; Condon, A.T.; Spielmeyer, W.; Dolferus, R. Breeding for improved water productivity in temperate cereals: Phenotyping, quantitative trait loci, markers and the selection environment. *Funct. Plant Biol.* 2010, 37(2), 85–97.

Richardson, K.L.; Vales, M.I.; Kling, J.G.; Mundt, C.C.; Hayes, P.M. Pyramid- ing and dissecting disease resistance QTL to barley stripe rust. *Theor. Appl. Genet.* 2006, 113, 485–495.

Robbins, M.D.; Masud, M.A.; Panthee, D.R.; Gardner, R.G.; Francis, D.M.; Stevens, M.R. Marker-assisted selection for coupling phase resistance to tomato spotted wilt virus and *Phytophthora infestans* (late blight) in tomato. *Hort. Sci.* 2010, 45(10), 1424–1428.

Rogers, W.J.; Payne, P.I.; Harinder, K. The HMW glutenin subunit and gliadin compositions of German-Grown wheat varieties and their relationship with bread-making quality. *Plant Breed.* 1989, 103, 89–100.

Ruangrak, E.; Su, X.; Huang, Z.; Wang, X.; Guo, Y.; Du, Y.; Gao, J. Fine mapping of a major QTL controlling early flowering in tomato using QTL-seq. *Canadian J. Plant Sci.* 2018, 98(3), 672–682.

Sahu, D.; Tripathy, S.K.; Sahoo, B.; Nayak, S.L.; Dalei, D.R. Role of molecular markers in vegetables improvement. *J. Pharmacogn. Phytochem.* 2019, 8(6), 456–460.

Salgon, S.; Jourda, C.; Sauvage, C.; Daunay, M.C.; Reynaud, B.; Wicker, E.; Dintinger, J. Eggplant resistance to the *Ralstonia solanacearum* species complex involves both broad-spectrum and strain-specific quantitative trait loci. *Front. Plant Sci.* 2017, 8, 828.

Sebolt, A.M.; Shoemaker, R.C.; Diers, B.W. Analysis of a quantitative trait locus allele from wild soybean that increases seed protein concentration in soybean. *Crop Sci.* 2000, 40, 1438–1444.

Servin, B.; Martin, O.C.; Mezard, M.; Hospital, F. Toward a theory of mark- er-assisted gene pyramiding. *Genetics* 2004, 168, 513–523.

Shi, A.; Chen, P.; Li, D.; Zheng, C.; Zhang, B.; Hou, A. Pyramiding multiple genes for resistance to soybean mosaic virus in soybean using molecular markers. *Mol. Breed.* 2009, 23, 113–124.

Singh, A.; Singh, V.K.; Singh, S.P.; Pandian, R.T.; Ellur, R.K.; Singh, D.; Bhowmick, P.K.; Gopala Krishnan, S.; Nagarajan, M.; Vinod, K.K.; Singh, U.D.; Prabhu, K.V.; Sharma, T.R.; Mohapatra, T.; Singh, A.K. Molecular breeding for the development of multiple disease resistance in Basmati rice. *AoB Plants* 2012, pls029. https://doi.org/10.1093/aobpla/pls029.

Singh, B.D.; Singh, A.K. *Marker-Assisted Plant Breeding: Principles and Practices*. Springer (India) Pvt. Ltd., 2015, pp. 1–542. DOI 10.1007/978-81-322-2316-0

Singh, S.K.; Majhi, P.K.; Anandan, A.; Korada, M.; Habde, S.V.; Khaire, A.R.; Singh, D.K.; Jayasudha, S.; Pandurang, A. Microsatellites based parental polymorphism survey for moisture stress in Rice (*Oryza sativa* L.) between the parental genotypes HUR-1309 and CR Dhan 801. *Emer. Life Sci. Res.* 2022, 8(2), 248–258.

Singh, S.; Sidhu, J.S.; Huang, N.; Vikal, Y.; Li, Z.; Brar, D.S.; Dhaliwal, H.S.; Khush, G.S. Pyramiding three bacterial blight resistance genes (xa5, xa13 and Xa21) using marker-assisted selection into indica rice cultivar PR106. *Theor. Appl. Genet.* 2001, 102, 1011–1015.

Sobrino, B.; Briona, M.; Carracedoa, A. SNPs in forensic genetics: A review on SNP typing methodologies. *Forensic Sci. Int.* 2005, 154, 181–194.

Somers, D.J.; Thomas, J.; DePauw, R.; Fox, S.; Humphreys, G.; Fedak, G. Assembling complex genotypes to resist Fusarium in wheat (*Triticum aestivum* L.). *Theor. Appl. Genet.* 2005, 111, 1623–1631.

Soriano, J.M.; Domingo, M.L.; Zuriaga, E.; Romero, C.; Zhebentyayeva, T.; Abbott, A.G.; Badenes, M.L. Identification of simple sequence repeat markers tightly linked to plum pox virus resistance in apricot. *Mol. Breed.* 2012, 30(2), 1017–1026.

Southern, E.M. Detection of specific sequences among DNA fragments sepa- rated by gel electrophoresis. *J. Mol. Biol.* 1975, 98, 503–517.

Stussey, T. (1990). *Plant Taxonomy*. Columbia University Press: New York.

Sudheesh, S.; Rodda, M.S.; Davidson, J.; Javid, M.; Stephens, A.; Slater, A.T.; Cogan, N.O.; Forster, J.W.; Kaur, S. SNP-based linkage mapping for validation of QTLs for resistance to ascochyta blight in lentil. *Front. Plant Sci.* 2016, 2(7), 1604.

Sun, Z.; Liu, Y.; Xiao, S.; Hu, J.; Pan, G.; He, J.; Xu, T.; Huang, J.; Qiu, Z.; Fan, D.; Zhang, L. Identification of quantitative trait loci for resistance to rice black-streaked dwarf virus disease and small brown planthopper in rice. *Mol. Breed.* 2017, 37(6), 1–9.

Tanksley, S.D.; Nelson, J.C. Advanced backcross QTL analysis: A method for the simultaneous discovery and transfer of valuable QTLs from unadapted germ-plasm into elite breeding lines. *Theor. Appl. Genet.* 1996, 92, 191–203.

Tanksley, S.D.; Orton, T.J.; (Eds.), (1983). *Isozymes in Plant Genetics and Breeding*. Elsevier: Amsterdam.

Taoutaou, A.; Berindean, I.V.; Csete, E.; Pamfil, D.; Botez, C. Development of a molecular marker for the resistance gene R11 of potato to late blight. *Romanian Agric. Res.* 2022, 39.

Tiwari, J.K.; Yerasu, S.R.; Rai, N.; Singh, D.P.; Singh, A.K.; Karkute, S.G.; Behera, T.K. Progress in marker-assisted selection to genomics-assisted breeding in tomato. *Critical Rev. Plant Sci.* 2022, 41, 321–350.

Tolmay, V.L.; Sydenham, S.L.; Sikhakhane, T.N.; Nhlapho, B.N.; Tsilo, T.J.; Elusive diagnostic markers for Russian wheat aphid resistance in bread wheat: Deliberating and reviewing the status quo. *Inter. J. Mol. Sci.* 2020, 21(21), 8271.

Verma, S.K.; Jasrotia, R.S.; Iquebal, M.A.; Jaiswal, S.; Angadi, U.B.; Rai, A.; Kumar, D. Deciphering genes associated with root wilt disease of coconut and development of its transcriptomic database (CnTDB). *Physiol. Mol. Plant Pathol.* 2017, 100, 255–263.

Vikal, Y.; Kaur, A.; Jindal, J.; Kaur, K.; Pathak, D.; Garg, T.; Singh, A.; Singh, P. Yadav, I. Identification of genomic regions associated with shoot fly resistance in maize and their syntenic relationships in the sorghum genome. *PLoS One* 2020, 15(6), e0234335.

Wan, J.; Jiang, L.; Zhai, H.; Wang, Q.; Zhou, Y. Molecular breeding of rice restorer lines and hybrids for brown planthopper resistance. *Rice* 2016, 9, Article 55. https://doi.org/10.1186/s12284-016-0113-3

Wang, B.; Chee, P.W. Application of advanced backcross quantitative trait locus (QTL) analysis in crop improvement. *J. Plant Breed. Crop Sci.* 2010, 2, 221–232.

Wang, X.; Jiang, G.L.; Green, M.; Scott, R.A.; Hyten, D.L.; Cregan, P.B. Quantitative trait locus analysis of saturated fatty acids in a population of recombinant inbred lines of soybean. *Mol. Breed.* 2012, 30, 1163–1179. DOI 10.1007/s11032-012-9704-0

Wang, Y.; Tan, J.; Wu, Z.; Vanden Langenberg, K.; Wehner, T.C.; Wen, C.; Weng, Y. Stay green, stay healthy: A loss-of-susceptibility mutation in the STAYGREEN gene provides durable, broad-spectrum disease resistances for over 50 years of US cucumber production. *New Phytol.* 2019, 221(1), 415–430.

Welsh, J.; McClelland, M. Fingerprinting genomes using PCR with arbitrary primers. *Nucleic Acids Res.* 1990. 18, 7213–7218.

Williams, J.G.K.; Kubelik, A.R.; Livak, K.J.; et al. DNA polymorphisms amplified by arbitrary primers are useful as genetic markers. *Nucleic Acids Res.* 1990. 18, 1631–1635

Xu, Y. *Molecular Plant Breeding, Chapter.* Marker Assisted Selection-Theory, CAB International: London, UK, 2010, pp. 286–327.

Yamamoto, T.; Kimura, T.; Soejima, J.; Sanada, T.; Ban, Y.; Hayashi, T. Identification of quince varieties using SSR markers developed from pear and apple. *Breed. Sci.* 2004, 54(3), 239–244.

Ye, G.; K.F. Smith. Marker-assisted gene pyramiding for inbred line development: Basic principles and practical guidelines. *Int. J. Plant Breed.* 2008, 2, 1–10.

Yi, C.; Guo, W.; Zhu, X.; Min, L.; Zhang, T. Pyramiding breeding by marker-assisted recurrent selection in upland cotton II. Selection effects on resistance to *Helicoverpa armigera*. *Sci. Agric. Sin.* 2004, 37, 801–807.

Young, N.D. A cautiously optimistic vision for marker-assisted breeding. *Mol. Breed.* 1999, 5, 505–510.

Youssef, M.; Ibrahim, R. Molecular markers associated with high vitamin-C content in guava. *J. Agric. Chem. Biotechnol.* 2016, 7(3), 49–55.

Zhang, J.; Panthee, D.R. Development of codominant SCAR markers to detect the Pto, Tm22, I3 and Sw5 genes in tomato (*Solanum lycopersicum*). *Plant Breed.* 2021, 140(2), 342–348.

Zhang, R.P.; Wu, J.; Li, X.G.; Khan, M.A.; Chen, H.; Korban, S.S.; Zhang, S.L. An AFLP, SRAP, and SSR genetic linkage map and identification of QTLs for fruit traits in pear (Pyrus L.). *Plant Mol. Biol. Rep.* 2013, 31(3), 678–687.

Zhao, X.; Tan, G.; Xing, Y.; Wei, L.; Chao, Q.; Zuo, W.; Lübberstedt, T.; Xu, M. Marker-assisted introgression of qHSR1 to improve maize resistance to head smut. *Mol. Breed.* 2012, 30, 1077–1088, DOI 10.1007/s11032-011-9694-3

Zhao, Y.; Liu, Y.; Zhang, Z.; Cao, Y.; Yu, H.; Ma, W.; Wang, L. Fine mapping of the major anthracnose resistance QTL AnRGO5 in *Capsicum chinense* 'PBC932'. *BMC Plant Biol.* 2020, 20(1), 1–8.

Zheng, J.; Zhao, H.; Ma, Y.; Jiang, M.; Zhan, Z.; Li, X.; Piao, Z. Marker-assisted pyramiding of genes for multilocular ovaries, self-compatibility, and clubroot resistance in Chinese cabbage (*Brassica rapa* L. ssp. *pekinensis*). *Horticulturae* 2022, 8(2), 139.

CHAPTER 9

Tissue Culture Techniques in Crop Improvement

AMOL D. GHOLAP[1*], PANKAJ R. KHUSPE[2], SADIKALI F. SAYYAD[3], MACHINDRA CHAVAN[4], NAVNATH T. HATVATE[5], MD. FAIYAZUDDIN[6], and MD JASIM UDDIN[7]

[1]*Department of Pharmaceutics, St. John Institute of Pharmacy and Research, Palghar, Maharashtra, India*

[2]*Department of Pharmaceutics, College of Pharmacy, Paniv, Solapur, Maharashtra, India*

[3]*Department of Pharmaceutics, Amrutvahini College of Pharmacy, Sangamner, Maharashtra, India*

[4]*Department of Pharmacognosy, Amrutvahini College of Pharmacy, Sangamner, Maharashtra, India*

[5]*Department of Pharmaceutical Technology, Institute of Chemical Technology, Marathwada Campus, Jalna, Maharashtra, India*

[6]*School of Pharmacy, Al-Karim University, Katihar, Bihar, India*

[7]*Department of Pharmaceutical Technology, Faculty of Pharmacy, Universiti Malaya, Kuala Lumpur, Malaysia*

*Corresponding author

ABSTRACT

The growing global population necessitates high yielding, sustainable crops, driving the need for innovative agricultural practices. This chapter explores the transformative impact of tissue culture techniques on crop

Technological Advancements in Sustainable Crop Improvement and Production.
Devendra Kumar, Alok Kumar Singh, and Nisha Malik (Eds.)

DOI: 10.1201/9781779643155-9

development strategies. It begins with a historical overview, followed by an in-depth examination of tissue culture principles, including genetic transformation and plant regeneration. The discussion covers medium composition, sterilization procedures, and cell culture techniques, providing a solid foundation for understanding tissue culture complexities. Organogenesis, somatic embryogenesis, and callus culture are highlighted as key pathways for plant regeneration. The chapter focuses on genetic transformation, exploring transgenic plant creation, gene transfer techniques, and marker genes. Applications in crop improvement include enhanced yield attributes, disease resistance, abiotic stress tolerance, and nutritional enhancement. However, challenges such as genetic stability, contamination, and regulatory issues are also evaluated. The chapter provides an outlook on future research areas considering biotechnology and agriculture advancements. Case studies offer valuable insights into effective crop development initiatives, serving as a resource for scholars and industry experts. By compiling information and providing case studies, the chapter aims to stimulate further study and promote sustainable agriculture breakthroughs, ultimately contributing to global food security and sustainable agricultural practices.

9.1 INTRODUCTION

The application of cutting-edge technology has become essential in the field of modern agriculture and plant breeding to improve crop yield, guarantee food security, and tackle the problems presented by environmental changes. Tissue culture is one of the methods that has become a potent instrument for crop development (Gholap, 2023b; Gholap et al., 2012f). To manipulate and propagate desired features, tissue culture entails the *in vitro* growing of plant cells, tissues, or organs in a controlled environment. The advancement and utilization of tissue culture methodologies in crop enhancement have fundamentally transformed conventional breeding approaches. This chapter explores the basic ideas and uses of tissue culture, with a focus on how it can be used to improve agricultural characteristics including yield, pest and disease resistance, and abiotic stress tolerance (Javaid et al., 2022). The chapter opens with a thorough historical overview, following the development of tissue culture from its early beginnings to its current position as a mainstay of contemporary agricultural research. Important turning points and discoveries in the area will be emphasized, illustrating how tissue culture techniques have evolved from experimental methods to widely used applications (Voss-Fels et al., 2019; Ahmad, 2023; Razzaq et al., 2021; Javaid et al., 2023).

The following sections explore the fundamental ideas of tissue culture, clarifying the process and essential elements needed to cultivate plant cells successfully. This involves talking about the hormones, environmental factors, and plant tissue culture media that are needed for the best possible growth and development. The chapter's main emphasis is on the useful uses of tissue culture in agricultural development. The presentation will include case studies and examples from a variety of crops to demonstrate the role that tissue culture has played in the advancement of superior varieties with enhanced features. Tissue culture provides a flexible platform for modifying plant genomes, enabling the rapid replication of elite genotypes, the production of genetic variety through somaclonal variation, or the inclusion of specific genes through genetic transformation. The chapter also discusses the drawbacks and difficulties of tissue culture methods, including genotype dependence, somaclonal variation, and implementation costs (Duta-Cornescu et al., 2023; Leva et al., 2012). It will be covered how to overcome these obstacles and enhance methods to make it easier for tissue culture to be successfully included into crop improvement initiatives. The chapter highlights how crucial tissue culture methods will be in determining the direction of crop breeding and agriculture in the future. The goal of this chapter is to give researchers, agronomists, and breeders the knowledge they need to fully utilize tissue culture for sustainable and effective crop improvement by offering a comprehensive understanding of its principles, applications, and challenges (Espinosa-Leal et al., 2018).

The use of tissue culture techniques in crop development has many goals and is essential to the advancement of agricultural practices. The issue culture is a powerful technique for the quick multiplication of desirable trait-rich elite plant genotypes, which makes high-quality planting material manufacturing efficient. This gives a way to quickly generate and share superior germplasm, addressing the growing need for enhanced crop types (Debnath and Arigundam, 2020; Türkoğlu et al., 2023). Second, tissue culture provides a controlled environment to select for and induce desired mutations, hence facilitating the creation of genetic variety through somaclonal variation. This offers breeding programs a priceless supply of genetic variation, facilitating the creation of crops with improved tolerance to abiotic stressors, diseases, and pests. Third, the exact integration of particular genes through genetic transformation is made possible in large part by tissue culture. This process results in transgenic crops that have enhanced yield, higher nutritional value, and more resilience to environmental stressors. Furthermore, tissue culture methods support sustainable agriculture and biodiversity conservation by enabling the rescue and regeneration of commercially significant or endangered plant species. Generally speaking, the goals of tissue culture in crop improvement include the targeted modification of plant genomes, the creation of genetic diversity, and the effective propagation of superior genotypes. These goals work together to help develop resilient, high-yielding, and sustainable crop varieties (Espinosa-Leal et al., 2018; Long et al., 2022; Hasnain et al., 2022).

This chapter provides a thorough examination of the origins, tenets, and real-world uses of tissue culture in the context of contemporary agriculture. The chapter begins with a historical overview and follows the evolution of tissue culture techniques from their inception to their current status as a mainstay of agriculture (Verma et al., 2020). It explores the fundamental ideas of tissue culture, clarifying the subtleties of hormone control, plant tissue culture medium, and the critical environmental factors required for successful *in vitro* growth. The main emphasis is on the practical uses of tissue culture, with case studies demonstrating how important it is in creating crop varieties with improved qualities including higher yield, disease resistance, and stress tolerance. The chapter also examines how tissue culture contributes to controlled mutagenesis for crop development by inducing genetic variety through somaclonal variation. It also emphasizes the critical function that tissue culture plays in genetic modification, which results in the creation of transgenic crops with certain desirable features (Thorpe, 2013; Brown and Thorpe, 1995). The chapter discusses the difficulties in using tissue culture methods and offers ways to improve them so that crop development initiatives can use them successfully. It also highlights tissue culture's importance in biodiversity conservation and sustainable agriculture, as well as how it helps conserve and regenerate economically significant or endangered plant species. To provide readers with a comprehensive understanding of the subject's scope, the final section offers insights into developing technologies and potential advancements in the field while reflecting on the future prospects of tissue culture in crop development (Wijerathna-Yapa and Hiti-Bandaralage, 2023; Murphy, 2007).

9.2 FUNDAMENTALS OF TISSUE CULTURE

The interesting voyage of tissue culture's historical growth highlights the advancement of this ground-breaking method in plant biology. The goal of comprehending plant growth and development at the cellular level was the primary driving force behind the early 20th-century emergence of tissue culture as a scientific field (Wijerathna-Yapa and Hiti-Bandaralage, 2023; Thorpe, 2007). The idea of cultivating plant tissues *in vitro* was made possible by the pioneering work of scientists like Gottlieb Haberlandt. It demonstrated that isolated plant tissues could thrive in a controlled, artificial environment when he successfully cultivated them in 1902. But tissue culture did not take off and become a potent tool for manipulating and propagating plants until the middle of the 20th century. The development of culture medium and the introduction of controlled-environment technology were crucial to the evolution of tissue culture techniques. The effective regeneration of plants from callus cultures in the 1950s signaled a paradigm shift and demonstrated the usefulness of tissue culture in plant breeding (Hussain et al., 2012; Sussex, 2008).

As tissue culture methods developed, researchers shifted their emphasis from fundamental concepts to real-world uses. Advances in the regeneration of entire plants from single cells during the 1970s gave rise to the technology known as micropropagation, which is now extensively used to produce large quantities of genetically identical plants. Tissue culture advanced to the forefront of plant biotechnology in the following decades as methods were improved and molecular biology technologies were incorporated (Long et al., 2022; Shashank et al., 2021).

Tissue culture is becoming an essential technique in crop improvement because it makes it possible to quickly replicate elite genotypes, introduce genetic variety, and carry out precise genetic transformations. The progression of tissue culture from Haberlandt's initial studies to its current uses illustrates the revolutionary influence it has had on our comprehension of plant biology and its usefulness in advancing agricultural practices (Wijerathna-Yapa and Hiti-Bandaralage, 2023).

9.3 CELL CULTURE TECHNIQUES

In several scientific fields, such as cell biology, medicine, and biotechnology, cell culture techniques are essential. These methods entail keeping cells in a controlled laboratory environment while cultivating them outside of their natural habitat. The primary goal is to create an environment that aids in the survival, growth, and study of cells. This will allow researchers to explore the behavior of cells, carry out tests, and create applications that can be used for everything from biopharmaceutical manufacturing to medical research (Wijerathna-Yapa and Hiti-Bandaralage, 2023; Haque et al., 2022; Khan et al., 2018).

Since their conception, cell culture techniques have undergone tremendous evolution, and the successful use of these techniques depends on a number of critical factors. First and foremost, selecting the right primary cell culture or cell line is essential. Certain cell types have distinct development requirements and behaviors; therefore, it is important to choose them carefully to make sure they are relevant to the goals of the research. Another important element is the culture medium, which contains growth stimulants, vitamins, and other minerals. This medium can be modified to meet the unique requirements of the cells being studied and is designed to offer an environment that is favorable to cell growth. It is essential to keep the cell culture environment sterile at all times to avoid contamination that can taint the results of the experiment. In cell culture laboratories, aseptic procedures that include the use of sterile equipment and laminar flow hoods are conventional procedures (Molina-Ruiz et al., 1984; Geraghty et al., 2014).

9.4 FORMULATION OF MEDIA

As it entails, creating a nutrient-rich environment that supports the growth, multiplication, and viability of cells outside of their native habitat, media formulation is an essential component of cell culture procedures. The culture medium's composition is carefully crafted to replicate the physiological circumstances necessary for cell viability and optimal performance. An efficient cell culture medium is formulated with the help of certain essential ingredients (Lin et al., 2017; Yao and Asayama, 2017).

The culture medium's base, the basal medium, supplies vital nutrients such as vitamins, carbohydrates, minerals, and amino acids. Roswell Park Memorial Institute Medium, Dulbecco's Modified Eagle Medium, and Minimum Essential Medium are examples of common basal media. The particular needs of the cell type being grown determine which basal media is best (Zhang and Wen, 2011; Yang and Xiong, 2012). The medium is frequently supplemented with fetal bovine serum or other animal-derived sera to supply a range of growth factors, hormones, and proteins essential for cell proliferation. For some applications, serum-free media formulations with predetermined additives or synthetic alternatives are also used to reduce variability and solve ethical issues with animal-derived goods (Subbiahanadar Chelladurai et al., 2021; Kolkmann et al., 2022). Buffering agents such as bicarbonate and HEPES (4-(2-hydroxyethyl)-1-piperazineethanesulfonic acid) are added to the solution to keep the pH within a physiological range. For metabolic activity and cell viability, proper pH regulation is crucial (Michl et al., 2019; Will et al., 2011). It is possible that some cell types need more hormones or growth factors to develop and differentiate to their full potential. Additives that affect cell activity in culture include insulin, basic fibroblast growth factor, and epidermal growth factor (Ahmad et al., 2023; Youssef et al., 2017). Iron, copper, and zinc are among the essential trace elements that are added to the medium to support cellular processes and preserve physiological equilibrium (Prabhu and Gadgil, 2021). Antibiotics (like penicillin-streptomycin) and antimycotics (like amphotericin B) are frequently added to the growth medium to avoid bacterial or fungal contamination (Campos et al., 2012). Supplements such as glutamine, nonessential amino acids, or certain lipid formulations may be added to improve cell growth and performance, depending on the kind of cell and the goals of the experiment (Yao and Asayama, 2017).

9.5 METHODS OF STERILIZATION

To avoid contamination and preserve the integrity of studies and cell cultures, sterilization is an essential step in microbiological and cell culture labs. Sterilization procedures are used for chemicals, media, and equipment. The choice of procedure relies on the particular needs of the materials to be sterilized, as each has pros and cons of its own. The type of material, how sensitive it is to heat, and the intended use all influence the sterilizing method selection. For thorough sterilization in laboratory settings, many techniques are frequently used. Sterilization procedures must be properly validated and monitored to guarantee their efficacy and preserve the integrity of experimental work (Jildeh et al., 2021; Mohapatra, 2017).

9.5.1 AUTOCLAVING

One popular technique for sterilizing under pressure using steam is autoclaving. The pressure and high temperature (usually between 121°C and 134°C) successfully kill spores, fungus, and bacteria. Glassware, media, culture solutions, and heat-resistant equipment can all be sterilized with autoclaving (Schoeb et al., 2017; Dockery, 2012).

9.5.2 DRY HEAT STERILIZATION

Materials are subjected to high temperatures (often between 160°C and 180°C) for a lengthy amount of time during the process of dry heat sterilizing. While it might take longer than autoclaving, it is appropriate for objects that can tolerate high temperatures. Certain powders, metal tools, and glassware are frequently sterilized using dry heat (Sadeque and Balachandran, 2020).

9.5.3 CHEMICAL STERILIZATION

Using chemical agents to eradicate bacteria is known as chemical sterilization. Hydrogen peroxide vapor and ethylene oxide gas are common chemical sterilants. These techniques work especially well for sterilizing heat-sensitive objects such as electronics, plastic ware, and some medical equipment (Rutala and Weber, 2015).

9.5.4 FILTRATION

The process of filtering eliminates germs from liquids or gasses by using porous membranes. Sterilizing heat-sensitive solutions, media, and serum is a popular use for it. Sterile liquids or gases can pass through membrane filters whose pores are small enough to hold germs (Coté, 1999).

9.5.5 ULTRAVIOLET IRRADIATION

Air and surfaces can be effectively sterilized by ultraviolet (UV) light. It causes harm to bacteria' DNA, which stops them from replicating.

Although it is frequently found in biosafety cabinets, laminar flow hoods, and other laboratory equipment, its limited penetration capabilities render it inappropriate for sterilizing opaque objects or liquids (Yen et al., 2014).

9.5.6 *RADIATION STERILIZATION*

Certain disposable objects including plastics, gloves, and medical gadgets are sterilized using ionizing radiation such as gamma rays and electron beams. By doing this, the microbes' DNA is damaged, making it impossible for them to reproduce (Singh et al., 2016).

9.5.7 *MICROWAVE STERILIZATION*

Certain materials can be effectively sterilized in a microwave, particularly those that can withstand heat. To achieve sterilization, it depends on the substance itself producing heat. Nevertheless, compared to other techniques, it is less frequently utilized in laboratory settings (Youssef and Amin, 2001).

9.6 PLANT REGENERATION

9.6.1 *ORGANOGENESIS*

Organogenesis, the process by which organs or particular tissues grow from undifferentiated cells, is an important component of plant regeneration. Organogenesis is essential to the regeneration of entire plants from explants that are tiny portions of plant tissues or organs that are cultivated *in vitro*, in the context of plant tissue culture as presented in Figure 9.1. This technique provides a regulated setting in which to alter a cell's course during development and bring about the genesis of new organs (Delporte et al., 2014; Bull and Michelmore, 2022).

9.6.1.1 THE PROCESS OF ORGANOGENESIS

The typically involves the key steps mentioned below.

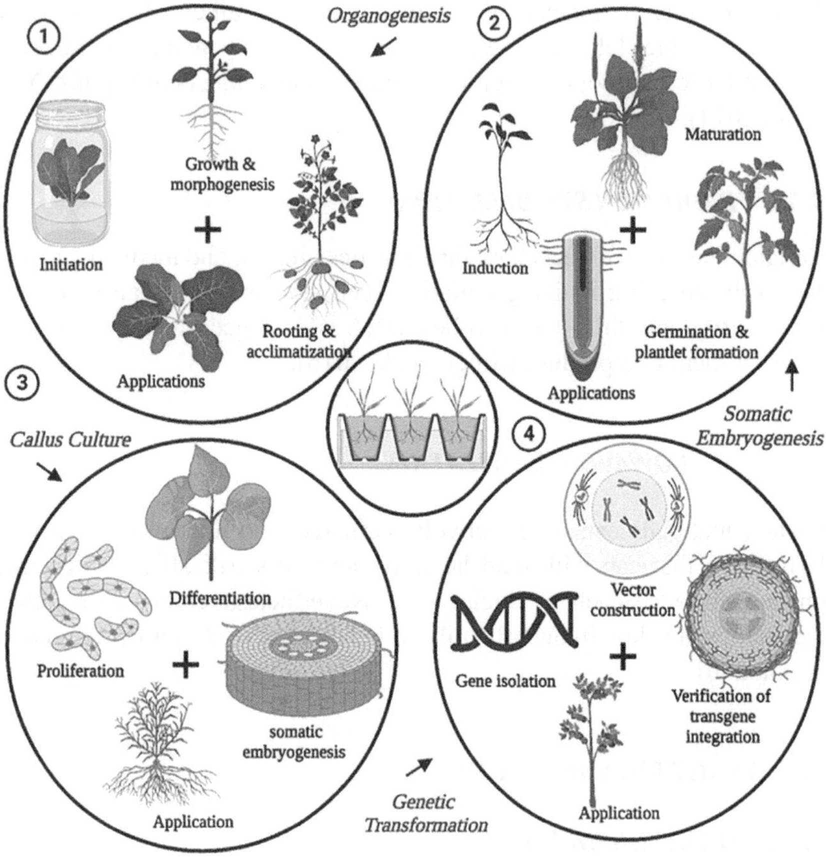

FIGURE 9.1 The summary of plant regeneration process involving organogenesis, somatic embryogenesis, callus culture, and genetic transformation.

9.6.1.1.1 Initiation Phase

The beginning of undifferentiated cells, which are frequently produced from explants like shoot tips, leaves, or embryos, is the initial stage of organogenesis. By being forced to dedifferentiate, these cells gain a meristematic state that allows them to potentially differentiate into different kinds of cells.

9.6.1.1.2 Determination and Differentiation

After initiation, cells go through a phase called determination whereby they commit to particular developmental pathways that result in the construction of organs. The differentiated cells then group together to form unique tissues, reflecting the developmental steps seen during typical plant growth (Bidabadi and Jain, 2020; Fehér, 2019).

9.6.1.1.3 Growth and Morphogenesis

The developing plantlet goes through morphogenesis as the differentiated tissues grow, taking on the distinctive shape and form of a mature plant. In this stage, a variety of tissues, including the shoot's apical meristem, leaves, stems, and roots, grow and organize in unison.

9.6.1.1.4 Rooting

The formation of roots is a crucial stage in numerous regeneration protocols, since it facilitates the proper acclimation and establishment of regenerated plants. To initiate roots, the culture media is frequently supplemented with certain hormones or growth regulators.

9.6.1.1.5 Acclimatization

The plantlets are acclimated to *ex vitro* conditions once they have formed roots and are able to support themselves. To get the plants ready for transplantation into soil, this entails progressively exposing them to ambient environmental conditions (Perianez-Rodriguez et al., 2014; Horstman et al., 2017).

9.6.1.2 APPLICATIONS AND SIGNIFICANCE

Numerous uses for organogenesis in plant tissue culture can be found in horticulture, agriculture, and plant biotechnology (Gholap et al., 2023a, 2024b, 2023c, 2023d, 2023e, 2023f, 2023g, 2023h, 2023, 2024a, 2024b, Pardeshi et al., 2023; Faiyazuddin et al., 2023). It makes it possible to produce large quantities of clonally identical plants with desired characteristics such as increased yield and resistance to disease (Gholap, 2012a, Gholap et al. 2012, 2023h; Gholap and Rajurkar, 2012a, 2012b, 2012c,

2012d; Vilas Bhagat, 2012; Rajurkar, 2012; Gholap et al., 2011a, 2011b; Gholap and Mahajan, 2011; Mahajan and Gholap, 2011; Gholap and Musmade, 2011; Mahajan and Gholap, 2011; Gholap, 2012; Khedkar and Tupe, 2012a, Gholap and Dighe, 2012a, 2012b; Khedkar and Tupe, 2012b; Tupe and Gholap, 2012). Organogenesis also plays a key role in the propagation of plants that are difficult to propagate using traditional techniques, which helps to preserve endangered species and produce superior plant types (Hasnain et al., 2022). To improve tissue culture procedures and increase plant regeneration efficiency, it is crucial to comprehend the physiological and molecular principles underpinning organogenesis. Organogenesis processes are still being investigated and improved upon by researchers, leading to advances in genetic engineering, crop development, and the sustainable production of plants for a variety of uses (Efferth, 2019; Ozyigit et al., 2023).

9.6.2 SOMATIC EMBRYOGENESIS

In plant tissue culture, a process known as somatic embryogenesis allows nonreproductive cells, or somatic cells, to mimic the embryogenic growth often associated with zygotic embryos. It is an intriguing and potent phenomenon. Somatic embryos are formed via this special regeneration pathway, and these embryos can eventually mature into fully grown plants. A flexible and effective technique for genetic transformation, clonal propagation, and germplasm preservation is somatic embryogenesis (Méndez-Hernández et al., 2019; Narváez et al., 2019).

9.6.2.1 SOMATIC EMBRYOGENESIS PROCESS

The process of somatic embryogenesis can be summarized in several key stages mentioned below.

9.6.2.1.1 Induction

Explant tissues, such as leaf or root tissues, are cultured under certain circumstances that include the injection of plant growth regulators, especially auxins and cytokinins, to induce somatic embryogenesis. Somatic cells dedifferentiate into embryogenic cells in response to these hormone cues.

9.6.2.1.2 Formation of Embryogenic Callus

The stimulated cells multiply and develop into an embryogenic callus, which is a mass of undifferentiated cells. The cells that make up this callus have the capacity to grow into somatic embryos.

9.6.2.1.3 Somatic Embryo Development

Somatic embryos start to form within the embryogenic callus. The process that these embryos go through, including the differentiation of embryonic organs like cotyledons, hypocotyls, and radicles, is similar to the stages of embryogenesis seen in zygotic embryos (Wang et al., 2023).

9.6.2.1.4 Maturation

The somatic embryos mature as they grow, gaining the capacity to sprout and mature into full-fledged, functioning plants. This stage frequently entails modifying the culture conditions such as the growth regulator concentration and the culture medium's makeup (Ramakrishna and Shasthree, 2016).

9.6.2.1.5 Germination and Plantlet Formation

Plantlets are created by inducing germination in mature somatic embryos. These plantlets can be grown further to produce shoots and roots, which will ultimately result in fully grown plants (Farhadi et al., 2021).

9.6.2.2 APPLICATIONS AND SIGNIFICANCE

Numerous real-world uses for somatic embryogenesis exist in plant biotechnology and agriculture. Plants with desired features can be mass-clonally propagated using somatic embryogenesis, which makes it possible to produce identical plantlets with high efficiency. Somatic embryos can be used for genetic transformation, which makes it easier to transfer particular genes into plants and create transgenic crops with enhanced characteristics (Efferth, 2019). By retaining the genetic material of endangered or commercially significant plant species in the form of somatic embryos, this procedure is beneficial for their conservation. A potent tool for crop improvement initiatives, somatic embryogenesis allows for the quick creation of plants with desired traits like increased yield, resistance to disease, or resilience to environmental stressors. It is essential to comprehend the complex physiological and molecular processes behind somatic embryogenesis to improve and optimize tissue culture procedures. Research in this field is important because it advances plant biotechnology, which in turn helps conservation efforts to preserve biodiversity and promote sustainable agriculture (Pais, 2019).

9.6.3 CALLUS CULTURE

A key component of plant tissue culture is callus culture, which involves growing the undifferentiated, proliferating masses of cells called callus. This technique provides a flexible platform for researching the behavior of plant cells, plant regeneration, and genetic modification. A variety of plant tissues can be used to induce callus, which is essential for the creation of regeneration protocols used in genetic engineering and agricultural improvement (Bidabadi and Jain, 2020).

9.6.3.1 CALLUS CULTURE PROCESS

The process of callus culture involves several key steps mentioned below.

9.6.3.1.1 Initiation

Explants are small portions of plant tissues that are often cultured on a nutrient-rich media supplemented with certain plant growth regulators, especially auxins like indole-3-acetic acid (IAA) or 2,4-dichlorophenoxyacetic acid (2,4-D). This process is known to trigger callus initiation. These hormones promote the creation of an undifferentiated mass of cells and cell dedifferentiation (Bhatia, 2015).

9.6.3.1.2 PROLIFERATION

Once the process is under way, the callus quickly divides into many cell masses, each of which is actively expanding. A robust and actively dividing callus can be maintained with regular subculturing on fresh medium (Ikeuchi et al., 2013).

9.6.3.1.3 DIFFERENTIATION

Depending on the makeup of the culture media, callus cells can develop into a variety of cell types and tissues. Differentiation pathways can be induced by varying the type and concentration of plant growth regulators.

9.6.3.1.4 MORPHOGENESIS

Callus cultures have the capacity to go through morphogenesis, which produces ordered structures like branches or roots. The regeneration of entire plants from callus cultures depends on this process (Yancheva and Kondakova, 2018).

9.6.3.2 APPLICATIONS AND SIGNIFICANCE

Organogenesis or somatic embryogenesis can be used to regenerate entire plants, and callus cultures are the first step in this process. Programs aimed at improving crops and clonal propagation will find this very helpful. Cultures of callus offer a target for genetic engineering. By introducing foreign DNA, they can help create transgenic plants that have desired characteristics like increased nutritional value or resistance to pests. Some callus cultures can create alkaloids, flavonoids, and medicinal substances as secondary metabolites. Because of this, callus cultures are useful for producing bioactive substances (Hussain et al., 2012; Bidabadi and Jain, 2020).

The production of disease-resistant cultivars and the research of plant-pathogen interactions are two applications for callus cultures. These interactions can be observed and manipulated in the controlled setting of tissue culture. By offering a way to conserve the genetic material of plant species—including those that are endangered or challenging to grow using conventional techniques—callus cultures support efforts to conserve germplasm. Optimizing tissue culture techniques and attaining good results in plant regeneration and biotechnological applications require an understanding of the mechanisms driving callus initiation, proliferation, and differentiation. The methods for cultivating callus are still being developed, and this is greatly advancing both plant science and sustainable agriculture (Hasnain et al., 2022; Hussain et al., 2012).

9.6.4 SHOOT AND ROOT FORMATION

In plant tissue culture, the development of shoots and roots is an important stage that marks the differentiation and growth of ordered structures from undifferentiated cells or tissues. This process is essential to plant regeneration methods and has important ramifications for genetic engineering, clonal propagation, and crop enhancement (Hasnain et al., 2022).

9.6.4.1 SHOOT AND ROOT FORMATION PROCESS

9.6.4.1.1 Shoot Formation (Organogenesis)

1. Induction: Usually, the plant growth regulators in the culture media are adjusted to start the induction of shoot production. It is common practice to use a balanced ratio of auxins (such as IAA) and cytokinins (like benzyladenine). Auxins aid in the formation of calluses, while cytokinins encourage the growth of shoots.
2. Dedifferentiation: Upon dedifferentiating, the callus's cells return to their meristematic stage and get ready to differentiate into shoot structures.
3. Shoot primordia formation: Cell clusters within the callus differentiate into miniature structures known as shoot primordia in response to particular hormonal stimuli. The early phases of shoot development are represented by these primordia.

9.6.4.1.2 Root Formation (Organogenesis)

1. Induction: Changing the kind and concentration of auxins in the culture media causes the formation of roots. Auxins such as naphthaleneacetic acid and indole-3-butyric acid are frequently employed.

2. Dedifferentiation: The cells adopt a meristematic state that may lead to the differentiation of root structures, much like in shoot induction.
3. Root primordia formation: Cell clusters within the callus undergo differentiation to become root primordia, which are early stages of root growth. These structures grow longer and longer until they become roots (Hnatuszko-Konka et al., 2021; Kumlay and Ercisli, 2015).

9.6.4.1.3 Maturation and Acclimatization

1. Maturation: Tissue culture settings are frequently changed to encourage the maturation of shoot and root primordia once they have formed. This entails adjusting the hormonal balance and culture medium composition.
2. Acclimatization: Plantlets with fully formed roots and shoots after maturation are acclimated to the *in vitro* environment. To get them ready for transplantation to soil, this entails progressively exposing them to ambient environmental conditions (Long et al., 2022; Miri, 2020).

9.6.4.2 APPLICATIONS AND SIGNIFICANCE

The development of shoots and roots in tissue culture facilitates the clonal replication of plants with desired characteristics, guaranteeing genetic homogeneity in the regenerated plants. In crop improvement initiatives, the capacity to promote shoot and root growth in tissue culture is crucial because it speeds up the process of producing genetically identical plants with enhanced traits. An essential part of the genetic metamorphosis of plants is the production of shoots and roots. It is possible to regenerate transgenic plants from tissue cultures by introducing genes for desired features (Espinosa-Leal et al., 2018). Plant germplasm is preserved in part by tissue culture techniques that incorporate the development of shoots and roots, especially for species that are challenging to replicate using conventional techniques. It is crucial to comprehend the nuances of shoot and root development in tissue culture to optimize procedures and accomplish efficient plant regeneration. The field's ongoing research keeps improving these procedures, broadening their uses and boosting their effectiveness across a range of plant biotechnology and agricultural applications (Bidabadi and Jain, 2020).

9.6.5 GENETIC TRANSFORMATION

9.6.5.1 TRANSGENIC PLANT PRODUCTION

The process of introducing foreign genes into an organism's genome to produce new features is known as genetic transformation. When it comes to plants, transgenic plant production is a potent biotechnology technique that makes it possible to incorporate desired genes, which frequently bestow features like pest resistance, abiotic stress tolerance, or enhanced nutritional value. This procedure is essential to crop enhancement and modern agriculture (Low et al., 2018).

9.6.5.2 TRANSGENIC PLANT PRODUCTION PROCESS

9.6.5.2.1 Gene Isolation

1. Finding and isolating the particular gene or genes of interest that encode the desired feature is the first stage in the identification process.
2. Gene cloning: To produce numerous copies for later transformation, the isolated gene is cloned, usually utilizing molecular biology techniques.

9.6.5.2.2 Vector Construction

1. Insertion into vector: The cloned gene is put into a vector, which is frequently a bacterial plasmid. The foreign gene is carried by the vector.
2. Regulatory elements such as a promoter, which drives gene expression, and a terminator, which signals the end of transcription, are included in the vector (Adrio and Demain, 2010).

9.6.5.2.3 Transformation

1. Introduction of the vector into the plant cells: The plant cells are exposed to the created vector. This is accomplished via a number of techniques such as electroporation, particle bombardment (gene cannon), and transformation mediated by *Agrobacterium*.
2. Integration into plant genome: The foreign gene becomes permanently integrated into the DNA of the host plant by integrating into the plant genome.

9.6.5.2.4 Selection and Regeneration

1. Selectable marker: To recognize and choose out transformed cells, the vector frequently contains a selectable marker gene such as a gene for herbicide or antibiotic resistance.
2. Regeneration of transgenic plants: To create complete plants from transformed cells, tissue culture is used to cultivate the transformed cells. Plant regeneration techniques are then applied (Rafeeq et al., 2023).

9.6.5.2.5 Verification of Transgene Integration

1. Molecular analysis: To verify that the transgene has been integrated into the plant genome and is present, polymerase chain reaction or other molecular techniques are employed.

9.6.5.2.6 Acclimatization and Growth

1. Acclimatization: After being exposed to *ex vitro* environments, the transgenic plants progressively adjust to their new surroundings.
2. Growth and development: Transgenic plants are cultivated in controlled environments to track the expression of the imported trait after they have acclimated (Low et al., 2018).

9.6.5.3 APPLICATIONS AND SIGNIFICANCE

By expressing genes that give plants resistance to pests, transgenic plants can lessen their need on chemical pesticides. Plant resilience can be increased by introducing genes that provide tolerance to environmental challenges like salinity or drought. By adding genes for higher vitamin or mineral levels, genetic transformation can be utilized to raise the nutritional content of plants. By modifying plants to withstand particular herbicides, weeds can be controlled in agriculture more successfully (Ahmad et al., 2012; Parmar et al., 2017). By introducing genes that provide resistance to particular illnesses, agricultural losses caused by pathogens can be decreased. The development of transgenic plants is a useful tool for studying the physiology of plants and the action of genes. Transgenic techniques have the potential to support plant germplasm conservation, especially for commercially significant or endangered species. Transgenic plant development and genetic modification provide novel approaches to a range of agricultural problems. The technology is very promising, but it is also under public and governmental scrutiny, which emphasizes how crucial ethical and environmental factors are when developing and implementing transgenic crops (Kobayashi et al., 2023).

9.7 METHODS OF GENE TRANSFER

The processes used to incorporate foreign genes into an organism's genome are referred to as gene transfer procedures. These techniques allow the insertion of particular genes into plant cells in the context of plant genetic engineering, leading to the creation of transgenic plants with desired features (Narusaka et al., 2012). Numerous techniques for gene transfer have been devised, each having pros and cons.

In plant biotechnology, the following are some typical gene transfer techniques.

9.7.1 AGROBACTERIUM-MEDIATED TRANSFORMATION

Soil bacteria *Agrobacterium tumefaciens* is frequently employed as a vector to introduce genes into plant cells. The altered *Agrobacterium* is then utilized to infect plant tissues after the desired genes are added to a Ti (tumor-inducing) plasmid. Because of its effectiveness and capacity to transfer genes directly into the plant genome, this technique is widely employed. It is frequently used with plants that are dicotyledonous (Hwang et al., 2017; Gelvin, 2003).

9.7.2 PARTICLE BOMBARDMENT (BIOLISTICS OR GENE GUN METHOD)

A gene gun is used to fire DNA-coated microprojectiles into plant cells. The DNA is delivered into the plant cell by the fast-moving particles that pierce the cell walls. Because of its versatility, particle bombardment can be applied to both monocots and dicots, among other plant species. It permits the transformation of particular tissues and cell types (Kikkert et al.).

9.7.3 ELECTROPORATION

To create transient holes in the cell membrane of plant protoplasts or cells containing foreign DNA, electric pulses are employed. This permits DNA to enter the cells. Applying electroporation to a variety of plant species is a quick and easy process. However, in comparison to other approaches, it frequently performs less well (Potter, 2003).

9.7.4 MICROINJECTION

The direct injection of foreign DNA into plant cells is accomplished with a micropipette. This labor-intensive technique is usually used on single cells or protoplasts. Although it is not commonly used, microinjection offers fine control over DNA delivery, making it appropriate for some applications (Zhang and Yu, 2008).

9.7.5 PROTOPLAST FUSION (SOMATIC HYBRIDIZATION)

This process creates hybrid cells by fusing the protoplasts of various plant species. Genes from both parent species may be incorporated into and expressed by these hybrid cells. Combining desired features from multiple species is made possible via protoplast fusion; however, the resultant plants may run into regulatory issues (Ranaware et al., 2023).

9.7.6 VIRUS-MEDIATED GENE TRANSFER

To introduce foreign genes into plant cells, viral vectors are employed. The desired gene is inserted into the viral genome, and when the virus infects the plant, the gene is transferred during the infection process. Although this approach makes effective gene delivery possible, the use of viral vectors may raise regulatory issues (Mahmood et al., 2023).

9.7.7 NANOTECHNOLOGY-MEDIATED GENE TRANSFER

Foreign DNA is encapsulated and delivered into plant cells using liposomes, which are lipid-based vesicles. DNA is released into the cell when the liposomes and cell membrane merge. Liposome-mediated gene transfer is a flexible method that can be used to numerous plant species (Balazs and Godbey, 2011).

The plant species, the kind of tissue to be changed, and the particular needs of the application or experiment all influence the choice of gene transfer technique. Plant genetic engineering advancements are leading to the development of genetically engineered crops and plant types with better features by improving the accuracy and efficiency of these techniques (Dong and Ronald, 2019).

9.8 MARKER GENES AND SELECTION SYSTEMS

In genetic engineering, selection mechanisms and marker genes are crucial elements that facilitate the recognition and separation of effectively altered cells. The development of transgenic organisms, including plants, depends heavily on these components. Marker genes serve as markers of successful gene transfer, delivered with the gene of interest. However, selection methods offer a way to identify and isolate only those cells that have successfully integrated the foreign genes (Gelvin, 2003).

9.8.1 MARKER DNA

Researchers can physically or biochemically identify altered cells thanks to marker genes, which often encode proteins or enzymes that are easy to detect (Li et al., 2020).

Frequently employed marker genes are mentioned below.

9.8.1.1 GREEN FLUORESCENT PROTEIN (GFP)

When exposed to UV light, GFP glows green, making it possible to see altered cells under a fluorescence microscope. It is widely used to indicate the successful transfer and expression of genes visually (Remington, 2011).

9.8.1.2 B-GLUCURONIDASE (GUS)

When GUS interacts with particular substrates, it produces a blue hue that makes it possible to identify altered cells by histochemical staining. It is extensively used as a biological marker in research on gene expression (Basu et al., 2004).

9.8.1.3 LUCIFERASE

The enzyme luciferase catalyzes the oxidation of luciferin, resulting in the emission of light. Glowing signifies a successful metamorphosis. It is utilized to track the expression of genes in real time as presented in Figure 9.2 (Greer and Szalay, 2002).

9.8.2 SELECTION SYSTEMS

Genes that confer resistance to specific selected agents, such as antibiotics or herbicides, are used in selection systems. It is possible to isolate transformed cells because only cells that have successfully integrated and expressed the inserted genes are resistant to exposure to these chemicals (Dormatey et al., 2021).

Typical selection markers are mentioned below.

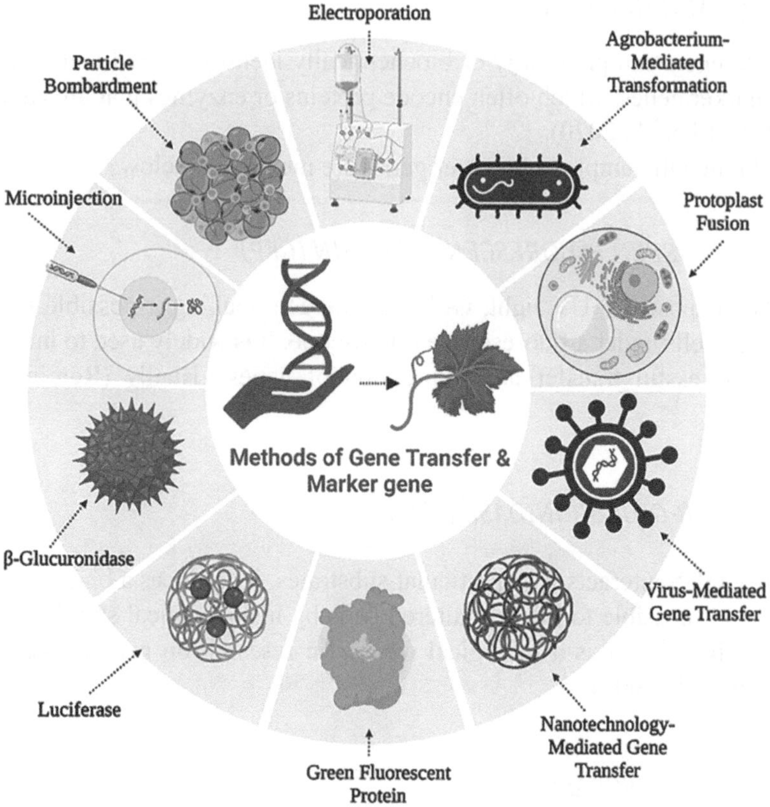

FIGURE 9.2 Methods of gene transfer and marker gene used in genetic transformation.

9.8.2.1 ANTIBIOTIC RESISTANCE GENES

Specific antibiotic exposure does not harm cells that possess antibiotic resistance genes. Plant transformation frequently makes use of antibiotic resistance markers such as kanamycin or hygromycin resistance.

9.8.2.2 HERBICIDE RESISTANCE GENES

When exposed to certain herbicides, cells with these genes survive. Many transgenic crops use indicators of herbicide resistance like resistance to glyphosate (Roundup) (Schütte et al., 2017).

9.8.2.3 SELECTIVE NUTRIENT UTILIZATION GENES

These genes allow cells to thrive in conditions deficient in particular nutrients. In certain systems, they are used in place of herbicide or antibiotic resistance indicators (Manyi-Loh et al., 2018).

9.8.3 APPLICATIONS AND SIGNIFICANCE

The effectiveness of genetic engineering projects depends on marker genes and selection mechanisms by making it easier to quickly and accurately identify transformed cells; marker genes free up researchers' time to concentrate on effective transformations. Only those cells that have incorporated and expressed the relevant foreign genes can be isolated and propagated using selection techniques (Dormatey et al., 2021). Tracer genes, like GFP and GUS, are very useful for tracking the expression patterns of newly added genes. The production of transgenic plants and creatures with enhanced features or characteristics depends heavily on the use of selection methods and marker genes. Although selection techniques and marker genes have greatly advanced genetic engineering, their use is fraught with ethical and legal issues, demanding careful assessment in the creation of genetically engineered organisms (Amiteye, 2021).

9.8.4 APPLICATIONS IN CROP IMPROVEMENT

9.8.4.1 DISEASE RESISTANCE

The creation of plants resistant to disease is one of the major uses of genetic engineering in crop improvement. Using this method, crops are genetically modified to impart resistance against infections such as bacteria, viruses, fungus, and other dangerous microbes (Dong and Ronald, 2019). The potential of genetic manipulation to increase crop resistance to disease has significant implications for global food security, sustainable agriculture, and crop loss reduction (Yu et al., 2022).

9.8.4.2 RESISTANCE TO DISEASE

Crops lacking natural defenses against particular infections can now have resistance genes introduced into them through genetic engineering. These genes for resistance can be derived from other plant species that are recognized for their ability to withstand specific illnesses (Dong and Ronald, 2019).

9.8.4.2.1 Fungal and Bacterial Resistance

For instance, antimicrobial proteins or peptides can be designed into crops like rice, wheat, or tomatoes that are prone to bacterial or fungal infestations. These proteins improve the plant's resistance to diseases by preventing the growth of pathogens (Tang et al., 2023).

9.8.4.2.2 Resistance to Viruses

For instance, adding genes that produce viral coat proteins can be used to create crops that are resistant to viruses. By stopping the virus from spreading within the plant, this interference offers strong defense against viral infections (Nicaise, 2014).

9.8.4.2.3 Interference with RNA (RNAi)

By silencing particular genes in the pathogen, RNAi technology might lessen the virus's capacity to infect crops and cause disease. As an illustration, RNAi has been utilized to provide resistance to illnesses brought on by insects, nematodes, and certain viruses (Younis et al., 2014).

9.8.4.2.4 Stacking Resistance Genes

To give broad-spectrum resistance against a variety of infections, multiple resistance genes can be layered in a single crop. This strategy is especially useful in areas where agricultural output is threatened by several diseases (Crété et al., 2020).

9.8.4.2 APPLICATIONS AND SIGNIFICANCE

By lessening the influence of pathogens, disease-resistant crops help cut down on crop losses brought on by illnesses. Maintaining stable and high-yielding agricultural systems depends on this. Using genetically modified crops that are *naturally* resistant to disease can help cut down on the need of chemical pesticides, which encourages ecologically friendly and environmentally sustainable farming methods (van Esse et al., 2020; Abdul Aziz et al., 2022). By guaranteeing a more dependable and steady food supply, disease-resistant crops help to promote global food security. With new and developing diseases, farmers are better able to supply the increasing demand for food. Disease resistance through genetic engineering is in line with the ideals of sustainable agriculture, which encourages methods that are socially, economically, and environmentally appropriate. By lowering the need for chemical treatments that can negatively impact nontarget creatures, disease-resistant crops can contribute to the preservation of biodiversity. Although there are many advantages to using disease-resistant genetically modified crops, its implementation necessitates careful consideration of safety, environmental effect, and regulatory compliance. To address issues in global agriculture, biotechnology developments and ongoing research will help develop crop varieties that are more resilient and resistant to disease (Bawa and Anilakumar, 2013; Vincelli, 2016).

9.8.5 TOLERANCE TO ABIOTIC STRESS

Abiotic stress tolerance, or the capacity of plants to tolerate harsh environmental circumstances, is a key function of genetic engineering in crop development. Abiotic stressors that can greatly affect agricultural productivity include salinity, drought, high temperatures, and nutrient shortages. Through the introduction of particular genes that increase a plant's resistance to these stressors, genetic modification makes it possible to provide food security and sustainable agriculture in the face of changing climatic conditions (Yadav et al., 2020; Waqas et al., 2019). Some of the cases are presented as per follows.

9.8.5.1 TOLERANCE FOR DROUGHT

Genes linked to water-use efficiency, osmotic control, and antioxidant defense systems are among the genes that can be introduced by genetic engineering that are implicated in drought response pathways. Crops that can withstand drought might express genes that promote better water retention, more root development, or the production of osmoprotectants like proline (Mishra et al., 2023).

9.8.5.2 SALINITY TOLERANCE

To improve a plant's capacity to flourish in saline soils, genes related to ion homeostasis, osmotic adjustment, and the exclusion of harmful ions can be added. Genes encoding ion transporters, detoxifying enzymes for salt, or proteins involved in osmotic adjustment may be expressed by salt-tolerant crops (Huang et al., 2020).

9.8.5.3 TOLERANCE TO TEMPERATURE STRESS

Genes involved in protein folding, membrane integrity, and antioxidant defense are among those that can be targeted by genetic engineering to control how the body reacts to heat or cold. Examples include the expression of heat shock proteins in heat-tolerant crops and increased membrane fluidity and antifreeze protein synthesis in cold-tolerant crops (Hasanuzzaman et al., 2013; Mondal et al., 2023).

9.8.5.4 TOLERANCE TO NUTRIENT DEFICIENCIES

Genes that enhance nutrient absorption, transport, and utilization within the plant can be introduced by genetic manipulation. Crops can be modified to acquire and use vital nutrients like nitrogen, phosphorus, or iron more effectively (Sathee et al., 2022; Hefferon, 2015).

9.8.5.5 TOLERANCE TO OXYGEN DEPRIVATION (FLOODING)

To improve resistance to oxygen-deprived conditions during floods, genes related to ethylene signaling, aerenchyma production, and other adaptive responses can be inserted. Crops that can withstand flooding might have better root aeration and different ethylene reactions (Yamauchi et al., 2021).

9.8.6 APPLICATIONS AND SIGNIFICANCE

Crops that are genetically engineered to be resistant to abiotic stressors can better tolerate the harsh circumstances of agriculture in the face of climate change, erratic weather patterns, and extreme weather. Crops that are resistant to abiotic stressors have enhanced productivity and performance in stressful environments, leading to higher yields and superior crop quality. By making better use of nutrients, water, and other resources, tolerable crops help to manage resources sustainably and lessen the negative effects of agriculture on the environment. By guaranteeing steady and reliable yields even in areas vulnerable to environmental stressors, the creation of abiotic stress-tolerant crops helps to ensure global food security (López-Arredondo et al., 2015; Raza et al., 2019). Tolerant crops might need less water and fertilizer inputs, which would lessen agriculture's negative environmental effects and protect ecosystems. While there are many advantages to developing crops resistant to abiotic stress, it is important to take into account possible agronomic and ecological effects, legal frameworks, and public acceptance before using genetically modified crops. Sustained investigation and progress in biotechnology are essential for tackling intricate problems in abiotic stress mitigation and enhancing the adaptability of worldwide agricultural production networks (Acevedo et al., 2020; Karavolias et al., 2021).

9.8.7 IMPROVED YIELD TRAITS

A major factor in increasing food yields is genetic engineering, which introduces genes that improve different aspects of plant growth, development, and production. Global food security is greatly enhanced by improved yield features, which allows farmers to grow more food on a smaller amount of land. These genetic alterations target processes that affect crop output overall, including photosynthesis, nutrient use, and resistance to environmental stresses (Qaim and Kouser, 2013).

9.8.8 GENETIC ENGINEERING FOR CROP IMPROVEMENT

By enhanced photosynthesis through genetic engineering, genes that maximize photosynthesis the process of turning sunlight into energy can be added. This could entail raising the rate at which carbon is fixed or increasing the use of light energy. To boost photosynthetic efficiency, crops can express genes encoding enzymes like Rubisco or introduce C4 photosynthetic processes (Batista-Silva et al., 2020).

Plants can be engineered to use available nutrients more efficiently by targeting genes involved in nutrient absorption, transport, and utilization. Better phosphorus uptake or increased nitrogen use efficiency can be bred into crops, which will improve nutrient use (Govindasamy et al., 2023). Increasing a crop's resilience to diseases and pests through genetic modification helps shield it against conditions that lower output. This may entail the insertion of genes that result in the production of insecticidal proteins or that provide resistance to particular infections. To reduce production losses from pests and illnesses, crops can express genes from *Bacillus thuringiensis* (Bt) for disease resistance or insect resistance (Dong and Ronald, 2019; Abbas, 2018).

By introducing genes that increase resistance to abiotic stresses like salinity, drought, or extremely high or low temperatures, genetic engineering can guarantee that crops can flourish in challenging environments. Genes for enhanced heat tolerance, osmotic control, or water-use efficiency may be expressed by crops, increasing their resistance to environmental stress (Esmaeili et al., 2022; Shelake et al., 2022). To maximize a plant's reproductive phase, genetic manipulation can target genes that control blooming time, floral structure, and fruit production. Genes that affect pollination, fruit setting, and flower development can be expressed by crops, which increases the yield of fruit and seeds (Tran et al., 2023).

9.8.9 APPLICATIONS AND SIGNIFICANCE

Higher crop yields are a result of improved yield qualities, which also guarantee a more plentiful and steady supply of food to fulfill the expanding demands of the world's population. Genetic engineering improves the way that resources, including sunshine, water, and nutrients, are used. This lessens resource waste, which supports sustainable agriculture. In areas with dense populations and little arable land, crops with enhanced yield features are essential for tackling food security issues (Datta, 2013). Genetically modified crops support environmentally sustainable agricultural methods by increasing resource efficiency and lowering the requirement for chemical inputs. Farmers gain economically from better crop yields, which raises their earnings and enhances their standard of living. Especially for smallholder farmers in emerging nations, this is crucial. Although the creation of crops with enhanced yield features has great potential, responsible and ethical application of these crops requires careful consideration of ecological, regulatory, and societal considerations. Genetic engineering technology developments and ongoing research help create crops that can sustainably provide the world's expanding population with food (Hefferon, 2015; Kavhiza et al., 2022).

9.9 NUTRITIONAL ENHANCEMENT

By biofortification with micronutrients through genetic modification, genes necessary for the production and accumulation of vital micronutrients, like vitamins and minerals, can be inserted into crop sections that are edible. A few examples of biofortified crops are wheat with higher zinc levels, beans with higher iron content, and Golden Rice, a type of rice enhanced with vitamin A. Increased protein content in crops with a higher protein content can be produced by genetic engineering by targeting the genes involved in protein synthesis. It is possible to modify crops like soybeans or maize to express genes that improve the synthesis of necessary amino acids or raise the amount of protein in general (Dhaliwal et al., 2022; Ofori et al., 2022). Amino acid enrichment with the nutritional value of protein in crops can be increased by genetic manipulation and by introducing genes that improve the production of particular amino acids. It is possible to genetically modify crops like rice or maize to increase their production of the necessary amino acid lysine. By fatty acid modification through genetic engineering, fatty acid biosynthesis-related genes can be targeted, resulting in crops with altered lipid profiles (Das et al., 2020). It is possible to genetically modify crops like canola or soybeans to yield oils with better nutritional qualities, like higher concentrations of omega-3 fatty acids. By reducing antinutritional factors through genetic modification, antinutritional components in crops can be reduced or eliminated, increasing the important nutrients' bioavailability. It is possible to engineer specific crop kinds, such as cassava or legumes, to have lowered amounts of antinutritional agents, increasing the accessibility of the nutrients (Beauregard and Hefford, 2006; Lea et al., 2016).

9.9.1 APPLICATIONS AND SIGNIFICANCE

Genetic engineering is used to improve nutrition to treat deficiencies in vital micronutrients, such as zinc, iron, and vitamin A, which are problems that impact millions of people worldwide. Raising the amount of protein and improving the quality of amino acids are beneficial for improving nutrition, particularly in areas where diets low in protein are prevalent. Increasing the amounts of vital fatty acids, including omega-3s, which are critical for cardiovascular health, by altering the fatty acid composition of crops has positive health effects. Genetically engineered crops that are high in nutrients are critical for treating hidden hunger, a condition in which people may eat enough calories but not enough of certain nutrients to maintain good health (Ghimire et al., 2023). Improving nutrition has the potential to have a major influence on global health, particularly in areas where access to a variety of nutrient-dense meals is restricted. By encouraging the growth of nutrient-dense cultivars and lowering reliance on dietary supplements, raising the nutritional content of staple crops helps to ensure agricultural sustainability. Genetic engineering presents a viable avenue for nutritional development; nevertheless, regulatory, ethical, and consumer acceptability issues must be taken into account. Ongoing studies and cooperative endeavors in the field of biotechnology facilitate the creation of crops that can tackle nutritional deficiencies and enhance the general health and welfare of a heterogeneous populace (Zafar and Jianlong, 2023; Kishore and Shewmaker, 1999).

9.10 ROLE OF MITIGATION STRATEGIES AND REGULATORY CONSIDERATION

Create and put into effect best practices for managing genetically modified, conventional, and organic crops together. This entails creating buffer zones, calculating isolation distances, and choosing sensible planting intervals. To comprehend the dynamics of gene flow in various crop systems, undertake in-depth study. Studying pollen distribution patterns, gene flow-influencing variables, and the possibility of transgenic persistence in the environment are all included in this. Investigate and create cutting-edge genetic technologies to help reduce pollution. This could involve developing crops with integrated genetic confinement systems or using gene-editing methods that have a lower chance of unintentional transmission (Caradus, 2023; Ceddia et al., 2007).

Boost genetically modified organism (GMO) detection and monitoring techniques to guarantee precise and sensitive transgene detection in crops, seeds, and processed goods. Enforcing regulations effectively requires this. Encourage open and honest communication and education to allay public fears and guarantee a knowledgeable discussion on the advantages, dangers, and mitigating techniques related to genetic engineering. Acceptance and well-informed decision-making can be encouraged by public knowledge. Genetic engineering contamination challenges require a multidisciplinary approach involving researchers, decision-makers, farmers, and the general public. Ensuring the responsible and sustainable deployment of genetic engineering in agriculture requires ongoing research, technological developments, and strong regulatory frameworks (Fraiture et al., 2015; Broeders et al., 2012).

Because it affects the dependability and long-term viability of GMOs, genetic stability is a crucial factor in genetic engineering. For genetically modified crops to behave consistently, inserted genes and their expression must be stabilized throughout multiple generations. Genetic stability issues can be caused by a number of things, including environmental variables, epigenetic modifications, and genomic integration. Unpredictable interactions and instability can occasionally result from the integration of foreign genes into the host genome. Unintentional changes or genomic rearrangements could happen during the integration process. Unpredictable phenotypic results can result from genetic instability at the integration site, which can impact the expression of the inserted genes (Bawa and Anilakumar, 2013; Raman, 2017).

Gene expression patterns can be influenced by epigenetic alterations, including DNA methylation and histone changes. Epigenetic modifications that affect the stability of gene expression can be brought on by developmental phases and environmental stimuli. Gene expression levels can vary as a result of epigenetic modifications, which may have an impact on the desirable qualities that are conferred by genetic modifications (Handy et al., 2011). The expression of genes that have been introduced can be influenced by environmental factors such as temperature, light, and the availability of nutrients. Changes in the environment can cause oscillations in the stability of gene expression. The dependability of genetically modified crops in various agricultural contexts may be impacted by inconsistent gene expression under variable environmental circumstances. The possibility of introducing genes to other organisms, such as bacteria, through horizontal gene transfer raises concerns. This calls into question the genetic changes' stability as well as their propensity to propagate to uninvited parties. Concerns about the environment and biosafety could arise from unintentional gene transfer's potential ecological effects and effects on nontarget organisms (Gibney and Nolan, 2010; Kumari et al., 2022).

9.11 REGULATORY AND ETHICAL CONSIDERATIONS

Genetic engineering has a lot of potential applications in agriculture, but there are also moral questions that need to be answered. To address concerns about the effects of GMOs on the environment, human health, and socioeconomic aspects, regulatory frameworks are essential for guaranteeing the safe and responsible use of GMOs (Zilberman et al., 2018). Determining the possible effects of genetically modified crops on the environment is a difficult task that necessitates in-depth research to grasp elements such as gene flow, ecological interactions, and unintended consequences. Strict environmental risk evaluations are carried out by regulatory bodies to determine how GMOs might affect nontarget animals, biodiversity, and ecosystems. It is crucial to guarantee that genetically modified crops are safe for ingestion by humans. Comprehensive assessment is necessary to address issues with allergenicity, unexpected consequences, and long-term health repercussions. To evaluate whether GMOs are safe for human consumption, regulatory agencies carry out thorough safety evaluations that include toxicity research, compositional analysis, and allergenicity testing (Nkonyam and Riedel, 2013; Meyer, 2011). The introduction of genetically modified crops may have social and economic repercussions such as problems with farmer access to technology, livelihoods, and the concentration of seed markets. When establishing regulations, policymakers take socioeconomic issues into account. They also put these elements into practice by enacting measures that provide fair access to technology, safeguard farmers' rights, and avoid market monopolies (Garcia-Yi et al., 2014). It might be difficult to tell customers in a straightforward and accurate manner whether or not food items include genetically modified components. It is crucial to strike a balance between the need to prevent misinformation and the right to know. Certain areas have labeling laws in place to alert customers to the existence of genetically modified foods. Regulatory organizations strive to prevent stigmatization while maintaining label openness (Giller et al., 2021; Azadi et al., 2015).

9.12 CHALLENGES AND FUTURE PERSPECTIVES

Even while genetic engineering holds out hope for improving crops, there are a number of issues that need to be carefully thought through and addressed. One of the main obstacles to the deployment of genetic engineering technology is contamination. The unintentional spread of GMOs or their genetic material can cause contamination at different stages, which raises questions about the effects on the environment, biodiversity, and coexistence with conventional or organic agriculture (Muzhinji and Ntuli, 2021; Buiatti et al., 2013). Through pollen transfer, genes from genetically modified crops may inadvertently cross pollinate with those from conventional or wild cousins, resulting in unintentional hybridization and the presence of transgenes in non-GMO populations. This prompt worries about the possible spread of genetically modified plants to closely related plant species, which could have an impact on natural ecosystems' ecological balance and biodiversity. GMO and non-GMO crops may inadvertently mix as a result of contamination that happens during the processing, storing, or transportation of seeds and grains. Unintentionally introducing GMOs into non-GMO products can have negative economic effects for markets subject to stringent laws or consumer attitudes against GMOs (Grantina-Ievina et al., 2019; Ryffel, 2014). After genetically modified crops are harvested, seeds left in the soil may give rise to volunteer plants, which may linger and inadvertently carry transgenes into next crops. Transgene persistence in the environment can cause ecological issues and make it more difficult to manage the coexistence of various agricultural practices. The lengths at which pollen from genetically modified crops can travel vary, and it can be difficult to calculate the proper isolation distances to stop gene flow. Insufficient isolation distances can cause transgenes to accidentally spread to surrounding fields, affecting nearby organic or conventional crops (Chapman and Burke, 2006; Shipman et al., 2021).

9.13 CONCLUSION

In summary, research into tissue culture methods for crop enhancement has shown that they have a revolutionary effect on farming. This thorough exploration of historical viewpoints, basic concepts, and cutting-edge applications has brought to light the adaptability of tissue culture in tackling critical issues related to crop management, genetic engineering, and rapid plant multiplication. Case studies highlight the practical uses and beneficial effects of tissue culture in improving agricultural attributes. Examples of these include the success tales of Golden Rice and Bt cotton. Lessons learnt highlight the accuracy, flexibility, and crossdisciplinary cooperation that are intrinsic to tissue culture research, illuminating the vital components that make it successful. Tissue culture has wide-ranging effects on crop improvement. This technology offers quick and accurate ways to manage illnesses, provide targeted genetic alterations, and multiply plants. Illustrative cases of genetically modified crops, such as Bt cotton and disease-resistant cultivars, highlight the beneficial effects on farming methods, promoting higher productivity, better quality, and the preservation of biodiversity. The ramifications go beyond the field of science to include biosecurity, ethics, and the delicate balancing act between innovation and social concerns. Looking ahead, a number of suggestions for additional tissue culture research come to light. These include developing crop varieties resistant to climate change, improving the nutritional value of existing kinds, and investigating more sophisticated genetic editing methods in conjunction with tissue culture. To provide comprehensive answers, efforts should also be focused on the conservation of biodiversity, addressing moral and social issues, investigating automation and scaling techniques, and encouraging interdisciplinary cooperation. Tissue culture is essentially the cornerstone upon which modern crop development is built. Tissue culture has shaped agriculture's course throughout history, from the creation of genetically modified crops to its historical roots in the manipulation of plant tissue. The trip via case studies, historical insights, and lessons gained highlights the accomplishments as well as the continuous effort needed for ethical and sustainable crop enhancement innovation. Tissue culture research holds great potential for robust, nutrient-dense, and sustainable agriculture in the future. This promise is fueled by ongoing investigation, ethical deliberation, and the combined efforts of the scientific community.

KEYWORDS

- **crop improvement**
- **embryogenesis**
- **genetic transformation**
- **tissue culture**
- **plant regeneration**

REFERENCES

Abbas, M.S.T. Genetically engineered (Modified) crops (Bacillus Thuringiensis Crops) and the world controversy on their safety. *Egypt. J. Biol. Pest Control* 2018, 28, 52, https://doi.org/10.1186/s41938-018-0051-2.

Abdul Aziz, M.; Brini, F.; Rouached, H.; Masmoudi, K. Genetically engineered crops for sustainably enhanced food production systems. *Front. Plant Sci.* 2022, 13, https://doi.org/10.3389/fpls.2022.1027828.

Acevedo, M.; Pixley, K.; Zinyengere, N.; Meng, S.; Tufan, H.; Cichy, K.; Bizikova, L.; Isaacs, K.; Ghezzi-Kopel, K.; Porciello, J. A Scoping review of adoption of climate-resilient crops by small-scale producers in low- and middle-income countries. *Nat. Plants* 2020, 6, 1231–1241, https://doi.org/10.1038/s41477-020-00783-z.

Adrio, J.-L.; Demain, A.L. Recombinant organisms for production of industrial products. *Bioeng. Bugs* 2010, 1, 116–131, https://doi.org/10.4161/bbug.1.2.10484.

Ahmad, M. Plant breeding advancements with "CRISPR-Cas" genome editing technologies will assist future food security. *Front. Plant Sci.* 2023, 14, https://doi.org/10.3389/fpls.2023.1133036.

Ahmad, P.; Ashraf, M.; Younis, M.; Hu, X.; Kumar, A.; Akram, N.A.; Al-Qurainy, F. Role of transgenic plants in agriculture and biopharming. *Biotechnol. Adv.* 2012, 30, 524–540, https://doi.org/10.1016/j.biotechadv.2011.09.006.

Ahmad, S.S.; Chun, H.J.; Ahmad, K.; Shaikh, S.; Lim, J.H.; Ali, S.; Han, S.S.; Hur, S.J.; Sohn, J.H.; Lee, E.J.; et al. The roles of growth factors and hormones in the regulation of muscle satellite cells for cultured meat production. *J. Anim. Sci. Technol.* 2023, 65, 16–31, https://doi.org/10.5187/jast.2022.e114.

Amiteye, S. Basic concepts and methodologies of DNA marker systems in plant molecular breeding. *Heliyon* 2021, 7, e08093, https://doi.org/10.1016/j.heliyon.2021.e08093.

Azadi, H.; Samiee, A.; Mahmoudi, H.; Jouzi, Z.; Rafiaani Khachak, P.; De Maeyer, P.; Witlox, F. Genetically modified crops and small-scale farmers: Main opportunities and challenges. *Crit. Rev. Biotechnol.* 2015, 1–13, https://doi.org/10.3109/07388551.2014.990413.

Balazs, D.A.; Godbey, W. Liposomes for use in gene delivery. *J. Drug Deliv.* 2011, 2011, 1–12, https://doi.org/10.1155/2011/326497.

Basu, C.; Kausch, A.P.; Chandlee, J.M. Use of β-glucuronidase reporter gene for gene expression analysis in turfgrasses. *Biochem. Biophys. Res. Commun.* 2004, 320, 7–10, https://doi.org/10.1016/j.bbrc.2004.05.128.

Batista-Silva, W.; da Fonseca-Pereira, P.; Martins, A.O.; Zsögön, A.; Nunes-Nesi, A.; Araújo, W.L. Engineering improved photosynthesis in the era of synthetic biology. *Plant Commun.* 2020, 1, 100032, https://doi.org/10.1016/j.xplc.2020.100032.

Bawa, A.S.; Anilakumar, K.R. Genetically modified foods: Safety, risks and public concerns—A review. *J. Food Sci. Technol.* 2013, 50, 1035–1046, https://doi.org/10.1007/s13197-012-0899-1.

Beauregard, M.; Hefford, M.A. Enhancement of essential amino acid contents in crops by genetic engineering and protein design. *Plant Biotechnol. J.* 2006, 4, 561–574, https://doi.org/10.1111/j.1467-7652.2006.00204.x.

Bhatia, S. Plant tissue culture. In: *Modern Applications of Plant Biotechnology in Pharmaceutical Sciences*. Elsevier, 2015, pp. 31–107.

Bidabadi, S.S.; Jain, S.M. Cellular, molecular, and physiological aspects of in vitro plant regeneration. *Plants* 2020, 9, 702, https://doi.org/10.3390/plants9060702.

Broeders, S.R.M.; De Keersmaecker, S.C.J.; Roosens, N.H.C. How to deal with the upcoming challenges in GMO detection in food and feed. *J. Biomed. Biotechnol.* 2012, 2012, 1–11, https://doi.org/10.1155/2012/402418.

Brown, D.C.W.; Thorpe, T.A. Crop improvement through tissue culture. *World J. Microbiol. Biotechnol.* 1995, 11, 409–415, https://doi.org/10.1007/BF00364616.

Buiatti, M.; Christou, P.; Pastore, G. The application of GMOs in agriculture and in food production for a better nutrition: Two different scientific points of view. *Genes Nutr.* 2013, 8, 255–270, https://doi.org/10.1007/s12263-012-0316-4.

Bull, T.; Michelmore, R. Molecular determinants of in vitro plant regeneration: Prospects for enhanced manipulation of lettuce (*Lactuca sativa* L.). *Front. Plant Sci.* 2022, 13, https://doi.org/10.3389/fpls.2022.888425.

Campos, C.O.; Bernuci, M.P.; Vireque, A.A.; Campos, J.R.; Silva-de-Sá, M.F.; Jamur, M.C.; Rosa-e-Silva, A.C.J.S. Preventing microbial contamination during long-term in vitro culture of human granulosa-lutein cells: An ultrastructural analysis. *ISRN Obstet. Gynecol.* 2012, 2012, 1–6, https://doi.org/10.5402/2012/152781.

Caradus, J.R. Intended and unintended consequences of genetically modified crops—Myth, fact and/or manageable outcomes? *N. Z. J. Agric. Res.* 2023, 66, 519–619, https://doi.org/10.1080/00288233.2022.2141273.

Ceddia, M.G.; Bartlett, M.; Perrings, C. Landscape gene flow, coexistence and threshold effect: The case of genetically modified herbicide tolerant oilseed rape (*Brassica napus*). *Ecol. Modell.* 2007, 205, 169–180, https://doi.org/10.1016/j.ecolmodel.2007.02.025.

Chapman, M.A.; Burke, J.M. Letting the gene out of the bottle: The population genetics of genetically modified crops. *New Phytol.* 2006, 170, 429–443, https://doi.org/10.1111/j.1469-8137.2006.01710.x.

Coté, R.J. Sterilization and filtration. *Curr. Protoc. Cell Biol.* 1999, 1, https://doi.org/10.1002/0471143030.cb0104s01.

Crété, R.; Pires, R.N.; Barbetti, M.J.; Renton, M. Rotating and stacking genes can improve crop resistance durability while potentially selecting highly virulent pathogen strains. *Sci. Rep.* 2020, 10, 19752, https://doi.org/10.1038/s41598-020-76788-7.

Das, P.; Adak, S.; Lahiri Majumder, A. Genetic manipulation for improved nutritional quality in rice. *Front. Genet.* 2020, 11, https://doi.org/10.3389/fgene.2020.00776.

Datta, A. Genetic engineering for improving quality and productivity of crops. *Agric. Food Secur.* 2013, 2, 15, https://doi.org/10.1186/2048-7010-2-15.

Debnath, S.C.; Arigundam, U. In vitro propagation strategies of medicinally important berry crop, lingonberry (*Vaccinium vitis-idaea* L.). *Agronomy* 2020, 10, 744, https://doi.org/10.3390/agronomy10050744.

Delporte, F.; Pretova, A.; du Jardin, P.; Watillon, B. Morpho-histology and genotype dependence of in vitro morphogenesis in mature embryo cultures of wheat. *Protoplasma* 2014, 251, 1455–1470, https://doi.org/10.1007/s00709-014-0647-7.

Dhaliwal, S.S.; Sharma, V.; Shukla, A.K.; Verma, V.; Kaur, M.; Shivay, Y.S.; Nisar, S.; Gaber, A.; Brestic, M.; Barek, V.; et al. Biofortification—A frontier novel approach to enrich micronutrients in field crops to encounter the nutritional security. *Molecules* 2022, 27, 1340, https://doi.org/10.3390/molecules27041340.

Dockery, G.D. Aseptic techniques. In: *Lower Extremity Soft Tissue and Cutaneous Plastic Surgery*. Elsevier, 2012, pp. 53–68.

Dong, O.X.; Ronald, P.C. Genetic engineering for disease resistance in plants: Recent progress and future perspectives. *Plant Physiol.* 2019, 180, 26–38, https://doi.org/10.1104/pp.18.01224.

Dormatey, R.; Sun, C.; Ali, K.; Fiaz, S.; Xu, D.; Calderón-Urrea, A.; Bi, Z.; Zhang, J.; Bai, J. PtxD/Phi as alternative selectable marker system for genetic transformation for bio-safety concerns: A review. *PeerJ* 2021, 9, e11809, https://doi.org/10.7717/peerj.11809.

Duta-Cornescu, G.; Constantin, N.; Pojoga, D.-M.; Nicuta, D.; Simon-Gruita, A. Somaclonal variation—Advantage or disadvantage in micropropagation of the medicinal plants. *Int. J. Mol. Sci.* 2023, 24, 838, https://doi.org/10.3390/ijms24010838.

Efferth, T. Biotechnology applications of plant callus cultures. *Engineering* 2019, 5, 50–59, https://doi.org/10.1016/j.eng.2018.11.006.

Esmaeili, N.; Shen, G.; Zhang, H. Genetic manipulation for abiotic stress resistance traits in crops. *Front. Plant Sci.* 2022, 13, https://doi.org/10.3389/fpls.2022.1011985.

Espinosa-Leal, C.A.; Puente-Garza, C.A.; García-Lara, S. In vitro plant tissue culture: means for production of biological active compounds. *Planta* 2018, 248, 1–18, https://doi.org/10.1007/s00425-018-2910-1.

Faiyazuddin, M.; Sophia, A.; Ashique, S.; Gholap, A.D.; Gowri, S.; Mohanto, S.; Karthikeyan, C.; Nag, S.; Hussain, A.; Akhtar, M.S.; et al. Virulence traits and novel drug delivery strategies for mucormycosis Post-COVID-19: A comprehensive review. *Front. Immunol.* 2023, 14, https://doi.org/10.3389/fimmu.2023.1264502.

Farhadi, H.; Hassanpouraghdam, M.B.; Aazami, M.A. The induction and development of somatic embryos from the in vitro culture of *Catharanthus roseus* (L.) G. Don. *Adv. Hortic. Sci.* 2021, 35, https://doi.org/10.36253/ahsc-9874.

Fehér, A. Callus, dedifferentiation, totipotency, somatic embryogenesis: What these terms mean in the era of molecular plant biology? *Front. Plant Sci.* 2019, 10, https://doi.org/10.3389/fpls.2019.00536.

Fraiture, M.-A.; Herman, P.; Taverniers, I.; De Loose, M.; Deforce, D.; Roosens, N.H. Current and new approaches in GMO detection: Challenges and solutions. *Biomed. Res. Int.* 2015, 2015, 1–22, https://doi.org/10.1155/2015/392872.

Garcia-Yi, J.; Lapikanonth, T.; Vionita, H.; Vu, H.; Yang, S.; Zhong, Y.; Li, Y.; Nagelschneider, V.; Schlindwein, B.; Wesseler, J. What are the socio-economic impacts of genetically modified crops worldwide? A systematic map protocol. *Environ. Evid.* 2014, 3, 24, https://doi.org/10.1186/2047-2382-3-24.

Gelvin, S.B. Agrobacterium-mediated plant transformation: The biology behind the "Gene-Jockeying" tool. *Microbiol. Mol. Biol. Rev.* 2003, 67, 16–37, https://doi.org/10.1128/MMBR.67.1.16-37.2003.

Geraghty, R.J.; Capes-Davis, A.; Davis, J.M.; Downward, J.; Freshney, R.I.; Knezevic, I.; Lovell-Badge, R.; Masters, J.R.W.; Meredith, J.; Stacey, G.N.; et al. Guidelines for the use of cell lines in biomedical research. *Br. J. Cancer* 2014, 111, 1021–1046, https://doi.org/10.1038/bjc.2014.166.

Ghimire, B.K.; Yu, C.Y.; Kim, W.-R.; Moon, H.-S.; Lee, J.; Kim, S.H.; Chung, I.M. Assessment of benefits and risk of genetically modified plants and products: Current controversies and perspective. *Sustainability* 2023, 15, 1722, https://doi.org/10.3390/su15021722.

Gholap, A.D. Translational nanomedicine: Status assessment and opportunities. *Int. J. Res. Pharm. Biomed. Sci.* 2012, 3, 1454–1463.

Gholap, A.D.; Dighe, S.B. Role of sebum in human health and acne. *Inven. Rapid Cosmeceuticals [ISSN 0976-3864]* 2012a, 1–3.

Gholap, A.; Pardeshi, S.; Giram, P. *Preparation, Antibacterial and Antiviral Activity Measurements and Detection Methods*. 2023a, pp. 33–64.

Gholap, A.D.; Rajurkar, V.G.; Dighe, S.B. Phonophoresis: Ultrasound-enhanced transdermal drug delivery. *Inven. Rapid NDDS [ISSN 0976-3791]* 2012, 1–3.

Gholap A.D.; Musmade, D.S. Advances in the field of nanooncology: An overview. *Inven. Impact NDDS* 2011, 1–3.

Gholap, A.D.; Rajurkar, V.G. Lipid nanocapsules: A novel promising tool as drug delivery system. *Inven. Rapid NDDS [ISSN 0976-3791]* 2012a, 1–4.

Gholap, A.D.; Rajurkar, V.G. New Threat of clinical toxicities evoked from nanocarrier systems. *Inven. Rapid NDDS [ISSN 0976-3791]* 2012b, 1–4.

Gholap, A.D.; Rajurkar, V.G. Potential applications of capillary electromigration methods for the improvement of separation efficiency. *Inven. Rapid Pharm Anal. Qual. Assur. [ISSN 0976-3813] 2012c*, 1–3.

Gholap, A.D.; Rajurkar, V.G. Potential applications of superparmagnetic nanoparticles technology. *Inven. Rapid NDDS* 2012d, 1–3.

Gholap, A.D.; Borude, S.S.; Mahajan, A.M. Smart crystals technology: A review. *Pharmacologyonline* 2011b, 238–243.

Gholap, A.D.; Bhowmik, D.D.; Deshmukh, A.Y.; Hatvate, N.T. *Quintessential Impact of Dendrimer Bioconjugates in Targeted Drug Delivery* 2023b, 103, 257–302.

Gholap, A.D.; Borude, S.S.; Mahajan, A.M.; Gholap, M.A.D. Aquasomes: A potential drug delivery carrier. *Pharmacologyonline* 2011a, 230–237.

Gholap, A.D.; Tupe, D. Artificial viral envelopes: Emerging vehicle for gene delivery. *Pharmatutor 1439* 2012a, 1–3.

Gholap, A.D.; Tupe, D. Potential applications of botanical extract. *Inven. Rapid Ethnopharmacol.* 2012b, 1–4.

Gholap, A.D.; Gupta, J.; Kamandar, P.; Bhowmik, D.D.; Rojekar, S.; Faiyazuddin, M.; Hatvate, N.T.; Mohanto, S.; Ahmed, M.G.; Subramaniyan, V.; et al. Harnessing nanovaccines for effective immunization—A special concern on COVID-19: Facts, fidelity, and future prospective. *ACS Biomater. Sci. Eng.* 2024a, 10, 271–297, https://doi.org/10.1021/acsbiomaterials.3c01247.

Gholap, A.D.; Gupta, J.S.; Kamandar, P.A.; Banchhod, G.V.; Hatvate, N.T. *Antibody-Drug Conjugates for Cancer Therapy: An up-to-Date Review on the Chemistry and Pharmacology*. 2023c.

Gholap, A.D.; Hatvate, N.T.; Khuspe, P.R.; Mandhare, T.A.; Kashid, P.; Gaikwad, V.D. Proteomics in oncology: Retrospect and prospects. In: *Novel Technologies in Biosystems, Biomedical and Drug Delivery*. Springer Nature Singapore: Singapore, 2023d, pp. 243–269.

Gholap, A.D.; Hatvate, N.T.; Sawant, S.D.; Khemka, P.N.; Vora, L.K. Unlocking the potential of nanobiohybrids to combat environmental pollution. In: *Nanobiohybrids for Advanced Wastewater Treatment and Energy Recovery*. IWA Publishing, 2023e, pp. 169–190.

Gholap, A.D.; Rojekar, S.; Kapare, H.S.; Vishwakarma, N.; Raikwar, S.; Garkal, A.; Mehta, T.A.; Jadhav, H.; Prajapati, M.K.; Annapure, U. Chitosan scaffolds: Expanding horizons in biomedical applications. *Carbohydr. Polym.* 2024b, 323, 121394, https://doi.org/10.1016/j.carbpol.2023.121394.

Gholap, A.D.; Dighe, S.B. Dandruff: Pathogenesis and treatment. *Inven. Rapid Cosmeceuticals [ISSN 0976-3864]* 2012b, 1–4.

Gholap, A.D.; Said, R.P.; Pawar, R.D.; Ambore, G.S.; Hatvate, N.T. *Importance of Carbohydrate-Drug Conjugates in Vaccine Development: A Detailed Review*. 2023f.

Gholap, A.D.; Sawant, S.D.; Sayyad, S.F.; Hatvate, N.T.; Chavan, M.; Rojekar, S.; Faiyazuddin, M. Food biopackaging for human benefits: Status and perspectives. In: *Biobased Packaging Materials*. Springer Nature Singapore: Singapore, 2023g, pp. 1–35.

Gholap, A.D.; Sayyad, S.F.; Hatvate, N.T.; Dhumal, V.V.; Pardeshi, S.R.; Chavda, V.P.; Vora, L.K. Drug delivery strategies for avobenzone: A case study of photostabilization. *Pharmaceutics* 2023h, 15, 1008, https://doi.org/10.3390/pharmaceutics15031008.

Gibney, E.R.; Nolan, C.M. Epigenetics and gene expression. *Heredity (Edinb).* 2010, 105, 4–13, https://doi.org/10.1038/hdy.2010.54.

Giller, K.E.; Delaune, T.; Silva, J.V.; Descheemaeker, K.; van de Ven, G.; Schut, A.G.T.; van Wijk, M.; Hammond, J.; Hochman, Z.; Taulya, G.; et al. The Future of farming: Who will produce our food? *Food Secur.* 2021, 13, 1073–1099, https://doi.org/10.1007/s12571-021-01184-6.

Govindasamy, P.; Muthusamy, S.K.; Bagavathiannan, M.; Mowrer, J.; Jagannadham, P.T.K.; Maity, A.; Halli, H.M.; Sujayananad, G. K.; Vadivel, R.; Das, T. K., et al. Nitrogen use efficiency—A key to enhance crop productivity under a changing climate. *Front. Plant Sci.* 2023, 14, https://doi.org/10.3389/fpls.2023.1121073.

Grantina-Ievina, L.; Ievina, B.; Evelone, V.; Berga, S.; Kovalcuka, L.; Bergspica, I.; Jakovele, A.; Malisevs, A.; Valcina, O.; Rodze, I.; et al. Potential risk evaluation for unintended entry of genetically modified plant propagating material in Europe through import of seeds and animal feed—The experience of latvia. *GM Crops Food* 2019, 10, 159–169, https://doi.org/10.1080/21645698.2019.1638721.

Greer, L.F.; Szalay, A.A. Imaging of light emission from the expression of luciferases in living cells and organisms: A review. *Luminescence* 2002, 17, 43–74, https://doi.org/10.1002/bio.676.

Handy, D.E.; Castro, R.; Loscalzo, J. Epigenetic modifications. *Circulation* 2011, 123, 2145–2156, https://doi.org/10.1161/CIRCULATIONAHA.110.956839.

Haque, M.I.; Singh, P.K.; Ghuge, S.; Kumar, A.; Chandra Rai, A.; Kumar, A.; Modi, A. A general introduction to and background of plant tissue culture: Past, current, and future aspects. In: *Advances in Plant Tissue Culture*, Elsevier, 2022, pp. 1–30.

Hasanuzzaman, M.; Nahar, K.; Alam, M.; Roychowdhury, R.; Fujita, M. Physiological, biochemical, and molecular mechanisms of heat stress tolerance in plants. *Int. J. Mol. Sci.* 2013, 14, 9643–9684, https://doi.org/10.3390/ijms14059643.

Hasnain, A.; Naqvi, S.A.H.; Ayesha, S.I.; Khalid, F.; Ellahi, M.; Iqbal, S.; Hassan, M.Z.; Abbas, A.; Adamski, R.; Markowska, D.; et al. Plants in vitro propagation with its applications in food, pharmaceuticals and cosmetic industries; Current scenario and future approaches. *Front. Plant Sci.* 2022, 13, 1009395, https://doi.org/10.3389/fpls.2022.1009395.

Hefferon, K. Nutritionally enhanced food crops; Progress and perspectives. *Int. J. Mol. Sci.* 2015, 16, 3895–3914, https://doi.org/10.3390/ijms16023895.

Hnatuszko-Konka, K.; Gerszberg, A.; Weremczuk-Jeżyna, I.; Grzegorczyk-Karolak, I. Cytokinin signaling and de novo shoot organogenesis. *Genes (Basel).* 2021, 12, 265, https://doi.org/10.3390/genes12020265.

Horstman, A.; Bemer, M.; Boutilier, K. A Transcriptional view on somatic embryogenesis. *Regeneration* 2017, 4, 201–216, https://doi.org/10.1002/reg2.91.

Huang, L.; Wu, D.; Zhang, G. Advances in studies on ion transporters involved in salt tolerance and breeding crop cultivars with high salt tolerance. *J. Zhejiang Univ. B* 2020, 21, 426–441, https://doi.org/10.1631/jzus.B1900510.

Hussain, A.; Ahmed, I.; Nazir, H.; Ullah, I. Plant tissue culture: Current status and opportunities. In *Recent Advances in Plant in vitro Culture*. InTech, 2012.

Hwang, H.-H.; Yu, M.; Lai, E.-M. Agrobacterium-mediated plant transformation: Biology and applications. *Arab. B.* 2017, 15, e0186, https://doi.org/10.1199/tab.0186.

Ikeuchi, M.; Sugimoto, K.; Iwase, A. Plant callus: Mechanisms of induction and repression. *Plant Cell* 2013, 25, 3159–3173, https://doi.org/10.1105/tpc.113.116053.

Javaid, M.; Haleem, A.; Khan, I.H.; Suman, R. Understanding the potential applications of artificial intelligence in agriculture sector. *Adv. Agrochem.* 2023, 2, 15–30, https://doi.org/10.1016/j.aac.2022.10.001.

Javaid, M.; Haleem, A.; Singh, R.P.; Suman, R. Enhancing smart farming through the applications of agriculture 4.0 technologies. *Int. J. Intell. Networks* 2022, 3, 150–164, https://doi.org/10.1016/j.ijin.2022.09.004.

Jildeh, Z.B.; Wagner, P.H.; Schöning, M.J. Sterilization of objects, products, and packaging surfaces and their characterization in different fields of industry: The status in 2020. *Phys. Status Solidi* 2021, 218, https://doi.org/10.1002/pssa.202000732.

Karavolias, N.G.; Horner, W.; Abugu, M.N.; Evanega, S.N. Application of gene editing for climate change in agriculture. *Front. Sustain. Food Syst.* 2021, 5, https://doi.org/10.3389/fsufs.2021.685801.

Kavhiza, N.J.; Zargar, M.; Prikhodko, S.I.; Pakina, E.N.; Murtazova, K.M.-S.; Nakhaev, M.R. Improving crop productivity and ensuring food security through the adoption of genetically modified crops in Sub-Saharan Africa. *Agronomy* 2022, 12, 439, https://doi.org/10.3390/agronomy12020439.

Khan, M.I.; Shin, J.H.; Kim, J.D. The promising future of microalgae: Current status, challenges, and optimization of a sustainable and renewable industry for biofuels, feed, and other products. *Microb. Cell Fact.* 2018, 17, 36, https://doi.org/10.1186/s12934-018-0879-x.

Khedkar, A., Tupe, D.; A.G. Pharmacosomes: An emerging vesicular system. *Inven. Rapid NDDS [ISSN 0976-3791]* 2012a, 1–3.

Khedkar, A.; Tupe, D.; Gholap, A. Cubosomes: Emerging drug delivery system. *Inven. Rapid NDDS [ISSN 0976-3791]* 2012b, 1–3.

Kikkert, J.R.; Vidal, J.R.; Reisch, B.I. Stable transformation of plant cells by particle Bombardment/Biolistics. In: *Transgenic Plants*. Humana Press: New Jersey, pp. 61–78.

Kishore, G.M.; Shewmaker, C. Biotechnology: Enhancing human nutrition in developing and developed worlds. *Proc. Natl. Acad. Sci.* 1999, 96, 5968–5972, https://doi.org/10.1073/pnas.96.11.5968.

Kobayashi, K., Wang, X.; Wang, W. Genetically modified rice is associated with hunger, health, and climate resilience. *Foods* 2023, 12, 2776, https://doi.org/10.3390/foods12142776.

Kolkmann, A.M.; Van Essen, A.; Post, M.J.; Moutsatsou, P. Development of a chemically defined medium for in vitro expansion of primary bovine satellite cells. *Front. Bioeng. Biotechnol.* 2022, 10, https://doi.org/10.3389/fbioe.2022.895289.

Kumari, P.; Khan, S.; Wani, I.A.; Gupta, R.; Verma, S.; Alam, P.; Alaklabi, A. Unravelling the role of epigenetic modifications in development and reproduction of angiosperms: A critical appraisal. *Front. Genet.* 2022, 13, https://doi.org/10.3389/fgene.2022.819941.

Kumlay, A.M.; Ercisli, S. Callus induction, shoot proliferation and root regeneration of potato (*Solanum tuberosum* L.) stem node and leaf explants under long-day conditions. *Biotechnol. Biotechnol. Equip.* 2015, 29, 1075–1084, https://doi.org/10.1080/13102818.2015.1077685.

Leva, A.R.; Petruccelli, R.; Rinaldi, L.M.R. Somaclonal variation in tissue culture: A case study with olive. In: *Recent Advances in Plant in vitro Culture*. InTech, 2012.

Li, H.; Yang, Y.; Hong, W.; Huang, M.; Wu, M.; Zhao, X. Applications of genome editing technology in the targeted therapy of human diseases: Mechanisms, advances and prospects. *Signal Transduct. Target. Ther.* 2020, 5, 1, https://doi.org/10.1038/s41392-019-0089-y.

Lin, H.; Leighty, R.W.; Godfrey, S.; Wang, S.B. Principles and approach to developing mammalian cell culture media for high cell density perfusion process leveraging established fed-batch media. *Biotechnol. Prog.* 2017, 33, 891–901, https://doi.org/10.1002/btpr.2472.

Long, Y.; Yang, Y.; Pan, G.; Shen, Y. New insights into tissue culture plant-regeneration mechanisms. *Front. Plant Sci.* 2022, 13, 926752, https://doi.org/10.3389/fpls.2022.926752.

López-Arredondo, D.; González-Morales, S.I.; Bello-Bello, E.; Alejo-Jacuinde, G.; Herrera, L. Engineering food crops to grow in harsh environments. *F1000Research* 2015, 4, 651, https://doi.org/10.12688/f1000research.6538.1.

Low, L.-Y.; Yang, S.-K.; Kok, D.-X.A.; Ong-Abdullah, J.; Tan, N.-P.; Lai, K.-S. Transgenic plants: Gene constructs, vector and transformation method. In: *New Visions in Plant Science*. InTech, 2018.

Magar, T.V.; Gholap, A.D.; Musmade, D.S. Elastic vesicles as drugs carriers through the skin. *Inven. Rapid NDDS [ISSN 0976-3791]* 2011, 1–2.

Mahajan, A.M.; Gholap, A.D.; Borude, S.S. POEMS syndrome: A review. In: *Pharmacologyonline* 2011, pp. 265–274.

Mahajan, A.M.; Gholap, A.D.; Borude, S.S. Recent updates on Q fever: Diagnosis and treatment. In: *Pharmacologyonline*, pp. 254–264.

Mahmood, M.A.; Naqvi, R.Z.; Rahman, S.U.; Amin, I.; Mansoor, S. Plant virus-derived vectors for plant genome engineering. *Viruses* 2023, 15, 531, https://doi.org/10.3390/v15020531.

Manyi-Loh, C.; Mamphweli, S.; Meyer, E.; Okoh, A. Antibiotic use in agriculture and its consequential resistance in environmental sources: Potential public health implications. *Molecules* 2018, 23, 795, https://doi.org/10.3390/molecules23040795.

Méndez-Hernández, H.A.; Ledezma-Rodríguez, M.; Avilez-Montalvo, R.N.; Juárez-Gómez, Y.L.; Skeete, A.; Avilez-Montalvo, J.; De-la-Peña, C.; Loyola-Vargas, V.M. Signaling overview of plant somatic embryogenesis. *Front. Plant Sci.* 2019, 10, 77, https://doi.org/10.3389/fpls.2019.00077.

Meyer, H. Systemic risks of genetically modified crops: The need for new approaches to risk assessment. *Environ. Sci. Eur.* 2011, 23, 7, https://doi.org/10.1186/2190-4715-23-7.

Michl, J.; Park, K.C.; Swietach, P. Evidence-based guidelines for controlling PH in mammalian live-cell culture systems. *Commun. Biol.* 2019, 2, 144, https://doi.org/10.1038/s42003-019-0393-7.

Miri, S.M. Micropropagation, callus induction and regeneration of ginger (Zingiber Officinale Rosc.). *Open Agric.* 2020, 5, 75–84, https://doi.org/10.1515/opag-2020-0008.

Mishra, N.; Jiang, C.; Chen, L.; Paul, A.; Chatterjee, A.; Shen, G. Achieving abiotic stress tolerance in plants through antioxidative defense mechanisms. *Front. Plant Sci.* 2023, 14, https://doi.org/10.3389/fpls.2023.1110622.

Mohapatra, S. Sterilization and disinfection. In: *Essentials of Neuroanesthesia*. Elsevier, 2017, pp. 929–944.

Molina-Ruiz, F.J.; Introna, C.; Bombau, G.; Galofre, M.; Canals, J.M. Standardization of cell culture conditions and routine genomic screening under a quality management system leads to reduced genomic instability in HPSCs. *Cells* 2022, 11, 1984, https://doi.org/10.3390/cells11131984.

Mondal, S.; Karmakar, S.; Panda, D.; Pramanik, K.; Bose, B.; Singhal, R.K. Crucial plant processes under heat stress and tolerance through heat shock proteins. *Plant Stress* 2023, 10, 100227, https://doi.org/10.1016/j.stress.2023.100227.

Murphy, D.J. Agricultural improvement in modern times. In: *People, Plants and Genes*. Oxford University Press: Oxford, 2007, pp. 261–278.

Muzhinji, N.; Ntuli, V. Genetically modified organisms and food security in Southern Africa: Conundrum and discourse. *GM Crops Food* 2021, 12, 25–35, https://doi.org/10.1080/21645698.2020.1794489.

Narusaka, Y.; Narusaka, M.; Yamasaki, S.; Iwabuchi, M. Methods to transfer foreign genes to plants. In: *Transgenic Plants—Advances and Limitations*. InTech, 2012.

Narváez, I.; Martín, C.; Jiménez-Díaz, R.M.; Mercado, J.A.; Pliego-Alfaro, F. Plant regeneration via somatic embryogenesis in mature wild olive genotypes resistant to the defoliating pathotype of verticillium dahliae. *Front. Plant Sci.* 2019, 10, 1471, https://doi.org/10.3389/fpls.2019.01471.

Nicaise, V. Crop immunity against viruses: Outcomes and future challenges. *Front. Plant Sci.* 2014, 5, https://doi.org/10.3389/fpls.2014.00660.

Nkonyam, D.; Riedel, H.; Semtansk, I. Social and economic issues – Genetically modified food. In: *Food Industry*. InTech, 2013.

Ofori, K.F.; Antoniello, S.; English, M.M.; Aryee, A.N.A. Improving nutrition through biofortification–A systematic review. *Front. Nutr.* 2022, 9, 1043655, https://doi.org/10.3389/fnut.2022.1043655.

Ozyigit, I.I.; Dogan, I.; Hocaoglu-Ozyigit, A.; Yalcin, B.; Erdogan, A.; Yalcin, I.E.; Cabi, E.; Kaya, Y. Production of secondary metabolites using tissue culture-based biotechnological applications. *Front. Plant Sci.* 2023, 14, 1132555, https://doi.org/10.3389/fpls.2023.1132555.

Pais, M.S. Somatic embryogenesis induction in woody species: The future after OMICs data assessment. *Front. Plant Sci.* 2019, 10, https://doi.org/10.3389/fpls.2019.00240.

Pardeshi, S.; Gholap, A.; More, M.; Togre, N.; Rebello, N.; Giram, P. *Dendrimers Based Antibacterial and Antiviral Materials*. In: 2023, pp. 139–169.

Parmar, N.; Singh, K.H.; Sharma, D.; Singh, L.; Kumar, P.; Nanjundan, J.; Khan, Y.J.; Chauhan, D.K.; Thakur, A.K. Genetic engineering strategies for biotic and abiotic stress tolerance and quality enhancement in horticultural crops: A comprehensive review. *3 Biotech* 2017, 7, 239, https://doi.org/10.1007/s13205-017-0870-y.

Perianez-Rodriguez, J.; Manzano, C.; Moreno-Risueno, M.A. Post-embryonic organogenesis and plant regeneration from tissues: Two sides of the same coin? *Front. Plant Sci.* 2014, 5, 219, https://doi.org/10.3389/fpls.2014.00219.

Potter, H. Transfection by electroporation. *Curr. Protoc. Mol. Biol.* 2003, 62, https://doi.org/10.1002/0471142727.mb0903s62.

Prabhu, A.; Gadgil, M. Trace metals in cellular metabolism and their impact on recombinant protein production. *Process Biochem.* 2021, 110, 251–262, https://doi.org/10.1016/j.procbio.2021.08.006.

Qaim, M.; Kouser, S. Genetically modified crops and food security. *PLoS One* 2013, 8, e64879, https://doi.org/10.1371/journal.pone.0064879.

Rafeeq, H.; Afsheen, N.; Rafique, S.; Arshad, A.; Intisar, M.; Hussain, A.; Bilal, M.; Iqbal, H.M.N. Genetically engineered microorganisms for environmental remediation. *Chemosphere* 2023, 310, 136751, https://doi.org/10.1016/j.chemosphere.2022.136751.

Rajurkar, V., Gholap, A. Recent technologies in nanomedicine for treatment of cancer. *Inven. Rapid Pharm. Tech.* 2012, 1–3. [ISSN 0976-3783]

Ramakrishna, D.; Shasthree, T. High efficient somatic embryogenesis development from leaf cultures of *Citrullus colocynthis* (L.) Schrad for generating true type clones. *Physiol. Mol. Biol. Plants* 2016, 22, 279–285, https://doi.org/10.1007/s12298-016-0357-z.

Raman, R. The impact of Genetically Modified (GM) crops in modern agriculture: A review. *GM Crops Food* 2017, 8, 195–208, https://doi.org/10.1080/21645698.2017.1413522.

Ranaware, A.S.; Kunchge, N.S.; Lele, S.S.; Ochatt, S.J. Protoplast technology and somatic hybridisation in the family apiaceae. *Plants* 2023, 12, 1060, https://doi.org/10.3390/plants12051060.

Raza, A.; Razzaq, A.; Mehmood, S.; Zou, X.; Zhang, X.; Lv, Y.; Xu, J. Impact of climate change on crops adaptation and strategies to tackle its outcome: A review. *Plants* 2019, 8, 34, https://doi.org/10.3390/plants8020034.

Razzaq, A.; Kaur, P.; Akhter, N.; Wani, S.H.; Saleem, F. Next-generation breeding strategies for climate-ready crops. *Front. Plant Sci.* 2021, 12, https://doi.org/10.3389/fpls.2021.620420.

Remington, S.J. Green fluorescent protein: A perspective. *Protein Sci.* 2011, 20, 1509–1519, https://doi.org/10.1002/pro.684.

Rutala, W.A.; Weber, D.J. Disinfection, sterilization, and control of hospital waste. In: *Mandell, Douglas, and Bennett's Principles and Practice of Infectious Diseases*. Elsevier, 2015, pp. 3294–3309.e4.

Ryffel, G.U. Transgene flow: Facts, speculations and possible countermeasures. *GM Crops Food* 2014, 5, 249–258, https://doi.org/10.4161/21645698.2014.945883.

Sadeque, M.; Balachandran, S.K. Overview of medical device processing. In: *Trends in Development of Medical Devices*. Elsevier, 2020, pp. 177–188.

Sathee, L.; Jagadhesan, B.; Pandesha, P.H.; Barman, D.; Adavi, B.S.; Nagar, S.; Krishna, G.K.; Tripathi, S.; Jha, S.K.; Chinnusamy, V. Genome editing targets for improving nutrient use efficiency and nutrient stress adaptation. *Front. Genet.* 2022, 13, https://doi.org/10.3389/fgene.2022.900897.

Schoeb, T.R.; Rahija, R.J.; Boyd, C.; Orcutt, R.P.; Eaton, K.A. Principles of establishing and operating a gnotobiotic facility. In: *Gnotobiotics*. Elsevier, 2017, pp. 21–63.

Schütte, G.; Eckerstorfer, M.; Rastelli, V.; Reichenbecher, W.; Restrepo-Vassalli, S.; Ruohonen-Lehto, M.; Saucy, A.-G.W.; Mertens, M. Herbicide resistance and biodiversity: Agronomic and environmental aspects of genetically modified herbicide-resistant plants. *Environ. Sci. Eur.* 2017, 29, 5, https://doi.org/10.1186/s12302-016-0100-y.

Shashank M., P.; V. B. Chandana, Kumari; Sumana, K.; Sujay, S.; Tejaswini, M.; Shirahatti, P.S.; Ramith, R. Sustainable development of plant tissue culture industry: The Indian scenario. *J. Appl. Biol. Biotechnol.* 2021, https://doi.org/10.7324/JABB.2021.9202.

Shelake, R.M.; Kadam, U.S.; Kumar, R.; Pramanik, D.; Singh, A.K.; Kim, J.-Y. Engineering drought and salinity tolerance traits in crops through CRISPR-mediated genome editing: Targets, tools, challenges, and perspectives. *Plant Commun.* 2022, 3, 100417, https://doi.org/10.1016/j.xplc.2022.100417.

Shipman, E.N.; Yu, J.; Zhou, J.; Albornoz, K.; Beckles, D.M. Can gene editing reduce postharvest waste and loss of fruit, vegetables, and ornamentals? *Hortic. Res.* 2021, 8, 1, https://doi.org/10.1038/s41438-020-00428-4.

Singh, R.; Singh, D.; Singh, A. Radiation sterilization of tissue allografts: A review. *World J. Radiol.* 2016, 8, 355, https://doi.org/10.4329/wjr.v8.i4.355.

Subbiahanadar Chelladurai, K.; Selvan Christyraj, J.D.; Rajagopalan, K.; Yesudhason, B.V.; Venkatachalam, S.; Mohan, M.; Chellathurai Vasantha, N.; Selvan Christyraj, J.R.S. Alternative to FBS in animal cell culture—An overview and future perspective. *Heliyon* 2021, 7, e07686, https://doi.org/10.1016/j.heliyon.2021.e07686.

Sussex, I.M. The scientific roots of modern plant biotechnology. *Plant Cell* 2008, 20, 1189–1198, https://doi.org/10.1105/tpc.108.058735.

Tang, R.; Tan, H.; Dai, Y.; Li, L.; Huang, Y.; Yao, H.; Cai, Y.; Yu, G. Application of antimicrobial peptides in plant protection: Making use of the overlooked merits. *Front. Plant Sci.* 2023, 14, 1139539, https://doi.org/10.3389/fpls.2023.1139539.

Thorpe, T.A. History of plant cell culture. In: *Plant Tissue Culture*. Elsevier, 2013, pp. 1–22.

Thorpe, T.A. History of plant tissue culture. *Mol. Biotechnol.* 2007, 37, 169–180, https://doi.org/10.1007/s12033-007-0031-3.

Tien Lea, D.; Duc Chua, H.; Quynh Lea, N. Improving nutritional quality of plant proteins through genetic engineering. *Curr. Genomics* 2016, 17, 220–229, https://doi.org/10.2174/1389202917666160202215934.

Tran, L.T.; Sugimoto, K.; Kasozi, M.; Mitalo, O.W.; Ezura, H. Pollination, pollen tube growth, and fertilization independently contribute to fruit set and development in tomato. *Front. Plant Sci.* 2023, 14, https://doi.org/10.3389/fpls.2023.1205816.

Tupe, D.; Gholap, A.D. The threat of photoaging: Clinical conditions, evaluation and treatment. *Inven. Rapid Cosmeceuticals [ISSN 0976-3864]* 2012, 1–3.

Türkoğlu, A.; Bolouri, P.; Haliloğlu, K.; Eren, B.; Demirel, F.; Işık, M.İ.; Piekutowska, M.; Wojciechowski, T.; Niedbała, G. Modeling callus induction and regeneration in hypocotyl explant of fodder pea (*Pisum sativum* var. Arvense L.) using machine learning algorithm method. *Agronomy* 2023, 13, 2835, https://doi.org/10.3390/agronomy13112835.

van Esse, H.P.; Reuber, T.L.; van der Does, D. Genetic modification to improve disease resistance in crops. *New Phytol.* 2020, 225, 70–86, https://doi.org/10.1111/nph.15967.

Verma, A.; Verma, M.; Singh, A. Animal tissue culture principles and applications. In: *Animal Biotechnology*. Elsevier, 2020, pp. 269–293.

Vikas G Rajurkar, A.D.G. Recent technologies in nanomedicine for treatment of cancer. *Inven. Rapid Pharm Tech* 2012, 1–3.

Vilas Bhagat, A.G. Gold nanorods: A promising drug delivery vehicle. *Inven. Rapid NDDS [ISSN 0976-3864]* 2012, 1–4.

Vincelli, P. Genetic engineering and sustainable crop disease management: Opportunities for case-by-case decision-making. *Sustainability* 2016, 8, 495, https://doi.org/10.3390/su8050495.

Voss-Fels, K.P.; Stahl, A.; Hickey, L.T. QandA: Modern crop breeding for future food security. *BMC Biol.* 2019, 17, 18, https://doi.org/10.1186/s12915-019-0638-4.

Wang, J.; Xia, X.; Qin, G.; Tang, J.; Wang, J.; Zhu, W.; Qian, M.; Li, J.; Cui, G.; Yang, Y.; et al. Somatic embryogenesis and plant regeneration from stem explants of pomegranate. *Horticulturae* 2023, 9, 1038, https://doi.org/10.3390/horticulturae9091038.

Waqas, M.A.; Kaya, C.; Riaz, A.; Farooq, M.; Nawaz, I.; Wilkes, A.; Li, Y. Potential mechanisms of abiotic stress tolerance in crop plants induced by thiourea. *Front. Plant Sci.* 2019, 10, 1336, https://doi.org/10.3389/fpls.2019.01336.

Wijerathna-Yapa, A.; Hiti-Bandaralage, J. Tissue culture—A sustainable approach to explore plant stresses. *Life* 2023, 13, 780, https://doi.org/10.3390/life13030780.

Will, M.A.; Clark, N.A.; Swain, J.E. Biological PH buffers in IVF: Help or hindrance to success. *J. Assist. Reprod. Genet.* 2011, 28, 711–724, https://doi.org/10.1007/s10815-011-9582-0.

Yadav, S.; Modi, P.; Dave, A.; Vijapura, A.; Patel, D.; Patel, M. Effect of abiotic stress on crops. In: *Sustainable Crop Production*. IntechOpen, 2020.

Yamauchi, T.; Noshita, K.; Tsutsumi, N. Climate-smart crops: Key root anatomical traits that confer flooding tolerance. *Breed. Sci.* 2021, 71, 51–61, https://doi.org/10.1270/jsbbs.20119.

Yancheva, S.; Kondakova, V. *Plant Tissue Culture Technology: Present and Future Development*. In: 2018, pp. 39–63.

Yang, Z.; Xiong, H.-R. Culture conditions and types of growth media for mammalian cells. In: *Biomedical Tissue Culture*. InTech, 2012.

Yao, T.; Asayama, Y. Animal-cell culture media: History, characteristics, and current issues. *Reprod. Med. Biol.* 2017, 16, 99–117, https://doi.org/10.1002/rmb2.12024.

Yen, S.; Sokolenko, S.; Manocha, B.; Patras, A.; Daynouri-Pancino, F.; Blondeel, E.J.M.; Sasges, M.; Aucoin, M.G. Treating cell culture media with UV irradiation against adventitious agents: Minimal impact on CHO performance. *Biotechnol. Prog.* 2014, 30, 1190–1195, https://doi.org/10.1002/btpr.1942.

Younis, A.; Siddique, M.I.; Kim, C.-K.; Lim, K.-B. RNA Interference (RNAi) induced gene silencing: A promising approach of Hi-tech plant breeding. *Int. J. Biol. Sci.* 2014, 10, 1150–1158, https://doi.org/10.7150/ijbs.10452.

Youssef, A.; Aboalola, D.; Han, V.K.M. The roles of insulin-like growth factors in mesenchymal stem cell niche. *Stem Cells Int.* 2017, 2017, 1–12, https://doi.org/10.1155/2017/9453108.

Youssef, E.M.A.; Amin, G.A. Microwave sterilization of tissue culture media. *Acta Hortic.* 2001, 513–516, https://doi.org/10.17660/ActaHortic.2001.560.104.

Yu, H.; Wang, Y.; Fu, F.; Li, W. Transgenic improvement for biotic resistance of crops. *Int. J. Mol. Sci.* 2022, 23, 14370, https://doi.org/10.3390/ijms232214370.

Zafar, S.; Jianlong, X. Recent advances to enhance nutritional quality of rice. *Rice Sci.* 2023, 30, 523–536, https://doi.org/10.1016/j.rsci.2023.05.004.

Zhang, X.; Wen, Y.; Yang, S.T. Modes of culture/animal cells. In: *Comprehensive Biotechnology*. Elsevier, 2011, pp. 285–302.

Zhang, Y.; Yu, L.-C. Microinjection as a tool of mechanical delivery. *Curr. Opin. Biotechnol.* 2008, 19, 506–510, https://doi.org/10.1016/j.copbio.2008.07.005.

Zilberman, D.; Holland, T.; Trilnick, I. Agricultural GMOs—What we know and where scientists disagree. *Sustainability* 2018, 10, 1514, https://doi.org/10.3390/su10051514.

CHAPTER 10

Application of Nanotechnology for Sustainable Developments of Agriculture

AMOL D. GHOLAP[1*], PANKAJ R. KHUSPE[2], SADIKALI F. SAYYAD[3], MACHINDRA CHAVAN[4], NAVNATH T. HATVATE[5], RAMESH D. BHUSAL[1], MD. FAIYAZUDDIN[6], and MD JASIM UDDIN[7]

[1]Department of Pharmaceutics, St. John Institute of Pharmacy and Research, Palghar, Maharashtra, India

[2]Department of Pharmaceutics, College of Pharmacy, Paniv, Solapur, Maharashtra, India

[3]Department of Pharmaceutics, Amrutvahini College of Pharmacy, Sangamner, Maharashtra, India

[4]Department of Pharmacognosy, Amrutvahini College of Pharmacy, Sangamner, Maharashtra, India

[5]Department of Pharmaceutical Technology, Institute of Chemical Technology, Marathwada Campus, Jalna, Maharashtra, India

[6]School of Pharmacy, Al-Karim University, Katihar, Bihar, India

[7]Department of Pharmaceutical Technology, Faculty of Pharmacy, Universiti Malaya, Kuala Lumpur, Malaysia

*Corresponding author

Technological Advancements in Sustainable Crop Improvement and Production.
Devendra Kumar, Alok Kumar Singh, and Nisha Malik (Eds.)

DOI: 10.1201/9781779643155-10

ABSTRACT

The manipulation of materials at the nanoscale, or nanotechnology, offers a revolutionary frontier for the development of sustainable agriculture. An extensive examination of nanotechnological applications that have the potential to transform agriculture is given in this chapter. It starts with fundamental concepts and explores the various uses of nanomaterials, like how nanoparticles can increase agricultural yields and how nanoencapsulation might improve the distribution of agrochemicals. The potential for real-time crop monitoring is investigated in relation to the integration of nanosensors into precision agriculture and their interaction with the Internet of Things. Furthermore, nanodevices such as smart delivery systems and nanorobotics become important assets in the advancement of sustainable farming methods and resource efficiency. With careful attention to detail, environmental and safety issues are covered, illuminating ecotoxicological issues and risk assessment techniques. The need for responsible innovation is emphasized throughout the chapter as a means of guaranteeing the benefits of nanotechnology while reducing associated hazards. Throughout, ethical elements are interwoven, highlighting societal and cultural ramifications, and adding to the foundation for deployment that is both responsible and equitable. Case studies from the real world offer useful insights by showcasing effective implementations and lessons discovered. The chapter concludes with a summary of the main conclusions, providing a thorough knowledge of the implications of nanotechnology for sustainable agriculture and directing future research and innovation toward resilient and environmentally conscious farming practices.

10.1 INTRODUCTION

The rapidly developing discipline of nanotechnology has had a profound impact on the landscape of scientific innovation in recent years, transforming various sectors in the process. Leading the way in this dramatic paradigm shift is the agricultural sector, which is essential to the world's sustenance and is currently on the verge of a technological renaissance thanks to the application of nanotechnology (Malik et al., 2023; Prasad et al., 2017). This introduction provides context for a thorough and nuanced investigation of the significant influence that nanotechnology can have on the course of sustainable agricultural development (Betz et al., 2023).

With its ability to manipulate and engineer materials at the nanoscale, nanotechnology has quickly developed into a disruptive development catalyst. These artificial materials, which function at nanoscale scales, have special chemical and physical characteristics that set them apart from bulk materials (Gholap et al., 2023a, 2023b, 2023c, 2023d; Gholap, 2012a; Mahajan et al., 2011a). A wide range of applications, including electronics, medicine, energy, and most importantly, agriculture, have been made possible by this capability (Khan et al., 2022; Chaturvedi et al., 2012; Baig et al., 2021). The underlying ideas of quantum physics and molecular interactions, which form the basis of nanotechnology, provide hitherto unthinkable levels of efficiency and precision. The capacity to customize materials at the molecular level enables researchers and scientists to tackle problems with an unprecedented level of detail, offering a plethora of creative solutions that have the potential to become sustainable farming methods (Barrat et al., 2024; Szefler, 2018).

Agriculture, which has always been characterized by human labor and conventional methods, is at a turning point in its history where incorporating cutting-edge technologies is essential. The growing world population necessitates a greater output of food, yet conventional farming practices are constrained by issues with space, resources, and environmental impact. As a ray of hope, nanotechnology presents the possibility of bridging these gaps by transforming the methods we use to manage, monitor, and maintain our agricultural ecosystems (Muhie, 2022; Talaviya ct al., 2020).

The application of nanotechnology in agriculture is a revolution rather than just an evolution. It represents a shift from traditional methods to a more advanced, data-driven, and environmentally conscious approach. The capacity to develop solutions at the nanoscale provides agriculture with instruments that maximize the use of available resources, improve crop yields, and lessen the impact on the environment (Saritha et al., 2022; Ghidan and Al Antary, 2020). The investigation that follows explores the various aspects of nanotechnological applications, their ramifications, and the revolutionary potential they have for sustainable agricultural growth as we stand on the cusp of this technological revolution (Yadav et al., 2023). The ensuing thorough investigation is directed by a diverse range of goals, all of which are intended to reveal the levels of intricacy involved in the application of nanotechnology to agriculture (Gholap, 2011; Mahajan et al. 2011b,, Khedkar and Tupe, 2012; Rajurkar, 2012; Tupe and Gholap,

2012; Gholap, 2012b, 2012c, 2012d). First and foremost, this investigation aims to clarify the fundamental ideas behind nanotechnology, creating the framework for a more thorough comprehension of its uses in agriculture. Building blocks for the debate that follows are the peculiarities of nanomaterials, their chemical details, and quantum effects (Ghidan and Al Antary, 2020; Shang et al., 2019). The second point of emphasis is on the particular uses of nanotechnology in agriculture. The story expands to cover the field of nanomaterials, examining how crop management, soil health, and pest control are being revolutionized by nanoparticles, nanoencapsulation, and nanocomposites. Nanotechnology can be used for effective crop protection, crop improvement, stress tolerance, development of nanofertilizers, precision farming, and social health enhancement as presented in Figure 10.1. The intention is to shed light on the various ways that traditional agricultural methods are being altered by nanotechnology (Ghidan and Al Antary, 2020; Shang et al., 2019). Additionally, the investigation delves into the incorporation of nanosensors and nanodevices, which are essential elements within the precision agricultural domain. A close examination of the relationship between nanotechnology and the Internet of Things (IoT) demonstrates how real-time data collecting and analysis are improving farming decision-making (Prasad et al., 2017; Javaid et al., 2021). But there are obstacles in the way of nanotechnology's revolutionary potential. The investigation also investigates the safety and environmental implications of using nanotechnology in agriculture. To guarantee a fair awareness of the risks and rewards, a comprehensive evaluation is conducted, encompassing everything from ecotoxicological implications to risk assessment methodologies (Bhagat, 2012; Gholap et al., 2023e, Gholap and Rajurkar, 2012; Rajurkar, 2012; Gholap, 2012e, 2012f, 2012g, 2012h; Bothiraja et al., 2014; Gholap et al., 2011a). The conversation includes ethical issues at every turn, highlighting the necessity of responsible innovation when applying nanotechnological solutions in agriculture (Yadav et al., 2023). The investigation seeks to negotiate the tricky terrain of moral frameworks, societal ramifications, and the wider influence on communities (Prasad et al., 2017; Iavicoli et al., 2017). Essentially, the goals of this investigation are based on offering a comprehensive and knowledgeable viewpoint regarding the significant influence of nanotechnology on the development of sustainable agriculture. This investigation aims to add to the body of knowledge that directs scholars, decision-makers, and practitioners toward a future in which nanotechnology and agriculture work together to improve environmental sustainability and global food security by removing obstacles, defining ethical issues, and removing layers of complexity (Tyagi et al., 2023; Mittal et al., 2020).

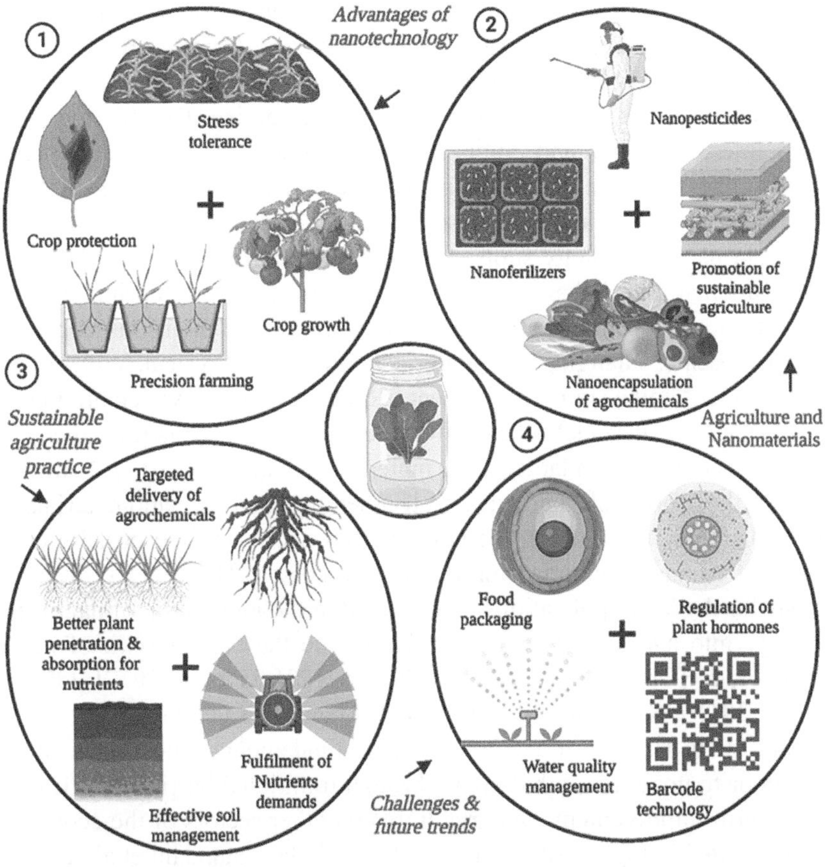

FIGURE 10.1 The intervention of nanotechnology in agriculture encompasses the following key aspects: (1) advantages of nanotechnology, (2) need of nanomaterials in agriculture, (3) sustainable agriculture practice, (4) and challenges and future trend of nanotechnology in agriculture.

10.2 FUNDAMENTALS OF NANOTECHNOLOGY

The study of nanoscale science and engineering, which focuses on the complexities of modifying matter at the nanometer scale, is where nanotechnology investigation starts. Materials have distinct characteristics at this tiny scale that frequently differ from those of their bulk counterparts (Bayda et al., 2019). Understanding and utilizing events that occur at dimensions smaller than 100 nm, when quantum effects become more prominent, is the focus of nanoscale science. This domain provides an interdisciplinary foundation for innovation by encompassing fields such as nanobiology, nanochemistry, and nanophysics. Conversely, the intentional design and construction of structures, apparatus, and systems at this minuscule scale is the focus of nanoscale engineering. By utilizing these principles, scientists and engineers may engineer materials with previously unheard-of accuracy, leading to the creation of innovative solutions for a variety of industries (Nasrollahzadeh et al., 2019). Fundamental ideas in nanotechnology support its transformational potential. The way that nanomaterials behave depends on their size is one such principle. The surface-area-to-volume ratio significantly rises as dimensions approach the nanoscale, changing the material's chemical, physical, and optical properties. Another important aspect is quantum effects, which are highly visible at the nanoscale and can affect electronic characteristics and enable functions not found in larger-scale materials (Khan et al., 2019). Additionally, a third principle known as self-assembly enables nanomaterials to spontaneously arrange into certain structures, offering the possibility of productive and economical production processes. These ideas form the basis for the later applications investigated in the integration of nanotechnology into agriculture, and they also define the basic toolkit of nanotechnology (Gholap et al., 2023c, 2023g, 2023f; Gholap, 2012I; Gholap et al., 2011b; Khedkar and Tupe, 2012; Magar and Gholap, 2011; Singh et al., 2016; Gholap et al., 2024a). This fundamental knowledge opens the door to novel ideas that break through traditional barriers and lead to long-term improvements in agriculture and other sectors of the economy, as presented in Figure 10.1 (Taha et al., 2022; Jeevanandam et al., 2018).

The foundation of the cutting-edge field of nanotechnology, which manipulates and designs materials and systems at dimensions typically ranging from 1 to 100 nm, is nanoscale science and engineering. At this size, a material's distinctive qualities become apparent, opening the door to revolutionary new directions in research and development (Bayda et al., 2019; Mahendra Kumar et al., 2023). The basic science of materials assumes a unique character in the field of nanoscale science. At the nanoscale, quantum effects become more and more prominent, introducing behaviors that are very different from those found in bulk materials.

Transformations occur in properties including chemical reactivity, optical properties, and electrical conductivity, providing a rich tapestry for investigation in fields such as nanochemistry and nanophysics (Mansoori, 2017; Chehelgerdi et al., 2023).

Another branch of nanoscale science called nanobiology studies how to apply the concepts of nanotechnology to the field of biology. Because biological molecules and structures may be accurately altered at this scale, advances in imaging, diagnostics, and medication administration are possible. The interaction of biology and nanoscale research could lead to revolutionary discoveries in the fields of medicine and healthcare (Saritha et al., 2022; Yadav et al., 2023). This scientific investigation is enhanced by nanoscale engineering, which offers the means to work with and create materials at the nanoscale. This field entails the purposeful, controlled design and construction of apparatuses, systems, and structures. Engineers can now create materials with specific features thanks to techniques such as molecular self-assembly and nanolithography, which is creating new opportunities for materials science, electronics, and other fields (Muhie, 2022; Talaviya et al., 2020).

Because nanoscale research and engineering are interdisciplinary, cooperation across conventional scientific boundaries is encouraged. To realize the potential of the nanoscale, scientists take ideas from engineering, biology, chemistry, and physics. This interdisciplinary convergence fosters groundbreaking discoveries and technical breakthroughs, providing the groundwork for the plethora of domains in which nanotechnology finds application, including agriculture (Porter and Youtie, 2009). Nanoscale science and engineering provide methods to handle issues ranging from precision farming to controlled release of agrochemicals in the context of sustainable agricultural development. This small-scale material manipulation technology holds the potential to completely transform agricultural methods and bring in a new era of resource efficiency, higher crop yields, and ecologically friendly farming (Gholap et al., 2024a, 2024b, 2023h, Pardeshi et al., 2023; Faiyazuddin et al., 2023). This basic knowledge of nanoscale science and engineering will pave the way for a more technologically advanced and sustainable future as we explore the uses of nanotechnology in agriculture (Prasad et al., 2017; Betz et al., 2023).

The revolutionary potential of nanotechnology is defined by numerous fundamental ideas that support the area. These ideas provide unique prospects for innovation and advancement in the manipulation and engineering of materials at the nanoscale. The size-dependent behavior of materials at the nanoscale is one of the basic principles (Haleem et al., 2023). Materials exhibit a significant rise in surface area-to-volume ratio when their dimensions decrease. Nanomaterials differ from their bulk counterparts due to the significant changes in the physical, chemical, and optical properties that result from this scale modification. Designing customized nanotechnological solutions requires leveraging these size-dependent phenomena. At the nanoscale, quantum effects play a bigger role and affect a material's electrical characteristics. At this level, the behavior of particles is governed by quantum mechanics, which also introduces concepts such as quantum confinement and tunneling. Due to these processes, distinct electronic structures are created, which makes it possible to create nanomaterials with new and frequently improved qualities, including conductivity, optical qualities, and reactivity (Szczyglewska et al., 2023).

The phenomenon of self-assembly, in which nanomaterials have the innate capacity to spontaneously assemble into particular structures, is another important fundamental. Fundamental forces, such as hydrogen bonds and van der Waals forces, are what propel this self-organization. By using self-assembly, complicated gadgets and structures can be made with little help from outside sources. This idea is essential for efficient and economical manufacturing procedures in the field of nanotechnology (Yadav et al., 2020). A large amount of the material is at or near the surface at the nanoscale. The overall characteristics of nanoparticles are largely determined by surface effects. Enhanced surface energy, reactivity, and the prominence of surface states are some of these impacts. Comprehending and regulating surface impacts is vital in customizing the functionality of nanomaterials for diverse uses (Mehraeen, 2020).

As materials get smaller in size, their physical and chemical properties alter. Nanotechnology abides by these rules. These principles control changes in optical characteristics, mechanical strength enhancement, and melting point depression, among other things. A theoretical foundation for forecasting and comprehending the behavior of nanomaterials, directing their design and implementation, is provided by adherence to scaling rules (Shrivastava and Dash, 2009; Joudeh and Linke, 2022). Together, these fundamental ideas make up the nanotechnology toolbox. Through the application of these concepts, materials can be engineered by scientists and engineers with previously unheard-of precision and control, leading to new discoveries in a variety of sectors, including electronics, medicine, and agriculture. As we investigate the revolutionary potential of nanotechnology in solving issues and promoting sustainable solutions in numerous fields, it is imperative that we comprehend and put these ideas into practice (Bayda et al., 2019).

10.3 AGRICULTURE AND NANOMATERIALS

A frontier of innovation, the use of nanomaterials in agricultural processes has the potential to transform several aspects of crop management, soil health, and pesticide delivery. Because of their size at the nanoscale, nanomaterials have special physical and chemical qualities that can be used to develop agriculture in a sustainable and effective manner. The purpose of nanoparticles, which include nanopesticides and nanofertilizers, is to improve nutrient uptake and shield crops from pests and illnesses (Fraceto et al., 2016). Nanofertilizers enable maximum plant nutrition, minimize environmental impact, and maximize agricultural yields through controlled nutrient release. Targeted delivery is another feature of nanopesticides, which lowers the amount of agrochemicals needed and lessens harmful effects on species that are not the intended targets (Saritha et al., 2022).

Encasing active substances in nanoscale carriers is known as nanoencapsulation. This method is used in agriculture to release agrochemicals such as insecticides and herbicides in a regulated manner, as presented in Figure 10.1. By shielding the active chemicals from deterioration, lowering the frequency of application, and avoiding environmental contamination, nanoencapsulation improves the effectiveness of pesticide administration. The use of nanocomposites, which are made up of nanoparticles scattered across a matrix, holds potential to improve soil. Soil structure and water retention can be improved by adding nanomaterials, such as nanoclays or nanoscale organic matter, to soil amendments (Chaud et al., 2021). This promotes sustainable agriculture operations by improving the availability of nutrients for plants and reducing soil erosion. Nanomaterials have uses in agriculture that go beyond these main classifications. Resources can be used more precisely and effectively by customizing nanoscale delivery methods for crops and environmental circumstances. In line with the objectives of precision agriculture, the increased surface area and reactivity of nanomaterials also improve the efficiency of the use of water and nutrients (Vega-Vásquez et al., 2020; An et al., 2022).

However, the use of nanomaterials in agriculture requires a thorough comprehension of the dangers and effects on the environment. To ensure the ethical and sustainable integration of nanotechnology into agricultural practices, research is being done to assess the long-term effects of nanomaterials on plant growth, soil health, and ecosystem dynamics. Applications of nanomaterials in agriculture have the potential to greatly improve both ecologically friendly farming practices and global food security as they continue to advance (Iavicoli et al., 2017; Mwaanga, 2018).

The application of nanoparticles in agriculture represents a novel strategy intended to enhance overall agricultural output, optimize nutrient consumption, and transform crop management. Because of their minuscule size on the nanoscale, nanoparticles provide unique and customized answers to modern agriculture's problems. Nanofertilizers, which represent a paradigm shift in the way nutrients are delivered to crops, are one important application. By precisely controlling the delivery of nutrients, engineered nanoparticles solve the inefficiencies and environmental issues related to conventional fertilization techniques (Mittal et al., 2020). Targeted nutrient delivery is made possible by nanofertilizers, which guarantee that plants receive the right amount of nutrients at the right time. This regulated release lessens the environmental impact on neighboring ecosystems by improving nutrient uptake and minimizing nutrient leakage (Yadav et al., 2023).

The field of crop protection is being transformed using nanoparticles in nanopesticides. Conventional pesticides frequently have overuse problems, which contaminate the environment and affect organisms that are not their intended targets. A more focused approach is provided by nanopesticides, which release pesticidal chemicals in a regulated manner thanks to nanoparticle technology (Wang et al., 2022). With this focused application, the number of insecticides needed is reduced, off-target impacts are lessened, and overall pest management effectiveness is increased. The outcome is a crop protection strategy that is more ecologically friendly and sustainable. Furthermore, plant stress tolerance is improved by nanoparticles. It is possible to engineer nanomaterials to function as stress mitigators, assisting plants in withstanding harsh environments such as salt, drought, or severe temperatures. This use is especially important in light of climate change, as global agriculture faces difficulties due to erratic environmental stresses (Asif et al., 2022; Yousef et al., 2023).

Even though using nanoparticles to improve crops has a lot of potential advantages, it is crucial to carefully assess any potential environmental effects and regulatory frameworks before deploying them. The goal of ongoing study is to better understand how nanoparticles move through soil and plants so that the application of nanotechnology in agriculture adheres to sustainable development and responsible innovation principles. To put it simply, crop enhancement with nanoparticles is a game-changing instrument in the agricultural toolbox. We are on the verge of developing a more effective, focused, and sustainable approach to crop management by utilizing the special qualities of nanomaterials, promoting global food security in an ecologically responsible way (Kumah et al., 2023; Wahab et al., 2023).

As a precision-driven method for the regulated release of agrochemicals, nanoencapsulation is at the forefront of agricultural innovation. By encapsulating active ingredients in nanoscale carriers, this technology offers numerous advantages in terms of increased agricultural sustainability, less environmental impact, and efficiency. A major benefit of nanoencapsulation is the precise and regulated release of agrochemicals, such as herbicides and pesticides. The technique of encapsulation protects the active ingredients from elements that could reduce their effectiveness, such as moisture and sunshine (Sampathkumar et al., 2020). The long-term stability of the agrochemicals is ensured by this protective layer, enabling a gradual and regulated release. Consequently, less application frequency is needed, which lessens the environmental impact of using traditional, nonencapsulated formulations. The slow release of agrochemicals is made possible by the nanoscale carriers, which are frequently made of biodegradable materials. This slows down the agrochemicals' quick dispersion and leaching into nearby ecosystems. In addition to increasing the agrochemicals' efficacy, this regulated release lessens the chance of contamination in water bodies and lessens their detrimental effects on organisms other than the intended targets (Fincheira et al., 2023; Machado et al., 2022).

The challenges associated with the targeted administration of agrochemicals to plant tissues are further addressed by nanoencapsulation. The nanoparticles' small size facilitates better plant penetration and absorption, guaranteeing that the active ingredients precisely reach their intended target. By reducing the total amount of agrochemicals required for efficient weed or insect control, this targeted delivery supports sustainable agriculture practices. Furthermore, formulations can be tailored to meet the unique needs of different crops, soil types, and environmental circumstances thanks to the encapsulation process as presented in Figure 10.1 (Grillo et al., 2021). This customization of pesticide delivery systems improves nanoencapsulation's flexibility and adaptability in a range of agricultural environments. Despite the significant potential advantages of nanoencapsulation in agriculture, current research efforts concentrate on resolving issues with scalability, affordability, and long-term environmental effects. Through responsible navigation of these obstacles, nanoencapsulation becomes apparent as a potential technology capable of revolutionizing pesticide uses and promoting a more efficient and sustainable method of crop protection in modern agriculture (Shang et al., 2019; Kookana et al., 2014).

The incorporation of nanocomposites into soil management techniques presents a viable path for improving soil health, encouraging effective nutrient usage, and reducing environmental effects in the context of sustainable agriculture. A new approach to soil enhancement techniques is provided by nanocomposites, which are made up of nanoparticles scattered throughout a matrix. They provide customized solutions to deal with the intricacies of contemporary agricultural ecosystems. Improved soil structure can be achieved using nanocomposites, which are frequently made of materials such as nanoclays or nanoscale organic matter (Prasad et al., 2017). By adding nanoparticles, soil aggregation is improved, and stable aggregates that are resistant to compaction and erosion are formed. Better water infiltration, root penetration, and general soil aeration are made possible by this improved structure, which promotes the development of a favorable environment for plant growth. One of the most important aspects of agricultural productivity is the soil's ability to hold water. Because of their capacity to alter the structure of soil, nanocomposites aid in improved water retention. This is especially helpful in areas where there is a shortage of water. Moreover, the presence of nanoparticles in the composite matrix affects the capacity for cation exchange, which enhances nutrient availability and retention for plants and maximizes nutrient usage efficiency (Yadav et al., 2023; Kim et al., 2024).

It is possible to design nanocomposites so that nutrients are controlled in their encapsulation and release. By coordinating nutrient availability with plant demand, this regulated release reduces nutrient runoff and leaching. The slow release of nutrients from nanocomposites improves plant uptake, lowering the frequency of traditional fertilizer treatments and lowering the resulting environmental effect. Heavy metals and other pollutants found in soil can be bound and rendered immobile by certain nanoparticles found in nanocomposites (Yadav et al., 2023). This remediation capability is a promising solution to problems with soil pollution, especially in places where industrial activity has historically occurred. An efficient and sustainable method of reducing the harmful impacts of soil pollutants on plant health and ecosystem integrity is provided by nanocomposites. Because of their adaptability, soil amendments can be tailored to the needs of certain crops and soil types. This flexibility makes it possible to precisely address problems and inadequacies in the soil, offering a focused and effective method of improving the soil (Nongbet et al., 2022; Omar et al., 2024). Although there is a lot of promise for using nanocomposites in agriculture, more study is needed to fully comprehend how these materials may affect soil microbiology, plant health, and ecosystem dynamics in the long run. Incorporating nanocomposites into soil management techniques in a responsible manner will aid in the creation of sustainable agricultural systems that strike a balance between environmental preservation and productivity (Dimkpa, 2014; Rajput et al., 2023).

10.4 NANOMATERIALS IN AGRICULTURE

10.4.1 NANOSENSORS FOR PRECISION AGRICULTURE

Nanosensors are emerging as game-changing instruments in the era of precision agriculture, providing real-time nanoscale data collecting and monitoring capabilities (Kaushal and Wani, 2017). Because they offer previously unheard-of insights into crop health, environmental conditions, and resource consumption, these tiny gadgets have the potential to completely transform the way farmers manage their crops. The incorporation of nanosensors into agricultural activities is an essential step toward maximizing sustainability and farm management (Tong et al., 2023).

10.4.2 REAL-TIME CROP OBSERVATION

Continuous and real-time monitoring of critical factors, including soil moisture, temperature, and nutrient levels—all essential for crop growth is made possible by nanosensors. Farmers are able to make well-informed decisions quickly thanks to this dynamic data stream, which guarantees that crops develop in the best possible conditions. Crop monitoring in real-time reduces waste, maximizes resource efficiency, and raises total agricultural productivity (Dhanaraju et al., 1745; Senapaty et al., 2023).

10.4.3 ACCURATE NUTRIENT MANAGEMENT

Because they provide precise information about the levels of nutrients in the soil, nanosensors are essential to precision nutrient management. When it comes to adjusting fertilizer treatments to the unique requirements of crops at various growth stages, this data is crucial. Nanosensors reduce environmental problems associated with excess nutrient runoff and help conserve resources by preventing over-fertilization and optimizing nutrient delivery (Yadav et al., 2023; Singh et al., 2020).

10.4.4 IOT INTEGRATION FOR SMART AGRICULTURE

The IoT and nanosensors work together effortlessly to create a network of connected devices that exchange data and interact (Dhanaraju et al., 1745; Chelliah et al., 2021). With the help of this synergy, a framework for smart farming practices is established. Using data from nanosensors and other sensors and devices, numerous elements of farm management may be automated and optimized. Nanosensors improve the precision and efficiency of farming operations in a variety of ways, including irrigation systems and insect management (Shi et al., 2022).

10.4.5 EARLY PEST AND DISEASE IDENTIFICATION IN CROPS

The use of nanosensors with chemical or biological markers allows for the early detection of pest infestations or agricultural illnesses. Because of their sensitivity, nanosensors can detect minute alterations in plant physiology or the presence of biomolecules that are suggestive of stress. Early diagnosis minimizes the impact on beneficial organisms and reduces the need for broad-spectrum medicines by enabling early interventions (Kulabhusan et al., 2022; Swierczewska et al., 2012).

10.4.6 ENVIRONMENTAL SURVEILLANCE

Nanosensors have uses in agriculture beyond crop-specific ones, such as environmental monitoring. They can identify and measure contaminants, evaluate the quality of the air and water, and offer information on the general environmental well-being of the agricultural ecosystem. This all-encompassing approach to monitoring promotes adherence to environmental standards and is consistent with sustainable farming methods. Nanosensors' importance in precision agriculture is expected to grow as they continue to advance and overcome issues with scalability, cost-effectiveness, and data security. Farmers are better equipped to make decisions, maximize resource use, and contribute to the creation of resilient and sustainable agricultural systems because of the insights produced by nanosensors (Neme et al., 2021; Sharma et al., 2021).

10.4.7 NANOSENSORS FUNCTION IN CROP MONITORING

The use of nanosensors in crop monitoring is essential because they provide a high-resolution, real-time method for obtaining vital information on the environmental factors, nutrient levels, and general health of the crops. With the help of this precision-driven technology, farmers can now make data-driven decisions that improve crop yield, resource efficiency, and sustainability, thereby revolutionizing modern agriculture (Javaid et al., 2021; El-Chaghaby and Rashad, 2024).

10.4.8 MONITORING THE ENVIRONMENT IN REAL TIME

Temperature, humidity, and light intensity are just a few of the environmental parameters that nanosensors are excellent at tracking in real time for crop growth. Farmers can react quickly to changes in weather patterns or unforeseen swings in environmental conditions because nanosensors provide constant and reliable data. By providing the ideal growing conditions for crops, this real-time monitoring supports adaptive farming techniques (El Hachimi et al., 2022).

10.4.9 MONITORING OF NUTRIENT STATUS

With remarkable accuracy, nutrient levels in the soil can be tracked thanks to nanosensors. These sensors can identify and measuring levels of important nutrients such as potassium, phosphorus, and nitrogen. Accurate nutrient management is made possible by the real-time data produced by nanosensors, which enables farmers to modify fertilizer inputs according to the unique requirements of crops at various growth stages. This focused strategy reduces fertilizer misuse, lessens its negative effects on the environment, and maximizes resource use (Yadav et al., 2023; Hamad et al., 2020).

10.4.10 WATER RESOURCES MANAGEMENT

In agriculture, water is a vital resource, and nanosensors help with effective water management. Nanosensors assist farmers in implementing irrigation systems that align with crop water requirements by continuously measuring soil moisture levels. This encourages water-use efficiency and guarantees crops receive the right quantity of moisture for maximum growth by preventing both water scarcity and waterlogging (Dasgupta et al., 2017).

10.4.11 EARLY DISEASE AND STRESS DETECTION

Early crop disease and stress detection is greatly aided by nanosensors fitted with chemical or biomolecular markers. These sensors can identify alterations in plant physiology as well as the presence of biomarkers linked to illnesses or stress. Timely intervention through early detection reduces the spread of diseases and makes tailored treatment techniques possible. This proactive strategy encourages sustainable pest management techniques while lowering the need for broad-spectrum pesticides (Rajendrasozhan et al., 2023; Mahlein, 2016).

10.4.12 COMBINING PRECISION AGRICULTURE TECHNOLOGIES

When combined with other precision agricultural technology, nanosensors create a seamless whole for farm management. A comprehensive picture of the farm can be created by combining the data gathered by nanosensors with other sensor data and Geographic Information System mapping. Precision farming techniques that maximize yields while reducing environmental impact can be implemented by farmers thanks to this integration, which facilitates data-driven decision-making (Yadav et al., 2023; Garg et al., 2023).

10.4.13 INTEGRATION OF IOT AND REMOTE SENSING

Farmers can monitor crops over wide areas by deploying nanosensors for remote sensing applications. Nanosensors, when combined with the IoT, provide a network that makes data processing and sharing easier. Smart farming techniques are supported by this networked technology, which gives farmers the ability to automate processes, remotely monitor and manage crops, and react instantly to changing circumstances (Garg et al., 2023). Nanosensors have a variety of applications in crop monitoring, giving farmers the means to maximize resource utilization, boost output, and support resilient and sustainable farming systems. Precision farming is anticipated to be heavily reliant on the incorporation of nanosensor technologies as they develop into farming methods (Mondal et al., 2022).

10.4.14 IOT INTEGRATION USING NANOSENSORS

The IoT and nanosensors work together to create a revolutionary synergy that is revolutionizing data collecting, analysis, and decision-making processes in a variety of fields, including agriculture. The capabilities of both systems are improved by this integration, which also creates a dynamic and interconnected ecosystem for automation, precision management, and real-time monitoring. The combination of IoT and nanosensors in agriculture presents a lot of potential for effective and sustainable agricultural management (Cruz Alvarado and Bazán, 2018; Alabdulatif et al., 2023).

10.4.15 REAL-TIME GATHERING OF DATA

Because they can record accurate information at the molecular level, nanosensors offer real-time information about the state of the environment, soil, and crops. Farmers can remotely monitor their fields thanks to the IoT integration, which allows this data to be smoothly relayed to centralized platforms. The capacity to gather data in real-time enables prompt responses to situations that change, improving farming methods' adaptability (Lew et al., 2020; Rehman et al., 2022).

10.4.16 REMOTE CONTROL AND MONITORING

Remote monitoring and management of agricultural activities are made possible by the integration of nanosensors with IoT systems. From any location with an internet connection, farmers may get up-to-date information on crop health, fertilizer levels, and soil moisture. Without having to be physically present in the fields, producers can use this remote accessibility to make well-informed decisions, modify irrigation schedules, and apply precision farming techniques (Dhanaraju et al., 1745; Zhu and Shang, 2022).

10.4.17 DECISION SUPPORT AND DATA ANALYTICS

IoT solutions efficiently manage and analyze the massive volume of data produced by nanosensors. Algorithms for data analytics can find trends, extract insightful information, and help farmers make decisions. Utilizing data from nanosensors, predictive analytics helps with resource allocation optimization, crop disease prevention, and yield forecasting as presented in Figure 10.2 (Han et al., 2023).

10.4.18 INTELLIGENT FARMING AND AUTOMATION

Automation in agriculture is facilitated by the combination of nanosensors with IoT. When certain circumstances or anomalies are detected by nanosensors, linked devices can automatically respond. For instance, a nanosensor can trigger an automated irrigation system if it senses low soil moisture. This degree of automation contributes to the idea of smart farming, lowers the need for manual intervention, and increases operational efficiency (Khelifa et al., 2015).

10.4.19 RESOURCE OPTIMIZATION AND ENERGY EFFICIENCY

Nanosensors with IoT capabilities aid in resource optimization and energy efficiency. The system can optimize energy utilization for irrigation, regulate fertilizer application based on real-time nutrient data, and modify lighting or ventilation in greenhouses by continuously monitoring environmental conditions. This degree of accuracy guarantees that resources are used wisely, minimizing waste and lessening the negative effects of farming methods on the environment (Khudoyberdiev et al., 2021).

10.4.20 INTERCONNECTIVITY AND SCALABILITY

Large agricultural landscapes can be equipped with several nanosensors due to the scalability of IoT platforms. A comprehensive picture of the entire farm is possible because of interconnectivity, which makes it possible to aggregate data from several sensors. In addition to facilitating the management of a variety of crops, its scalability and interconnection allow farmers to monitor and optimize operations at a scale that was previously difficult (Jayaraman et al., 2016).

10.4.21 ENVIRONMENTAL SURVEILLANCE AND OBSERVATION

IoT platforms that are combined with nanosensors enable thorough environmental monitoring. This includes monitoring the quality of the air and water, evaluating how agricultural practices affect ecosystems, and making sure that environmental laws are followed.

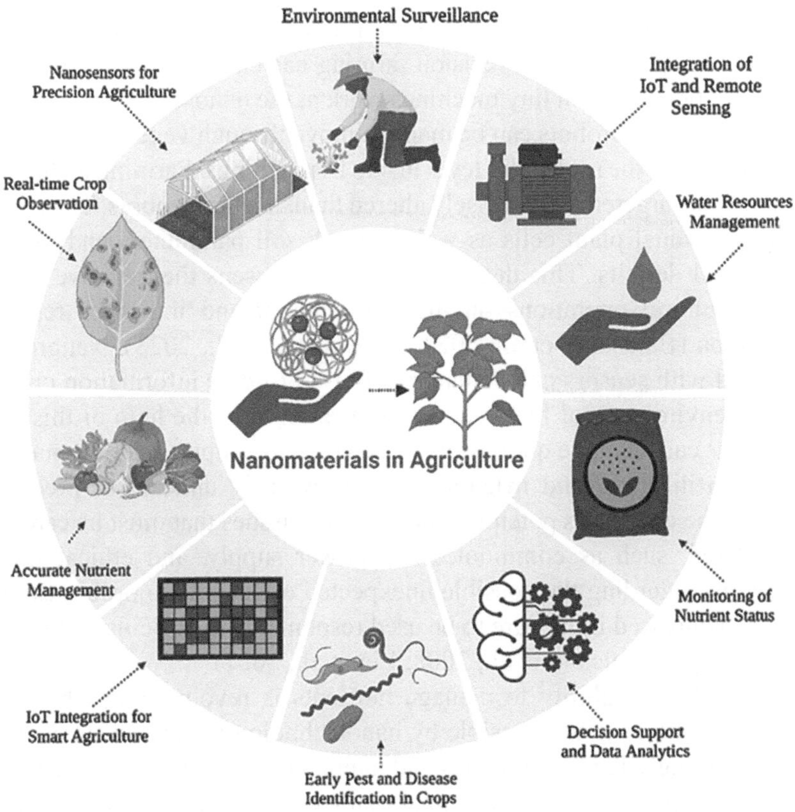

FIGURE 10.2 Applications of nanomaterials in agriculture through technological advancements such as IoT, data analytics, and nanosensors for the improvement of agriculture and crop production.

Sustainable and ethical agriculture methods benefit from the openness that real-time environmental data offers. Agriculture's use of IoT and nanosensor integration is a developing field with enormous promise. This combination will probably be very important in determining how precision agriculture develops in the future as technology progresses, helping farmers to overcome obstacles, make the most use of available resources, and support the production of food in a sustainable manner (Miteu et al., 2023; Tovar-Lopez, 2023).

10.5 PRECISION AGRICULTURE USING NANOSENSORS

The advancement of sustainable farming practices is greatly aided by nanodevices, which provide creative solutions that improve environmental stewardship, efficiency, and precision. Nanorobotics for precision farming and smart delivery systems for nutrient management are two prominent features of nanodevices. Precision farming has entered a new phase with nanorobotics, wherein tiny machines work at the nanoscale to do complex tasks. These nanorobots can be made to move through vegetation, through soil, or even at the molecular level inside living things. Farming techniques can now be targeted and precisely altered thanks to nanorobots. They could treat individual plant cells as well as track soil parameters and identify nutritional deficits. This degree of accuracy lessens the negative effects of agricultural operations on the environment and improves resource utilization (Balusamy et al., 2023; El-Ramady et al., 2023). Nanorobots, outfitted with sensors and actuators, offer up-to-date information on crop health, environmental factors, and soil health. With the help of this data, decisions can be made quickly and based on data, improving pest management, fertilization, and irrigation with previously unheard-of precision. Despite the enormous potential, there are still issues that must be carefully considered, such as communication, power supply, and ethical issues. Concerns regarding the possible unexpected effects of nanorobotics must also be addressed if they are to be used responsibly in agriculture (Mandal et al., 2020; Cavalcanti et al., 2008; Liu et al., 2021).

Agriculture's ability to manage nutrients is revolutionized by smart delivery systems made possible by nanotechnology. These systems make use of cutting-edge technologies and nanomaterials to precisely distribute nutrients to plants, maximizing uptake and reducing environmental impact. Nutrients are encapsulated in nanoparticles, which permits a regulated and progressive release. By coordinating fertilizer delivery with the plant's growth stages, this regulated release reduces nutrient runoff and avoids overfertilization. Nutrients can be delivered precisely to plant roots with the help of intelligent delivery systems. This focused strategy lowers waste and improves the overall efficiency of nutrient utilization by ensuring that nutrients reach the designated plants (Saritha et al., 2022). Smart delivery systems help to maintain environmental sustainability by minimizing nutrient runoff and lowering total fertilizer usage. This is in line with precision agriculture's guiding principles, which advocate for commercially and environmentally sound techniques. Precision agriculture and smart delivery systems work together smoothly, sharing sensor and monitoring data. This connection promotes a dynamic and flexible approach to agriculture by enabling adaptive nutrient management strategies based on real-time information. In conclusion, cutting-edge technologies such as nanorobotics and smart delivery systems, or nanodevices, have the potential to make agriculture a more efficient and sustainable sector of the economy. These advancements not only improve accuracy and resource efficiency but also make a substantial contribution to environmental preservation and a more sustainable farming industry in the future (Rayhan Shaheb et al., 2022; Bongiovanni and Lowenberg-Deboer, 2004).

10.6 NANOTECHNOLOGY FOR ECOLOGICAL AGRICULTURE

A multitude of potential exist to improve farming operations' sustainability through the use of nanodevices, which are distinguished by their small size and precise functionality at the nanoscale. By boosting resource efficiency, lowering environmental impact, and enhancing the overall resilience of agricultural ecosystems, these cutting-edge technologies can be specifically designed to solve difficulties in the agricultural sector. Precision agriculture relies heavily on nanodevices because they make it easier to target the application of agrochemicals such as pesticides and fertilizers. Nanodevices guarantee that these inputs reach the desired destination through nanoencapsulation and regulated release processes, minimizing waste and lowering the total amount of chemicals needed. By using a precision method, the environmental impact of traditional agrochemical applications is reduced, and resource efficiency is increased. Because they make it possible to monitor soil moisture levels in real time, nanodevices aid in the development of smart irrigation systems (Rayhan Shaheb et al., 2022; Bongiovanni and Lowenberg-Deboer, 2004).

Water delivery may be precisely and automatically controlled by integrating these devices with irrigation systems. Nanodevices help to conserve water by ensuring that crops receive the right amount of water depending on their real demands. This helps to lessen the detrimental effects of over-irrigation on soil health and water supplies. Crop monitoring capacities are improved by nanodevices, especially nanosensors. Real-time data on nutrient levels, crop health, and environmental variables can be detected and transmitted by these sensors. Nanosensor-gathered data facilitates data-driven decision-making, enabling farmers to optimize farming operations for increased yields and sustainability and to react quickly to changing conditions (Gutierrez et al., 2014; Bhavsar et al., 2023). In order to evaluate and preserve the health of the soil, nanodevices are essential. These instruments offer important insights into soil conditions by measuring soil properties at the nanoscale, such as pH, organic matter content, and nutrient levels. With the use of this knowledge, tailored soil management techniques can be put into place, which will lessen the need for overfertilization and increase soil resilience and fertility over the long run. Plants are able to absorb nutrients more efficiently thanks in part to nanodevices (Magnabosco et al., 2023; Nanoscience and Plant–Soil Systems, 2017). Nanodevices help plants absorb nutrients more efficiently by developing nanofertilizers and nanomaterials that increase nutrient availability and solubility. By minimizing nitrogen loss through leaching, this focused nutrient delivery maximizes fertilizer use and lessens environmental impact (Silva et al., 2023). Plant disease early detection is made possible by nanodevices fitted with biosensors. These tools are able to recognize particular pathogens or biomarkers that indicate plant stress or diseases. Early detection reduces the need for broad-spectrum chemical therapies and enables prompt intervention to stop the spread of disease. This focused strategy is in line with environmentally responsible and sustainable pest management techniques. Nanodevices help farms monitor their environmental conditions by identifying pollutants and evaluating the quality of the air and water. Furthermore, environmental remediation using nanodevices can help lessen the negative effects of pollutants on soil and water resources (Saritha et al., 2022; Silva et al., 2023). Both adherence to environmental standards and sustainable farming methods are supported by this dual capability. There is great promise for addressing the issues facing contemporary agriculture and promoting sustainability through the use of nanodevices in farming techniques. With further development, these technologies have the potential to become vital instruments for enhancing effective resource management, mitigating environmental effects, and bolstering resilience within agricultural systems (Khiyami et al., 2014; Dyussembayev et al., 2021).

Precision farming could undergo a substantial transformation if nanorobotics, the application of nanotechnology to the design and manufacturing of robots at the nanoscale, is successful. These tiny robots, also known as nanobots, can carry out a variety of activities with previously unheard-of precision, supporting sustainable agricultural methods (Koleoso et al., 2020). The following are some important ways that nanorobotics can improve precision farming. It is possible to build nanorobots that precisely supply nutrients to plants. Nanorobots make sure that plants get the necessary components precisely where and when they need them by delivering and releasing nutrients at the root level. This precise supply of nutrients promotes plant development, reduces fertilizer waste, and helps ensure effective nutrient usage (Soto et al., 2020). Sensing and acting nanorobots can be useful in intelligent pest control. These nanobots are capable of early illness or pest detection and can provide focused treatments, such as localized medications or the release of natural predators. This method lessens the need for broad-spectrum insecticides, reducing their negative effects on the environment and fostering environmentally friendly pest management. By locating and removing weeds with precision, nanorobots can aid with automated weed control. Nanorobots can minimize the use of herbicides and lessen their impact on nontarget plants by using precise mechanical or chemical approaches (Urso et al., 2023). This focused approach reduces the environmental impact of weed management techniques while also improving crop health overall. By examining soil composition, moisture content, and nutrient content at the nanoscale, nanorobots can keep an eye on the health of the soil (Zhang et al., 2022). Nanorobots can release additives from the soil, increase microbial activity, or alleviate soil compaction based on the data they acquire. Practices for sustainable soil management are aided by this real-time monitoring and treatment (Melanie et al., 2022).

Nanorobots could be developed to help in pollination in light of the disappearance of natural pollinators. With the help of pollen-equipped small robots, crops might be efficiently and precisely pollinated. This strategy could increase agricultural yields and improve overall food security, especially in regions with pollination issues. By measuring soil moisture levels at the microscale and adjusting water distribution accordingly, nanorobots can help with precision irrigation. This degree of accuracy makes sure that crops get the right amount of water, which lowers water waste and improves agricultural water use efficiency. As environmental sensors, nanorobots may gather information on temperature, water and air quality, and other pertinent factors. This real-time data gathering helps to promote data-driven decision-making for sustainable farming methods by advancing our understanding of the agricultural ecosystem (Potts et al., 2018; Wu et al., 2022).

When fitted with diagnostic tools, nanorobots can detect plant illnesses early on. These nanorobots can provide specific therapies, including administering antifungal medicines or triggering plant defense mechanisms, once they are detected. By using a precise approach, the likelihood of developing resistance is decreased, and fewer broad-spectrum treatments are required (Aggarwal and Kumar, 2022). Precision farming with nanorobotics has a lot of promise, but there are some issues that need to be resolved, including scalability, cost-effectiveness, and ethical issues. The ethical integration of nanorobots into agricultural systems has the potential to transform farming methods and make them more ecologically friendly, sustainable, and productive as technology develops. An innovation in agricultural technology that improves accuracy and productivity in nutrient management is smart delivery systems (Zhang et al., 2022). Based on real-time monitoring and analysis, these systems use a variety of technologies, such as sensors, actuators, and data analytics, to accurately distribute nutrients to crops. The following are important features and advantages of intelligent delivery systems for agricultural fertilizer management: Utilizing sensors, smart delivery systems keep an eye on the nutrients, moisture content, and other pertinent aspects of the soil. The dynamic awareness of the plant's nutrient requirements provided by this real-time monitoring enables prompt modifications and accurate nutrient administration (Abaszadeh et al., 2023).

Data collection is aided by a variety of sensor technologies, including satellite photography, drones, and soil sensors. While aerial technologies offer a more comprehensive view of crop health and fertilizer distribution, soil sensors can measure nutrient levels in the root zone. A thorough evaluation of nutritional requirements at various scales is made possible by the integration of these sensors (Talaviya et al., 2020). The advanced analytics platforms are utilized to process the gathered data. The data is analyzed by machine learning algorithms and data models to provide insights into growth trends, possible stress factors, and nutrient requirements.

Subsequently, decision support systems offer practical suggestions for enhancing nutrient delivery tactics. Actuators accurately control the release of nutrients and are frequently integrated into smart delivery systems (Akhter and Sofi, 2022). This can include drip irrigation systems, automatic fertilizer applicators, or other devices that supply nutrients straight to the root zone. Delivering nutrients precisely guarantees that crops get the necessary nutrients at the right time while minimizing waste and its negative effects on the environment (Atalla et al., 2023). Variable rate application is supported by smart delivery systems, which makes it possible to modify fertilizer levels in response to geographical changes in soil conditions. By adjusting fertilizer supply to the unique requirements of several fields, this method maximizes resource utilization and enhances crop performance. A common component of the larger IoT ecosystem is smart delivery systems. Real-time data sharing and communication are facilitated by sensors, actuators, and connected devices (Bargaz et al., 2018). By enhancing their reactivity and adaptability, nutrient management techniques become more adaptable to changing conditions. Farmers can monitor fertilizer management activities from any location with internet access thanks to the remote monitoring features of smart delivery systems. By enabling prompt reactions to new problems, this remote control lessens the requirement for field presence and boosts operational effectiveness (Bijay-Singh and Craswell, 2021).

Smart delivery systems help ensure effective resource usage by accurately supplying nutrients where and when they are needed. This entails making the best use of fertilizers, cutting down on runoff and leaching, and avoiding the harm that comes from applying too many nutrients to the ecosystem. Smart distribution methods give important insights into the data for farmers. Historical patterns, relationships between crop performance and nutrient levels, and suggestions for ongoing development are some of these observations (Govindasamy et al., 2023; Solanki et al., 2015). The management of nutritional resources is improved overall when decisions are made based on data. Smart delivery systems' accuracy and effectiveness support agriculture's environmental sustainability. These technologies assist in lessening the negative effects of fertilizer misuse and nutrient runoff on water bodies and ecosystems. In conclusion, intelligent delivery systems for managing nutrients mark a substantial development in contemporary agriculture, encouraging accuracy, effectiveness, and sustainability. As these technologies advance, they will become increasingly important in tackling the problems of feeding a growing world population and reducing the environmental impact of agricultural operations (Bisht and Singh Chauhan, 2021).

10.7 SAFETY ISSUES AND THE ENVIRONMENT

The use of smart delivery systems for fertilizer management in agriculture presents safety and environmental issues that should be carefully considered, as is the case with any new technology.

Overapplication of nutrients may result in runoff that contaminates adjacent bodies of water (Basavegowda and Baek, 2021). Elevated levels of nutrients, particularly nitrogen and phosphorus, are linked to eutrophication, toxic algal blooms, and other ecological disorders (Penuelas et al., 2023). Although overapplication is minimized by precision nutrient delivery systems, runoff can still be avoided by routinely monitoring and modifying procedures. Over time, a persistent dependence on specific nutrients may have a detrimental impact on soil health by changing the composition of the soil and microbial activity. To preserve soil fertility and structure, intelligent delivery systems should be used in conjunction with complete soil health management approaches (Raimondi et al., 2021).

Increasing energy consumption could be a result of the energy needed to run sensors, data analytics, and precise application procedures. It is critical to support renewable energy sources, energy-efficient technology, and an evaluation of the life-cycle energy costs of intelligent delivery systems (Li, 2022). Local ecosystems may be impacted by the introduction of cutting-edge technologies such as drones and driverless cars. These issues can be resolved by implementing technologies with less ecological impact, taking wildlife corridors into account, and implementing behaviors that cause the least amount of disturbance (Zhang et al., 2018). Because smart delivery systems depend on data analysis and collection, there are worries over the security and privacy of private agricultural information. Information security requires the use of strong data encryption, safe cloud storage, and compliance with data protection laws (Gupta et al., 2020).

Reliance on technology increases the possibility of malfunctions, cyberattacks, or system breakdowns that could interfere with nutrient management procedures. To guarantee the dependability of smart delivery technologies, backup plans, routine maintenance, and cybersecurity precautions are essential. It may be necessary to modify nutrient management practices if persistent and consistent nutrient treatment leads to pest or crop resistance (Tariq et al., 2023). Sustainable agriculture requires the use of a variety of nutrient management techniques, crop rotation, and ongoing knowledge of resistance issues. Human health may be at risk from exposure to specific herbicides, fertilizers, or chemicals used in smart delivery systems (Riggs et al., 2023).

Health hazards can be reduced by putting in place appropriate safety procedures, training operators, and utilizing environmentally friendly inputs. Traditional farming methods may lose jobs as smart delivery systems become more widely used. These socioeconomic issues can be resolved by offering farmers programs of education and training that will help them adjust to technology advancements (Javaid et al., 2022). Smart delivery systems' compliance with ever-changing laws pertaining to data privacy, environmental effects, and safety may face difficulties. It's critical to maintain awareness of new legislation, comply with standards, and engage proactively with regulatory organizations. It takes a multi-stakeholder strategy involving farmers, technology developers, regulators, and academics to address these environmental and safety concerns (El Bilali and Allahyari, 2018).

The ethical and sustainable implementation of intelligent distribution systems necessitates ongoing observation, modification, and cooperation to guarantee the long-term advantages of these technologies without jeopardizing public health or safety (Javaid et al., 2022; Siebrecht, 2020). When assessing the environmental effects of agricultural methods, such as the use of smart delivery systems for fertilizer management, ecotoxicological issues are crucial. The study of ecotoxicology aims to comprehend the potential effects of chemicals, such as agrochemicals, on ecosystems and the creatures that inhabit them (Abideen et al., 2021).

Even when applied precisely, agrochemicals may unintentionally affect organisms other than the intended target, including birds, insects, and aquatic life. To assist in allaying these worries, comprehensive risk assessments, ecologically friendly input selection, and application techniques that reduce exposure to nontarget organisms are used. Certain agrochemicals may build up in living things' tissues and biomagnify their way up the food chain, endangering trophic levels higher up. To reduce these ecological concerns, buffer zones should be put in place, low-persistent agrochemicals should be used, and bioaccumulation potential should be monitored. Using some agrochemicals may have an impact on the microbial communities in the soil, upsetting vital soil functions (Kumar et al., 2021; Dhananjayan et al., 2020). Crop rotation, the use of organic methods, and the selection of inputs with little effect on soil microbiota can all contribute to the preservation of robust soil ecosystems. Aquatic habitats may be impacted by runoff from fields treated with agrochemicals that enter water bodies. The risk of aquatic contamination can be decreased by employing buffer strips, controlled-release formulations, and precision irrigation techniques. Target pests may become resistant to some insecticides if they are applied consistently and consistently. You can slow the emergence of resistance by using integrated pest management tactics, rotating pesticides, and using a variety of pest control techniques (Pathak et al., 2022).

The physiology and biochemistry of plants may be adversely affected by the accidental application of some agrochemicals. Choosing chemicals with advantageous profiles, keeping an eye on crop health, and carrying out in-depth research on the physiological effects of agrochemicals on crops are crucial. The use of agrochemicals may have cumulative and long-term impacts that are not immediately evident, which could have unanticipated ecological repercussions. To help with these issues, long-term monitoring programs, environmental impact analyses, and the adoption of adaptive management techniques can be put into place. Aquatic and animal species may be subject to endocrine disruption by certain agrochemicals. Reducing these risks can be achieved by selecting inputs with a lower potential for endocrine disruption, testing for toxicity, and following regulations (Kalyabina et al., 2021; Tudi et al., 2021). Even when applied precisely, agrochemicals may unintentionally harm pollinators such as butterflies and bees. Mitigation techniques include applying pollinator-friendly measures, employing low-toxicity pesticides to pollinators, and avoiding application during crucial pollination times. To identify possible hazards and track changes over time, it is crucial to continuously check the quality of the soil and water. Early detection and mitigation of environmental risks are facilitated by the implementation of monitoring programs, the use of cutting-edge sensor technology, and routine assessment (Nicolopoulou-Stamati et al., 2016).

10.8 ETHICAL CONSIDERATIONS IN THE APPLICATION OF NANOTECHNOLOGY

With its enormous potential for use in a wide range of industries, including agriculture, the application of nanotechnology raises several ethical issues that require careful study. First of all, there are ethical questions raised by the possible hazards to health and the environment posed by nanomaterials and nanodevices (Schulte and Salamanca-Buentello, 2007; Kumari et al., 2023). To reduce possible harm to human health and the environment during the development and deployment of nanotechnology, it is imperative that thorough risk assessments be carried out and that safety be given first priority (Biswas et al., 2022). Another ethical consideration that comes to mind, considering the general public's lack of knowledge about nanotechnology, is transparency and information availability. Transparent communication in research, development, and application is essential to ensuring ethical practices (Isigonis et al., 2019; Wang et al., 2023). Accessible and understandable information is provided to support well-informed decision-making and cultivate public confidence. Concerns about equity and access also surface because nanotechnology may make social and economic disparities worse. Ensuring fair distribution of benefits and access to nanotechnology breakthroughs is crucial for preventing societal inequities and supporting social justice principles. These factors are emphasized by ethical considerations (Ray, 2023). An important ethical component is privacy and data security, especially in fields such as nanomedicine and nanosurveillance. Strict data security protocols and observance of privacy standards are necessary to strike a balance between innovation and the defense of people's rights to privacy. An ethical imperative to reduce the environmental effects of nanotechnology by appropriate disposal, recycling methods, and the creation of eco-friendly nanomaterials is advocated by the ethical consideration of environmental impact (Jaime et al., 2023; Aldboush and Ferdous, 2023).

10.9 CONCLUSION

Upon analyzing the uses of nanotechnology in agriculture, several significant discoveries highlight its revolutionary potential. Most notably, nanotechnology makes precision agriculture possible by enabling exact control over irrigation, insect control, and nutrient delivery. Resource-efficient farming techniques that minimize waste and lessen the impact on the environment help to promote this shift toward precision agriculture. By strengthening plant health overall, disease resistance, and nutrient uptake, nanomaterials and nanofertilizers improve agricultural performance. Data-driven decision-making in agriculture is made possible by the integration of nanosensors and nanodevices, which provides real-time information for well-informed decisions. However, open communication, stakeholder involvement, and the creation of ethical frameworks are necessary to solve issues with public acceptance and ethical considerations. Cooperation between scientists, farmers, legislators, and regulatory agencies is necessary for successful implementation. The benefits of applying nanotechnology to agriculture go well beyond short-term improvements, as sustainable farming methods stand to gain greatly. Through precise control over the application of water, fertilizer, and pesticides, as well as waste reduction and efficiency promotion, nanotechnology enables efficient resource utilization. Targeted delivery methods and nanoencapsulation also support environmental preservation by preserving the quality of soil and water and fostering ecosystem health. In order to monitor and mitigate the effects of climate change and increase crop tolerance to environmental challenges, nanosensors and nanobiosensors are essential. By increasing crop yields and reducing losses, nanotechnology's increased productivity and efficiency promote global food security.

KEYWORDS

- **nanotechnology**
- **agriculture**
- **sustainable development**
- **nanomaterials**
- **environmental impacts**
- **risk assessment**
- **sustainable farming**
- **case studies**

REFERENCES

Abaszadeh, F.; Ashoub, M.H.; Khajouie, G.; Amiri, M. Nanotechnology development in surgical applications: Recent trends and developments. *Eur. J. Med. Res.* 2023, 28, 537, https://doi.org/10.1186/s40001-023-01429-4.

Abideen, A.Z.; Sundram, V.P.K.; Pyeman, J.; Othman, A.K.; Sorooshian, S. Food supply chain transformation through technology and future research directions—A systematic review. *Logistics* 2021, 5, 83, https://doi.org/10.3390/logistics5040083.

Aggarwal, M.; Kumar, S. The use of nanorobotics in the treatment therapy of cancer and its future aspects: A review. *Cureus* 2022, 14(9), e29366, https://doi.org/10.7759/cureus.29366.

Akhter, R.; Sofi, S.A. Precision agriculture using IoT data analytics and machine learning. *J. King Saud Univ. Comput. Inf. Sci.* 2022, 34, 5602–5618, https://doi.org/10.1016/j.jksuci.2021.05.013.

Alabdulatif, A.; Thilakarathne, N.N.; Lawal, Z.K.; Fahim, K.E.; Zakari, R.Y. Internet of Nano-Things (IoNT): A comprehensive review from architecture to security and privacy challenges. *Sensors* 2023, 23, 2807, https://doi.org/10.3390/s23052807.

Aldboush, H.H.H.; Ferdous, M. Building trust in fintech: An analysis of ethical and privacy considerations in the intersection of big data, AI, and customer trust. *Int. J. Financ. Stud.* 2023, 11, 90, https://doi.org/10.3390/ijfs11030090.

An, C.; Sun, C.; Li, N.; Huang, B.; Jiang, J.; Shen, Y.; Wang, C.; Zhao, X.; Cui, B.; Wang, C.; et al. Nanomaterials and nanotechnology for the delivery of agrochemicals: Strategies towards sustainable agriculture. *J. Nanobiotechnol.* 2022, 20, 11, https://doi.org/10.1186/s12951-021-01214-7.

Asif, M.; Islam, S.; A. Malik, M.; Mahdi Dar, Z.; Masood, A.; Shafi, S.; Rashid, B.; Sidique, S. Nano pesticides application in agriculture and their impact on environment. In: *Agricultural Development in Asia—Potential Use of Nano-Materials and Nano-Technology*. IntechOpen, 2022.

Atalla, S.; Tarapiah, S.; Gawanmeh, A.; Daradkeh, M.; Mukhtar, H.; Himeur, Y.; Mansoor, W.; Bin Hashim, K.F.; Daadoo, M. IoT-Enabled precision agriculture: Developing an ecosystem for optimized crop management. *Information* 2023, 14, 205, https://doi.org/10.3390/info14040205.

Baig, N.; Kammakakam, I.; Falath, W. Nanomaterials: A review of synthesis methods, properties, recent progress, and challenges. *Mater. Adv.* 2021, 2, 1821–1871, https://doi.org/10.1039/D0MA00807A.

Balusamy, S.R.; Joshi, A.S.; Perumalsamy, H.; Mijakovic, I.; Singh, P. Advancing sustainable agriculture: A critical review of smart and eco-friendly nanomaterial applications. *J. Nanobiotechnol.* 2023, 21, 372, https://doi.org/10.1186/s12951-023-02135-3.

Bargaz, A.; Lyamlouli, K.; Chtouki, M.; Zeroual, Y.; Dhiba, D. Soil microbial resources for improving fertilizers efficiency in an integrated plant nutrient management system. *Front. Microbiol.* 2018, 9, 1606, https://doi.org/10.3389/fmicb.2018.01606.

Barrat, J.-L.; Del Gado, E.; Egelhaaf, S.U.; Mao, X.; Dijkstra, M.; Pine, D.J.; Kumar, S.K.; Bishop, K.; Gang, O.; Obermeyer, A.; et al. Soft matter roadmap. *J. Phys. Mater.* 2024, 7, 012501, https://doi.org/10.1088/2515-7639/ad06cc.

Basavegowda, N.; Baek, K.-H. Current and future perspectives on the use of nanofertilizers for sustainable agriculture: The case of phosphorus nanofertilizer. *3 Biotech* 2021, 11, 357, https://doi.org/10.1007/s13205-021-02907-4.

Bayda, S.; Adeel, M.; Tuccinardi, T.; Cordani, M.; Rizzolio, F. The History of nanoscience and nanotechnology: From chemical–physical applications to nanomedicine. *Molecules* 2019, 25, 112, https://doi.org/10.3390/molecules25010112.

Betz, U.A.K.; Arora, L.; Assal, R.A.; Azevedo, H.; Baldwin, J.; Becker, M.S.; Bostock, S.; Cheng, V.; Egle, T.; Ferrari, N.; et al. Game changers in science and technology—Now and beyond. *Technol. Forecast. Soc. Change* 2023, 193, 122588, https://doi.org/10.1016/j.techfore.2023.122588.

Bhagat, V.; A.G. Gold nanorods: A promising drug delivery vehicle. *Inven. Rapid NDDS* 2012, 1–4.

Bhavsar, D.; Limbasia, B.; Mori, Y.; Imtiyazali Aglodiya, M.; Shah, M. A Comprehensive and systematic study in smart drip and sprinkler irrigation systems. *Smart Agric. Technol.* 2023, 5, 100303, https://doi.org/10.1016/j.atech.2023.100303.

Bijay-Singh; Craswell, E. Fertilizers and nitrate pollution of surface and ground water: An increasingly pervasive global problem. *SN Appl. Sci.* 2021, 3, 518, https://doi.org/10.1007/s42452-021-04521-8.

Bisht, N.; Singh Chauhan, P. Excessive and disproportionate use of chemicals cause soil contamination and nutritional stress. In: *Soil Contamination—Threats and Sustainable Solutions*. IntechOpen, 2021.

Biswas, R.; Alam, M.; Sarkar, A.; Haque, M.I.; Hasan, M.M.; Hoque, M. Application of nanotechnology in food: Processing, preservation, packaging and safety assessment. *Heliyon* 2022, 8, e11795, https://doi.org/10.1016/j.heliyon.2022.e11795.

Bongiovanni, R.; Lowenberg-Deboer, J. Precision agriculture and sustainability. *Precis. Agric.* 2004, 5, 359–387, https://doi.org/10.1023/B:PRAG.0000040806.39604.aa.

Bothiraja, C.; Gholap, A.D.; Shaikh, K.S.; Pawar, A.P. Investigation of ethyl cellulose microsponge gel for topical delivery of eberconazole nitrate for fungal therapy. *Ther. Deliv.* 2014, 5, 781–794, https://doi.org/10.4155/tde.14.43.

Cavalcanti, A.; Shirinzadeh, B.; Zhang, M.; Kretly, L. Nanorobot hardware architecture for medical defense. *Sensors* 2008, 8, 2932–2958, https://doi.org/10.3390/s8052932.

Chaturvedi, S.; Dave, P.N.; Shah, N.K. Applications of nano-catalyst in new era. *J. Saudi Chem. Soc.* 2012, 16, 307–325, https://doi.org/10.1016/j.jscs.2011.01.015.

Chaud, M.; Souto, E.B.; Zielinska, A.; Severino, P.; Batain, F.; Oliveira-Junior, J.; Alves, T. Nanopesticides in agriculture: Benefits and challenge in agricultural productivity, toxicological risks to human health and environment. *Toxics* 2021, 9, 131, https://doi.org/10.3390/toxics9060131.

Chehelgerdi, M.; Chehelgerdi, M.; Allela, O.Q.B.; Pecho, R.D.C.; Jayasankar, N.; Rao, D.P.; Thamaraikani, T.; Vasanthan, M.; Viktor, P.; Lakshmaiya, N.; et al. Progressing nanotechnology to improve targeted cancer treatment: Overcoming hurdles in its clinical implementation. *Mol. Cancer* 2023, 22, 169, https://doi.org/10.1186/s12943-023-01865-0.

Chelliah, R.; Wei, S.; Daliri, E.B.-M.; Rubab, M.; Elahi, F.; Yeon, S.-J.; Jo, K. H.; Yan, P.; Liu, S.; Oh, D.H. Development of nanosensors based intelligent packaging systems: Food quality and medicine. *Nanomaterials* 2021, 11, 1515, https://doi.org/10.3390/nano11061515.

Cruz Alvarado, M.A.; Bazán, P.A. Understanding the internet of nano things: Overview, trends, and challenges. *e-Ciencias la Inf.* 2018, 9(1), https://doi.org/10.15517/eci.v1i1.33807.

Dasgupta, N.; Ranjan, S.; Ramalingam, C. Applications of nanotechnology in agriculture and water quality management. *Environ. Chem. Lett.* 2017, 15, 591–605, https://doi.org/10.1007/s10311-017-0648-9.

Dhananjayan, V.; Jayanthi, P.; Jayakumar, S.; Ravichandran, B. Agrochemicals impact on ecosystem and bio-monitoring. In: *Resources Use Efficiency in Agriculture*. Springer Singapore: Singapore, 2020, pp. 349–388.

Dhanaraju, M.; Chenniappan, P.; Ramalingam, K.; Pazhanivelan, S.; Kaliaperumal, R. Smart farming: Internet of Things (IoT)-based sustainable agriculture. *Agriculture* 2022, 12, 1745, https://doi.org/10.3390/agriculture12101745.

Dimkpa, C.O. Can nanotechnology deliver the promised benefits without negatively impacting soil microbial life? *J. Basic Microbiol.* 2014, 54, 889–904, https://doi.org/10.1002/jobm.201400298.

Dyussembayev, K.; Sambasivam, P.; Bar, I.; Brownlie, J.C.; Shiddiky, M.J.A.; Ford, R. Biosensor technologies for early detection and quantification of plant pathogens. *Front. Chem.* 2021, 9, 636245, https://doi.org/10.3389/fchem.2021.636245.

El Bilali, H.; Allahyari, M.S. Transition towards sustainability in agriculture and food systems: Role of information and communication technologies. *Inf. Process. Agric.* 2018, 5, 456–464, https://doi.org/10.1016/j.inpa.2018.06.006.

El Hachimi, C.; Belaqziz, S.; Khabba, S.; Sebbar, B.; Dhiba, D.; Chehbouni, A. Smart weather data management based on artificial intelligence and big data analytics for precision agriculture. *Agriculture* 2022, 13, 95, https://doi.org/10.3390/agriculture13010095.

El-Chaghaby, G.A.; Rashad, S. Nanosensors in agriculture: Applications, prospects, and challenges. In: *Handbook of Nanosensors*. Springer Nature Switzerland: Cham, 2024, pp. 1–29.

El-Ramady, H.; Abdalla, N.; Sári, D.; Ferroudj, A.; Muthu, A.; Prokisch, J.; Fawzy, Z.F.; Brevik, E.C.; Solberg, S.Ø. Nanofarming: Promising solutions for the future of the global agricultural industry. *Agronomy* 2023, 13, 1600, https://doi.org/10.3390/agronomy13061600.

Faiyazuddin, M.; Sophia, A.; Ashique, S.; Gholap, A.D.; Gowri, S.; Mohanto, S.; Karthikeyan, C.; Nag, S.; Hussain, A.; Akhtar, M.S.; et al. Virulence traits and novel drug delivery strategies for mucormycosis post-COVID-19: A comprehensive review. *Front. Immunol.* 2023, 14, 1264502, https://doi.org/10.3389/fimmu.2023.1264502.

Fincheira, P.; Hoffmann, N.; Tortella, G.; Ruiz, A.; Cornejo, P.; Diez, M.C.; Seabra, A.B.; Benavides-Mendoza, A.; Rubilar, O. Eco-efficient systems based on nanocarriers for the controlled release of fertilizers and pesticides: Toward smart agriculture. *Nanomaterials* 2023, 13, 1978, https://doi.org/10.3390/nano13131978.

Fraceto, L.F.; Grillo, R.; de Medeiros, G.A.; Scognamiglio, V.; Rea, G.; Bartolucci, C. Nanotechnology in agriculture: Which innovation potential does it have? *Front. Environ. Sci.* 2016, 4, https://doi.org/10.3389/fenvs.2016.00020.

Garg, S.; Rumjit, N.P.; Roy, S. Smart agriculture and nanotechnology: Technology, challenges, and new perspective. *Adv. Agrochem* 2023, https://doi.org/10.1016/j.aac.2023.11.001.

Ghidan, A.Y.; Al Antary, T.M. Applications of nanotechnology in agriculture. In: *Applications of Nanobiotechnology*. IntechOpen, 2020.

Gholap, A.; Pardeshi, S.; Giram, P. *Preparation, Antibacterial and Antiviral Activity Measurements and Detection Methods*. 2023a, pp. 33–64.

Gholap, A.D. Translational nanomedicine: Status assessment and opportunities. *Int. J. Res. Pharm. Biomed. Sci.* 2012a, 3, 1454–1463.

Gholap, A.D.; Bhowmik, D.D.; Deshmukh, A.Y.; Hatvate, N.T. *Quintessential Impact of Dendrimer Bioconjugates in Targeted Drug Delivery*. 2023b.

Gholap, A.D.; Borude, S.S.; Mahajan, A.M. Smart crystals technology: A review. *Pharmacologyonline* 2011a, 238–243.

Gholap, A.D.; Borude, S.S.; Mahajan, A.M.; Gholap, M.A.D. Aquasomes: A potential drug delivery carrier. *Pharmacologyonline* 2011b, 230–237.

Gholap, A.D.; Musmade, D.S. Advances in the field of nanooncology: An overview. *Inven. Impact NDDS* 2011a, 1–3.

Gholap, A.D.; Tupe, D. Artificial viral envelopes: Emerging vehicle for gene delivery. *Pharmatutor-Art-1439* 2012a, 1–3.

Gholap, A.D.; Tupe, D Potential applications of botanical extract. *Inven. Rapid Ethnopharmacol.* 2012b, 1–4.

Gholap, A.D.; Gupta, J.; Kamandar, P.; Bhowmik, D.D.; Rojekar, S.; Faiyazuddin, M.; Hatvate, N.T.; Mohanto, S.; Ahmed, M.G.; Subramaniyan, V.; et al. Harnessing nanovaccines for effective immunization—A special concern on COVID-19: Facts, fidelity, and future prospective. *ACS Biomater. Sci. Eng.* 2024, 10, 271–297, https://doi.org/10.1021/acsbiomaterials.3c01247.

Gholap, A.D.; Gupta, J.S.; Kamandar, P.A.; Banchhod, G.V.; Hatvate, N.T. *Antibody-Drug Conjugates for Cancer Therapy: An Up-to-Date Review on the Chemistry and Pharmacology*. 2023c.

Gholap, A.D.; Hatvate, N.T.; Khuspe, P.R.; Mandhare, T.A.; Kashid, P.; Gaikwad, V.D. Proteomics in oncology: Retrospect and prospects. In: *Novel Technologies in Biosystems, Biomedical and Drug Delivery*. Springer Nature Singapore: Singapore, 2023d, pp. 243–269.

Gholap, A.D.; Hatvate, N.T.; Sawant, S.D.; Khemka, P.N.; Vora, L.K. Unlocking the potential of nanobiohybrids to combat environmental pollution. In: *Nanobiohybrids for Advanced Wastewater Treatment and Energy Recovery*. IWA Publishing, 2023e, pp. 169–190.

Gholap, A.D.; Rajurkar, V.G.; Dighe, S.B. Phonophoresis: Ultrasound-enhanced transdermal drug delivery. *Inven. Rapid NDDS* 2012, 1–3.

Gholap, A.D.; Rojekar, S.; Kapare, H.S.; Vishwakarma, N.; Raikwar, S.; Garkal, A.; Mehta, T.A.; Jadhav, H.; Prajapati, M.K.; Annapure, U. Chitosan scaffolds: Expanding horizons in biomedical applications. *Carbohydr. Polym.* 2024a, 323, 121394, https://doi.org/10.1016/j.carbpol.2023.121394.

Gholap, A.D.; Dighe, S.B. Dandruff: Pathogenesis and treatment. *Inven. Rapid Cosmeceuticals* 2012a, 1–4.

Gholap, A.D.; Dighe, S.B. Role of sebum in human health and acne. *Inven. Rapid Cosmeceuticals* 2012b, 1–3.

Gholap, A.D.; Said, R.P.; Pawar, R.D.; Ambore, G.S.; Hatvate, N.T. Importance of carbohydrate-drug conjugates in vaccine development: A detailed review. 2023f.

Gholap, A.D.; Sawant, S.D.; Sayyad, S.F.; Hatvate, N.T.; Chavan, M.; Rojekar, S.; Faiyazuddin, M. Food biopackaging for human benefits: Status and perspectives. In: *Biobased Packaging Materials*. Springer Nature Singapore: Singapore, 2023g, pp. 1–35.

Gholap, A.D.; Sayyad, S.F.; Hatvate, N.T.; Dhumal, V.V.; Pardeshi, S.R.; Chavda, V.P.; Vora, L.K. Drug delivery strategies for avobenzone: A case study of photostabilization. *Pharmaceutics* 2023h, 15, 1008, https://doi.org/10.3390/pharmaceutics15031008.

Gholap, A.D.; Rajurkar, V.G. Lipid nanocapsules: A novel promising tool as drug delivery system. *Inven. Rapid NDDS* 2012b, 1–4.

Gholap, A.D.; Rajurkar, V.G. New threat of clinical toxicities evoked from nanocarrier systems. *Inven. Rapid NDDS* 2012c, 1–4.

Gholap, A.D.; Rajurkar, V.G. Potential applications of capillary electromigration methods for the improvement of separation efficiency. *Inven. Rapid Pharm Anal. Qual. Assur.* 2012d, 1–3.

Gholap, A.D.; Rajurkar, V.G. Potential applications of superparmagnetic nanoparticles technology. *Inven. Rapid NDDS* 2012e, 1–3.

Govindasamy, P.; Muthusamy, S.K.; Bagavathiannan, M.; Mowrer, J.; Jagannadham, P.T.K.; Maity, A.; Halli, H.M.; G. K., S.; Vadivel, R.; Sujayananad G. K., Rajagopal Vadivel; Das T. K.; Raj, R.; Pooniya, V.; Babu, S.; Rathore, S.S.; Muralikrishnan L.; Tiwari, G. Nitrogen use efficiency—A key to enhance crop productivity under a changing climate. *Front. Plant Sci.* 2023, 14, 1121073, https://doi.org/10.3389/fpls.2023.1121073.

Grillo, R.; Mattos, B.D.; Antunes, D.R.; Forini, M.M.L.; Monikh, F.A.; Rojas, O.J. Foliage adhesion and interactions with particulate delivery systems for plant nanobionics and intelligent agriculture. *Nano Today* 2021, 37, 101078, https://doi.org/10.1016/j.nantod.2021.101078.

Gupta, M.; Abdelsalam, M.; Khorsandroo, S.; Mittal, S. Security and privacy in smart farming: Challenges and opportunities. *IEEE Access* 2020, 8, 34564–34584, https://doi.org/10.1109/ACCESS.2020.2975142.

Gutierrez, J.; Villa-Medina, J.F.; Nieto-Garibay, A.; Porta-Gandara, M.A. Automated irrigation system using a wireless sensor network and GPRS module. *IEEE Trans. Instrum. Meas.* 2014, 63, 166–176, https://doi.org/10.1109/TIM.2013.2276487.

Haleem, A.; Javaid, M.; Singh, R.P.; Rab, S.; Suman, R. Applications of nanotechnology in medical field: A brief review. *Glob. Heal. J.* 2023, 7, 70–77, https://doi.org/10.1016/j.glohj.2023.02.008.

Hamad, H.T.; Al-Sharify, Z.T.; Al-Najjar, S.Z.; Gadooa, Z.A. A Review on nanotechnology and its applications on fluid flow in agriculture and water recourses. *IOP Conf. Ser. Mater. Sci. Eng.* 2020, 870, 012038, https://doi.org/10.1088/1757-899X/870/1/012038.

Han, D.; Hosamo, H.; Ying, C.; Nie, R. A Comprehensive review and analysis of nanosensors for structural health monitoring in bridge maintenance: Innovations, challenges, and future perspectives. *Appl. Sci.* 2023, 13, 11149, https://doi.org/10.3390/app132011149.

Iavicoli, I.; Leso, V.; Beezhold, D.H.; Shvedova, A.A. Nanotechnology in agriculture: Opportunities, toxicological implications, and occupational risks. *Toxicol. Appl. Pharmacol.* 2017, 329, 96–111, https://doi.org/10.1016/j.taap.2017.05.025.

Isigonis, P.; Hristozov, D.; Benighaus, C.; Giubilato, E.; Grieger, K.; Pizzol, L.; Semenzin, E.; Linkov, I.; Zabeo, A.; Marcomini, A. Risk governance of nanomaterials: Review of criteria and tools for risk communication, evaluation, and mitigation. *Nanomaterials* 2019, 9, 696, https://doi.org/10.3390/nano9050696.

Jaime, F.J.; Muñoz, A.; Rodríguez-Gómez, F.; Jerez-Calero, A. Strengthening privacy and data security in biomedical microelectromechanical systems by IoT communication security and protection in smart healthcare. *Sensors* 2023, 23, 8944, https://doi.org/10.3390/s23218944.

Javaid, M.; Haleem, A.; Singh, R.P.; Rab, S.; Suman, R. Exploring the potential of nanosensors: A brief overview. *Sensors Int.* 2021, 2, 100130, https://doi.org/10.1016/j.sintl.2021.100130.

Javaid, M.; Haleem, A.; Singh, R.P.; Suman, R. Enhancing smart farming through the applications of agriculture 4.0 technologies. *Int. J. Intell. Networks* 2022, 3, 150–164, https://doi.org/10.1016/j.ijin.2022.09.004.

Jayaraman, P.; Yavari, A.; Georgakopoulos, D.; Morshed, A.; Zaslavsky, A. Internet of Things platform for smart farming: Experiences and lessons learnt. *Sensors* 2016, 16, 1884, https://doi.org/10.3390/s16111884.

Jeevanandam, J.; Barhoum, A.; Chan, Y.S.; Dufresne, A.; Danquah, M.K. Review on nanoparticles and nanostructured materials: History, sources, toxicity and regulations. *Beilstein J. Nanotechnol.* 2018, 9, 1050–1074, https://doi.org/10.3762/bjnano.9.98.

Joudeh, N.; Linke, D. Nanoparticle classification, physicochemical properties, characterization, and applications: A comprehensive review for biologists. *J. Nanobiotechnol.* 2022, 20, 262, https://doi.org/10.1186/s12951-022-01477-8.

Kalyabina, V.P.; Esimbekova, E.N.; Kopylova, K.V.; Kratasyuk, V.A. Pesticides: Formulants, distribution pathways and effects on human health – A review. *Toxicol. Rep.* 2021, 8, 1179–1192, https://doi.org/10.1016/j.toxrep.2021.06.004.

Kaushal, M.; Wani, S.P. Nanosensors: Frontiers in precision agriculture. In: *Nanotechnology*. Springer Singapore: Singapore, 2017, pp. 279–291.

Khan, I.; Saeed, K.; Khan, I. Nanoparticles: Properties, applications and toxicities. *Arab. J. Chem.* 2019, 12, 908–931, https://doi.org/10.1016/j.arabjc.2017.05.011.

Khan, Y.; Sadia, H.; Ali Shah, S.Z.; Khan, M.N.; Shah, A.A.; Ullah, N.; Ullah, M.F.; Bibi, H.; Bafakeeh, O.T.; Ben Khedher, N.; et al. Classification, synthetic, and characterization approaches to nanoparticles, and their applications in various fields of nanotechnology: A review. *Catalysts* 2022, 12, 1386, https://doi.org/10.3390/catal12111386.

Khedkar, A.; Tupe, D.; Gholap, A. Cubosomes: Emerging drug delivery system. *Inven. Rapid NDDS* 2012, 1–3.

Khedkar, A.; Tupe, D.; Gholap, A. Pharmacosomes: An emerging vesicular system. *Inven. Rapid NDDS* 2012, 1–3.

Khelifa, B.; Amel, D.; Amel, B.; Mohamed, C.; Tarek, B. Smart irrigation using internet of things. In: *Proceedings of the 2015 Fourth International Conference on Future Generation Communication Technology (FGCT)*; IEEE, July 2015, pp. 1–6.

Khiyami, M.A.; Almoammar, H.; Awad, Y.M.; Alghuthaymi, M.A.; Abd-Elsalam, K.A. Plant pathogen nanodiagnostic techniques: forthcoming changes? *Biotechnol. Biotechnol. Equip.* 2014, 28, 775–785, https://doi.org/10.1080/13102818.2014.960739.

Khudoyberdiev, A.; Ullah, I.; Kim, D. Optimization-assisted water supplement mechanism with energy efficiency in IoT based greenhouse. *J. Intell. Fuzzy Syst.* 2021, 40, 10163–10182, https://doi.org/10.3233/JIFS-200618.

Kim, D.-Y.; Patel, S.K.S.; Rasool, K.; Lone, N.; Bhatia, S.K.; Seth, C.S.; Ghodake, G.S. Bioinspired silver nanoparticle-based nanocomposites for effective control of plant pathogens: A review. *Sci. Total Environ.* 2024, 908, 168318, https://doi.org/10.1016/j.scitotenv.2023.168318.

Koleoso, M.; Feng, X.; Xue, Y.; Li, Q.; Munshi, T.; Chen, X. Micro/Nanoscale magnetic robots for biomedical applications. *Mater. Today Bio.* 2020, 8, 100085, https://doi.org/10.1016/j.mtbio.2020.100085.

Kookana, R.S.; Boxall, A.B.A.; Reeves, P.T.; Ashauer, R.; Beulke, S.; Chaudhry, Q.; Cornelis, G.; Fernandes, T.F.; Gan, J.; Kah, M.; et al. Nanopesticides: Guiding principles for regulatory evaluation of environmental risks. *J. Agric. Food Chem.* 2014, 62, 4227–4240, https://doi.org/10.1021/jf500232f.

Kulabhusan, P.K.; Tripathi, A.; Kant, K. Gold nanoparticles and plant pathogens: An overview and prospective for biosensing in forestry. *Sensors* 2022, 22, 1259, https://doi.org/10.3390/s22031259.

Kumah, E.A.; Fopa, R.D.; Harati, S.; Boadu, P.; Zohoori, F.V.; Pak, T. Human and environmental impacts of nanoparticles: A scoping review of the current literature. *BMC Public Health* 2023, 23, 1059, https://doi.org/10.1186/s12889-023-15958-4.

Kumar, J.K.S.; Monica, S.S.; Vinothkumar, B.; Suganthi, A.; Paramasivam, M. Impact of pesticide exposure on environment and biodiversity: A review. *Agric. Rev.* 2021, https://doi.org/10.18805/ag.R-2325.

Kumari, R.; Suman, K.; Karmakar, S.; Mishra, V.; Lakra, S.G.; Saurav, G.K.; Mahto, B.K. Regulation and safety measures for nanotechnology-based agri-products. *Front. Genome Ed.* 2023, 5, 1200987, https://doi.org/10.3389/fgeed.2023.1200987.

Lew, T.T.S.; Sarojam, R.; Jang, I.-C.; Park, B.S.; Naqvi, N.I.; Wong, M.H.; Singh, G.P.; Ram, R.J.; Shoseyov, O.; Saito, K.; et al. Species-independent analytical tools for next-generation agriculture. *Nat. Plants* 2020, 6, 1408–1417, https://doi.org/10.1038/s41477-020-00808-7.

Li, B. Effective Energy Utilization through economic development for sustainable management in smart cities. *Energy Rep.* 2022, 8, 4975–4987, https://doi.org/10.1016/j.egyr.2022.02.303.

Liu, C.; Zhou, H.; Zhou, J. The applications of nanotechnology in crop production. *Molecules* 2021, 26, 7070, https://doi.org/10.3390/molecules26237070.

Machado, T.O.; Grabow, J.; Sayer, C.; de Araújo, P.H.H.; Ehrenhard, M.L.; Wurm, F.R. Biopolymer-based nanocarriers for sustained release of agrochemicals: A review on materials and social science perspectives for a sustainable future of agri- and horticulture. *Adv. Colloid Interface Sci.* 2022, 303, 102645, https://doi.org/10.1016/j.cis.2022.102645.

Magar, T.V.; Gholap, A.D.; Musmade, D.S. Elastic vesicles as drugs carriers through the skin. *Inven. Rapid NDDS* 2011, 1–2.

Magnabosco, P.; Masi, A.; Shukla, R.; Bansal, V.; Carletti, P. Advancing the impact of plant biostimulants to sustainable agriculture through nanotechnologies. *Chem. Biol. Technol. Agric.* 2023, 10, 117, https://doi.org/10.1186/s40538-023-00491-8.

Mahajan, A.M.; Gholap, A.D.; Borude, S.S.. POEMS syndrome: A review. *Pharmacologyonline* 2011a, 265–274.

Mahajan, A.M.; Gholap, A.D.; Borude, S.S.. Recent updates on Q fever: Diagnosis and treatment. *Pharmacologyonline* 2011b, 3, 254–264.

Mahendra Kumar, S.; Rajni, Y.; Sandeep Prasad, T. Recent advances in nanotechnology. *Int. J. Nanomater. Nanotechnol. Nanomed.* 2023, 9, 015–023, https://doi.org/10.17352/2455-3492.000053.

Mahlein, A.-K. Plant disease detection by imaging sensors—Parallels and specific demands for precision agriculture and plant phenotyping. *Plant Dis.* 2016, 100, 241–251, https://doi.org/10.1094/PDIS-03-15-0340-FE.

Malik, S.; Muhammad, K.; Waheed, Y. Nanotechnology: A revolution in modern industry. *Molecules* 2023, 28, 661, https://doi.org/10.3390/molecules28020661.

Mandal, N.; Adhikary, S.; Rakshit, R. Nanobiosensors: Recent developments in soil health assessment. In: *Soil Analysis: Recent Trends and Applications*. Springer Singapore: Singapore, 2020, pp. 285–304.

Mansoori, G.A. *An Introduction to Nanoscience and Nanotechnology*. 2017, pp. 3–20.

Mehraeen, S. Introductory chapter: Self-assembly of nanostructures. In: *Self-Assembly of Nanostructures and Patchy Nanoparticles*. IntechOpen, 2020.

Melanie, M.; Miranti, M.; Kasmara, H.; Malini, D.M.; Husodo, T.; Panatarani, C.; Joni, I.M.; Hermawan, W. Nanotechnology-based bioactive antifeedant for plant protection. *Nanomaterials* 2022, 12, 630, https://doi.org/10.3390/nano12040630.

Miteu, G.D.; Emmanuel, A.A.; Addeh, I.; Ojeokun, O.; Olayinka, T.; Godwin, J.S.; Adeyemo, O.I.; Benneth, E.O. Nanoscience and technology as a pivot for sustainable agriculture and its one health approach awareness. *Sci. One Heal.* 2023, 2, 100020, https://doi.org/10.1016/j.soh.2023.100020.

Mittal, D.; Kaur, G.; Singh, P.; Yadav, K.; Ali, S.A. Nanoparticle-based sustainable agriculture and food science: Recent advances and future outlook. *Front. Nanotechnol.* 2020, 2, https://doi.org/10.3389/fnano.2020.579954.

Mondal, R.; Dam, P.; Chakraborty, J.; Paret, M.L.; Katı, A.; Altuntas, S.; Sarkar, R.; Ghorai, S.; Gangopadhyay, D.; Mandal, A.K.; et al. Potential of nanobiosensor in sustainable agriculture: The state-of-art. *Heliyon* 2022, 8, e12207, https://doi.org/10.1016/j.heliyon.2022.e12207.

Muhie, S.H. Novel approaches and practices to sustainable agriculture. *J. Agric. Food Res.* 2022, 10, 100446, https://doi.org/10.1016/j.jafr.2022.100446.

Mwaanga, P. Risks, Uncertainties, and ethics of nanotechnology in agriculture. In: *New Visions in Plant Science*. InTech, 2018.

Nanoscience and Plant–Soil Systems. In: Ghorbanpour, M.; Manika, K.; Varma, A.; (Eds.), *Soil Biology*. Springer International Publishing: Cham, 2017, Vol. 48. ISBN 978-3-319-46833-4.

Nasrollahzadeh, M.; Sajadi, S.M.; Sajjadi, M.; Issaabadi, Z. *An Introduction to Nanotechnology*. 2019, pp. 1–27.

Neme, K.; Nafady, A.; Uddin, S.; Tola, Y.B. Application of nanotechnology in agriculture, postharvest loss reduction and food processing: Food security implication and challenges. *Heliyon* 2021, 7, e08539, https://doi.org/10.1016/j.heliyon.2021.e08539.

Nicolopoulou-Stamati, P.; Maipas, S.; Kotampasi, C.; Stamatis, P.; Hens, L. Chemical pesticides and human health: The urgent need for a new concept in agriculture. *Front. Public Heal.* 2016, 4, 148, https://doi.org/10.3389/fpubh.2016.00148.

Nongbet, A.; Mishra, A.K.; Mohanta, Y.K.; Mahanta, S.; Ray, M.K.; Khan, M.; Baek, K.-H.; Chakrabartty, I. Nanofertilizers: A smart and sustainable attribute to modern agriculture. *Plants* 2022, 11, 2587, https://doi.org/10.3390/plants11192587.

Omar, R.A.; Talreja, N.; Ashfaq, M.; Chauhan, D. *Nanostructure-Based Smart Fertilizers and Their Interaction with Plants*. 2024, pp. 399–430.

Pardeshi, S.; Gholap, A.; More, M.; Togre, N.; Rebello, N.; Giram, P. *Dendrimers Based Antibacterial and Antiviral Materials*. 2023, pp. 139–169.

Pathak, V.M.; Verma, V.K.; Rawat, B.S.; Kaur, B.; Babu, N.; Sharma, A.; Dewali, S.; Yadav, M.; Kumari, R.; Singh, S.; et al. Current status of pesticide effects on environment, human health and it's eco-friendly management as bioremediation: A comprehensive review. *Front. Microbiol.* 2022, 13, 962619, https://doi.org/10.3389/fmicb.2022.962619.

Penuelas, J.; Coello, F.; Sardans, J. A better use of fertilizers is needed for global food security and environmental sustainability. *Agric. Food Secur.* 2023, 12, 5, https://doi.org/10.1186/s40066-023-00409-5.

Porter, A.L.; Youtie, J. How interdisciplinary is nanotechnology? *J. Nanoparticle Res.* 2009, 11, 1023–1041, https://doi.org/10.1007/s11051-009-9607-0.

Potts, S.G.; Neumann, P.; Vaissière, B.; Vereecken, N.J. Robotic bees for crop pollination: Why drones cannot replace biodiversity. *Sci. Total Environ.* 2018, 642, 665–667, https://doi.org/10.1016/j.scitotenv.2018.06.114.

Prasad, R.; Bhattacharyya, A.; Nguyen, Q.D. Nanotechnology in sustainable agriculture: Recent developments, challenges, and perspectives. *Front. Microbiol.* 2017, 8, 1014, https://doi.org/10.3389/fmicb.2017.01014.

Raimondi, G.; Maucieri, C.; Toffanin, A.; Renella, G.; Borin, M. Smart fertilizers: What should we mean and where should we go? *Ital. J. Agron.* 2021, 16, https://doi.org/10.4081/ija.2021.1794.

Rajendrasozhan, S.; Sherwani, S.; Ahmed, F.; Shaalan, N.; Alsukaibi, A.; Al-Motair, K.; Khan, M.W.A. Emerging nanomaterials biosensors in breathalyzers for detection of COVID-19: Future prospects. *Inorganics* 2023, 11, 483, https://doi.org/10.3390/inorganics11120483.

Rajput, V.D.; Kumari, A.; Upadhyay, S.K.; Minkina, T.; Mandzhieva, S.; Ranjan, A.; Sushkova, S.; Burachevskaya, M.; Rajput, P.; Konstantinova, E.; et al. Can nanomaterials improve the soil microbiome and crop productivity? *Agriculture* 2023, 13, 231, https://doi.org/10.3390/agriculture13020231.

Rajurkar, V.G, Gholap, A.D. Recent technologies in nanomedicine for treatment of cancer. *Inven. Rapid Pharm. Tech.* 2012, 1–3.

Ray, P.P. ChatGPT: A comprehensive review on background, applications, key challenges, bias, ethics, limitations and future scope. *Internet Things Cyber-Physical Syst.* 2023, 3, 121–154, https://doi.org/10.1016/j.iotcps.2023.04.003.

Rayhan Shaheb, M.; Sarker, A.; A. Shearer, S. Precision agriculture for sustainable soil and crop management. In: *Soil Science—Emerging Technologies, Global Perspectives and Applications*. IntechOpen, 2022.

Rehman, A.; Saba, T.; Kashif, M.; Fati, S.M.; Bahaj, S.A.; Chaudhry, H. A Revisit of internet of things technologies for monitoring and control strategies in smart agriculture. *Agronomy* 2022, 12, 127, https://doi.org/10.3390/agronomy12010127.

Riggs, H.; Tufail, S.; Parvez, I.; Tariq, M.; Khan, M.A.; Amir, A.; Vuda, K.V.; Sarwat, A.I. Impact, vulnerabilities, and mitigation strategies for cyber-secure critical infrastructure. *Sensors* 2023, 23, 4060, https://doi.org/10.3390/s23084060.

Sampathkumar, K.; Tan, K.X.; Loo, S.C.J. Developing nano-delivery systems for agriculture and food applications with nature-derived polymers. *iScience* 2020, 23, 101055, https://doi.org/10.1016/j.isci.2020.101055.

Saritha, G.N.G.; Anju, T.; Kumar, A. Nanotechnology—Big impact: How nanotechnology is changing the future of agriculture? *J. Agric. Food Res.* 2022, 10, 100457, https://doi.org/10.1016/j.jafr.2022.100457.

Schulte, P.A.; Salamanca-Buentello, F. Ethical and scientific issues of nanotechnology in the workplace. *Environ. Health Perspect.* 2007, 115, 5–12, https://doi.org/10.1289/ehp.9456.

Senapaty, M.K.; Ray, A.; Padhy, N. IoT-enabled soil nutrient analysis and crop recommendation model for precision agriculture. *Computers* 2023, 12, 61, https://doi.org/10.3390/computers12030061.

Shang, Y.; Hasan, M.K.; Ahammed, G.J.; Li, M.; Yin, H.; Zhou, J. Applications of nanotechnology in plant growth and crop protection: A review. *Molecules* 2019, 24, 2558, https://doi.org/10.3390/molecules24142558.

Sharma, P.; Pandey, V.; Sharma, M.M.M.; Patra, A.; Singh, B.; Mehta, S.; Husen, A. A Review on biosensors and nanosensors application in agroecosystems. *Nanoscale Res. Lett.* 2021, 16, 136, https://doi.org/10.1186/s11671-021-03593-0.

Shi, Q.; Yang, Y.; Sun, Z.; Lee, C. Progress of advanced devices and Internet of Things systems as enabling technologies for smart homes and health care. *ACS Mater. Au* 2022, 2, 394–435, https://doi.org/10.1021/acsmaterialsau.2c00001.

Shrivastava, S.; Dash, D. Applying nanotechnology to human health: Revolution in biomedical sciences. *J. Nanotechnol.* 2009, 2009, 1–14, https://doi.org/10.1155/2009/184702.

Siebrecht, N. Sustainable agriculture and its implementation gap—Overcoming obstacles to implementation. *Sustainability* 2020, 12, 3853, https://doi.org/10.3390/su12093853.

Silva, A.M.M.; de Araújo, V.L.V.P.; Cardoso, E.J.B.N. Revisiting the past to understand the present and future of soil health in Brazil. *Front. Soil Sci.* 2023, 3, https://doi.org/10.3389/fsoil.2023.1172436.

Singh, P.; Pandey, P.C.; Petropoulos, G.P.; Pavlides, A.; Srivastava, P.K.; Koutsias, N.; Deng, K.A.K.; Bao, Y. Hyperspectral remote sensing in precision agriculture: Present status, challenges, and future trends. In: *Hyperspectral Remote Sensing*. Elsevier, 2020, pp. 121–146.

Singh, T.; Kumar, V.R.; Gholap, A.D.; Sanap, G.; Anjaiah, B.; Roja, K. N vitro antipyrial activity of psidium guajava leaf extract. *Eur. J. Pharm. Med. Res.* 2016, 3, 312–315.

Solanki, P.; Bhargava, A.; Chhipa, H.; Jain, N.; Panwar, J. Nano-fertilizers and their smart delivery system. In: *Nanotechnologies in Food and Agriculture*. Springer International Publishing: Cham, 2015, pp. 81–101.

Soto, F.; Wang, J.; Ahmed, R.; Demirci, U. Medical Micro/Nanorobots in precision medicine. *Adv. Sci.* 2020, 7, 2002203, https://doi.org/10.1002/advs.202002203.

Swierczewska, M.; Liu, G.; Lee, S.; Chen, X. High-sensitivity nanosensors for biomarker detection. *Chem. Soc. Rev.* 2012, 41, 2641–2655, https://doi.org/10.1039/C1CS15238F.

Szczyglewska, P.; Feliczak-Guzik, A.; Nowak, I. Nanotechnology—General aspects: A chemical reduction approach to the synthesis of nanoparticles. *Molecules* 2023, 28, 4932, https://doi.org/10.3390/molecules28134932.

Szefler, B. Nanotechnology, from quantum mechanical calculations up to drug delivery. *Int. J. Nanomed.* 2018, 13, 6143–6176, https://doi.org/10.2147/IJN.S172907.

Taha, T.B.; Barzinjy, A.A.; Hussain, F.H.S.; Nurtayeva, T. Nanotechnology and computer science: Trends and advances. *Memories—Mater. Devices, Circuits Syst.* 2022, 2, 100011, https://doi.org/10.1016/j.memori.2022.100011.

Talaviya, T.; Shah, D.; Patel, N.; Yagnik, H.; Shah, M. Implementation of artificial intelligence in agriculture for optimisation of irrigation and application of pesticides and herbicides. *Artif. Intell. Agric.* 2020, 4, 58–73, https://doi.org/10.1016/j.aiia.2020.04.002.

Tariq, U.; Ahmed, I.; Bashir, A.K.; Shaukat, K. A Critical cybersecurity analysis and future research directions for the internet of things: A comprehensive review. *Sensors* 2023, 23, 4117, https://doi.org/10.3390/s23084117.

Tong, S.W.; Goh, W.P.; Jiang, C. Review—Recent advances in nanosensors for precision agriculture. *J. Electrochem. Soc.* 2023, 170, 127507, https://doi.org/10.1149/1945-7111/ad1306.

Tovar-Lopez, F.J. Recent progress in micro- and nanotechnology-enabled sensors for biomedical and environmental challenges. *Sensors* 2023, 23, 5406, https://doi.org/10.3390/s23125406.

Tudi, M.; Daniel Ruan, H.; Wang, L.; Lyu, J.; Sadler, R.; Connell, D.; Chu, C.; Phung, D.T. Agriculture development, pesticide application and its impact on the environment. *Int. J. Environ. Res. Public Health* 2021, 18, 1112, https://doi.org/10.3390/ijerph18031112.

Tupe, D.; Gholap, A.D.* The Threat of photoaging: Clinical conditions, evaluation and treatment. *Inven. Rapid Cosmeceuticals* 2012, 1–3.

Tyagi, P.K.; Arya, A.; Ramniwas, S.; Tyagi, S. Editorial: Recent trends in nanotechnology in precision and sustainable agriculture. *Front. Plant Sci.* 2023, 14, 1256319, https://doi.org/10.3389/fpls.2023.1256319.

Urso, M.; Ussia, M.; Pumera, M. Smart micro- and nanorobots for water purification. *Nat. Rev. Bioeng.* 2023, 1, 236–251, https://doi.org/10.1038/s44222-023-00025-9.

Vega-Vásquez, P.; Mosier, N.S.; Irudayaraj, J. Nanoscale drug delivery systems: From medicine to agriculture. *Front. Bioeng. Biotechnol.* 2020, 8, 79, https://doi.org/10.3389/fbioe.2020.00079.

Wahab, A.; Batool, F.; Muhammad, M.; Zaman, W.; Mikhlef, R.M.; Naeem, M. Current knowledge, research progress, and future prospects of phyto-synthesized nanoparticles interactions with food crops under induced drought stress. *Sustainability* 2023, 15, 14792, https://doi.org/10.3390/su152014792.

Wang, D.; Saleh, N.B.; Byro, A.; Zepp, R.; Sahle-Demessie, E.; Luxton, T.P.; Ho, K.T.; Burgess, R.M.; Flury, M.; White, J.C.; et al. Nano-enabled pesticides for sustainable agriculture and global food security. *Nat. Nanotechnol.* 2022, 17, 347–360, https://doi.org/10.1038/s41565-022-01082-8.

Wang, R.; Bush-Evans, R.; Arden-Close, E.; Bolat, E.; McAlaney, J.; Hodge, S.; Thomas, S.; Phalp, K. Transparency in persuasive technology, immersive technology, and online marketing: Facilitating users' informed decision making and practical implications. *Comput. Human Behav.* 2023, 139, 107545, https://doi.org/10.1016/j.chb.2022.107545.

Wu, S.; Liu, J.; Lei, X.; Zhao, S.; Lu, J.; Jiang, Y.; Xie, B.; Wang, M. Research progress on efficient pollination technology of crops. *Agronomy* 2022, 12, 2872, https://doi.org/10.3390/agronomy12112872.

Yadav, A.; Yadav, K.; Abd-Elsalam, K.A. Exploring the potential of nanofertilizers for a sustainable agriculture. *Plant Nano Biol.* 2023, 5, 100044, https://doi.org/10.1016/j.plana.2023.100044.

Yadav, A.; Yadav, K.; Abd-Elsalam, K.A. Nanofertilizers: Types, delivery and advantages in agricultural sustainability. *Agrochemicals* 2023, 2, 296–336, https://doi.org/10.3390/agrochemicals2020019.

Yadav, A.; Yadav, K.; Ahmad, R.; Abd-Elsalam, K.A. Emerging frontiers in nanotechnology for precision agriculture: Advancements, hurdles and prospects. *Agrochemicals* 2023, 2, 220–256, https://doi.org/10.3390/agrochemicals2020016.

Yadav, S.; Sharma, A.K.; Kumar, P. Nanoscale self-assembly for therapeutic delivery. *Front. Bioeng. Biotechnol.* 2020, 8, 127, https://doi.org/10.3389/fbioe.2020.00127.

Yousef, H.A.; Fahmy, H.M.; Arafa, F.N.; Abd Allah, M.Y.; Tawfik, Y.M.; El Halwany, K.K.; El-Ashmanty, B.A.; Al-anany, F.S.; Mohamed, M.A.; Bassily, M.E. Nanotechnology in pest management: Advantages, applications, and challenges. *Int. J. Trop. Insect Sci.* 2023, 43, 1387–1399, https://doi.org/10.1007/s42690-023-01053-z.

Zhang, Y.; Huang, T.; Bompard, E.F. Big data analytics in smart grids: A review. *Energy Inform.* 2018, 1, 8, https://doi.org/10.1186/s42162-018-0007-5.

Zhang, Y.; Zhang, Y.; Han, Y.; Gong, X. Micro/Nanorobots for medical diagnosis and disease treatment. *Micromachines* 2022, 13, 648, https://doi.org/10.3390/mi13050648.

Zhu, M.; Shang, J. Remote monitoring and management system of intelligent agriculture under the internet of things and deep learning. *Wirel. Commun. Mob. Comput.* 2022, 2022, 1–13, https://doi.org/10.1155/2022/1206677.

Index

D

E

H

W

Z

For Product Safety Concerns and Information please contact our EU representative GPSR@taylorandfrancis.com Taylor & Francis Verlag GmbH, Kaufingerstraße 24, 80331 München, Germany